MATHEMATICAL ANALYSIS IN ENGINEERING

MATHEMATICAL ANALYSIS IN ENGINEERING
How to Use the Basic Tools

CHIANG C. MEI
Massachusetts Institute of Technology

Published by the Press Syndicate of the University of Cambridge
The Pitt Building, Trumpington Street, Cambridge CB2 1RP
40 West 20th Street, New York, NY 10011-4211, USA
10 Stamford Road, Oakleigh, Melbourne 3166, Australia

© Cambridge University Press 1995

First published 1995

Printed in the United States of America

Library of Congress Cataloguing in Publication Data
Mei, Chiang C.
Mathematical analysis in engineering : how to use the basic tools / Chiang C. Mei.
p. cm.
Includes bibliographical references and index.
ISBN 0-521-46053-0
1. Engineering mathematics. I. Title.
TA330.M43 1995
620′.00151 – dc20   94-30063

A catalog record of this book is available from the British Library.

ISBN 0-521-46053-0 hardback

To My Parents and My Wife

Contents

Preface		*page* xiii
Acknowledgments		xvii
1	**Formulation of physical problems**	**1**
1.1	Transverse vibration of a taut string	1
1.2	Longitudinal vibration of an elastic rod	4
1.3	Traffic flow on a freeway	6
1.4	Seepage flow through a porous medium	7
1.5	Diffusion in a stationary medium	10
1.6	Shallow water waves and linearization	12
	1.6.1 Nonlinear governing equations	12
	1.6.2 Linearization for small amplitude	15
2	**Classification of equations with two independent variables**	**20**
2.1	A first-order equation	20
2.2	System of first-order equations	22
2.3	Linear second-order equations	25
	2.3.1 Constant coefficients	25
	2.3.2 Variable coefficients	28
3	**One-dimensional waves**	**33**
3.1	Waves due to initial disturbances	33
3.2	Reflection from the fixed end of a string	38
3.3	Specification of initial and boundary data	40
3.4	Forced waves in a long string	41
3.5	Uniqueness of the Cauchy problem	44
3.6	Traffic flow – a taste of nonlinearity	45
3.7	Green light at the head of traffic	46
3.8	Traffic congestion and jam	50

4	**Finite domains and separation of variables**	**57**
4.1	Separation of variables	57
4.2	One-dimensional diffusion	63
4.3	Eigenfunctions and base vectors	65
4.4	Partially insulated slab	66
4.5	Sturm–Liouville problems	71
4.6	Steady forcing	75
4.7	Transient forcing	76
4.8	Two-dimensional diffusion	78
4.9	Cylindrical polar coordinates	80
4.10	Steady heat conduction in a circle	82
5	**Elements of Fourier series**	**91**
5.1	General Fourier series	91
5.2	Trigonometric Fourier series	93
	5.2.1 Full Fourier series	93
	5.2.2 Fourier cosine and sine series	95
	5.2.3 Other intervals	96
5.3	Exponential Fourier series	98
5.4	Convergence of Fourier series	99
6	**Introduction to Green's functions**	**105**
6.1	The δ function	105
6.2	Static deflection of a string under a concentrated load	108
6.3	String under a simple harmonic point load	110
6.4	Sturm–Liouville boundary-value problem	112
6.5	Bending of an elastic beam on an elastic foundation	114
	6.5.1 Formulation of the beam problem	114
	6.5.2 Beam under a sinusoidal concentrated load	117
6.6	Fundamental solutions	121
6.7	Green's function in a finite domain	124
6.8	Adjoint operator and Green's function	125
7	**Unbounded domains and Fourier transforms**	**132**
7.1	Exponential Fourier transform	132
	7.1.1 From Fourier series to Fourier transform	132
	7.1.2 Transforms of derivatives	134
	7.1.3 Convolution theorem	134
7.2	One-dimensional diffusion	135
	7.2.1 General solution in integral form	135
	7.2.2 A localized source	137
	7.2.3 Discontinuous initial temperature	139

7.3	Forced waves in one dimension	141
7.4	Seepage flow into a line drain	143
7.5	Surface load on an elastic ground	145
	7.5.1 Field equations for plane elasticity	145
	7.5.2 Half plane under surface load	147
	7.5.3 Response to a line load	149
7.6	Fourier sine and cosine transforms	153
7.7	Diffusion in a semi-infinite domain	154
7.8	Potential problem in a semi-infinite strip	159
8	**Bessel functions and circular boundaries**	**165**
8.1	Circular region and Bessel's equation	165
8.2	Bessel function of the first kind	167
8.3	Bessel function of the second kind for integer order	171
8.4	Some properties of Bessel functions	174
	8.4.1 Recurrence relations	175
	8.4.2 Behavior for small argument	176
	8.4.3 Behavior for large argument	176
	8.4.4 Wronskians	178
	8.4.5 Partial wave expansion	179
8.5	Oscillations in a circular region	180
	8.5.1 Radial eigenfunctions and natural modes	180
	8.5.2 Orthogonality of natural modes	182
	8.5.3 Transient oscillations in a circular pond	184
8.6	Hankel functions and wave propagation	185
	8.6.1 Wave radiation from a circular cylinder	186
	8.6.2 Scattering of plane waves by a circular cylinder	190
8.7	Modified Bessel functions	192
8.8	Bessel functions with complex argument	194
8.9	Pipe flow through a vertical thermal gradient	195
	8.9.1 Formulation	195
	8.9.2 Solution for rising ambient temperature	199
8.10	Differential equations reducible to Bessel form	201
9	**Complex variables**	**210**
9.1	Complex numbers	210
9.2	Complex functions	212
9.3	Branch cuts and Riemann surfaces	215
9.4	Analytic functions	220
9.5	Plane seepage flows in porous media	223
9.6	Plane flow of a perfect fluid	225

9.7	Simple irrotational flows	227
9.8	Cauchy's theorem	229
9.9	Cauchy's integral formula and inequality	235
9.10	Liouville's theorem	237
9.11	Singularities	238
9.12	Evaluation of integrals by Cauchy's theorems	239
9.13	Jordan's lemma	246
9.14	Forced harmonic waves and the radiation condition	247
9.15	Taylor and Laurent series	250
9.16	More on contour integration	254
10	**Laplace transform and initial value problems**	**260**
10.1	The Laplace transform	260
10.2	Derivatives and the convolution theorem	263
10.3	Coupled pendula	265
10.4	One-dimensional diffusion in a strip	267
10.5	A string-oscillator system	269
10.6	Diffusion by sudden heating at the boundary	272
10.7	Sound diffraction near a shadow edge	275
10.8	*Temperature in a layer of accumulating snow	280
11	**Conformal mapping and hydrodynamics**	**289**
11.1	What is conformal mapping?	289
11.2	Relevance to plane potential flows	290
11.3	Schwarz–Christoffel transformation	291
11.4	An infinite channel	295
	11.4.1 Mapping onto a half plane	295
	11.4.2 Source in an infinite channel	296
11.5	A semi-infinite channel	298
11.6	An estuary	301
11.7	Seepage flow under an impervious dam	302
11.8	Water table above an underground line source	307
12	**Riemann–Hilbert problems in hydrodynamics and elasticity**	**318**
12.1	Riemann–Hilbert problem and Plemelj's formulas	318
12.2	Solution to the Riemann–Hilbert problem	320
12.3	Linearized theory of cavity flow	322
12.4	Schwarz's principle of reflection	327
12.5	*Complex formulation of plane elasticity	330
	12.5.1 Airy's stress function	330
	12.5.2 Stress components	331

	12.5.3 Displacement components	332
	12.5.4 Half-plane problems	333
12.6	*A strip footing on the ground surface	335
	12.6.1 General solution to the boundary-value problem	335
	12.6.2 Vertically pressed flat footing	336
13	**Perturbation methods – the art of approximation**	**343**
13.1	Introduction	343
13.2	Algebraic equations	345
	13.2.1 Regular perturbations	345
	13.2.2 Singular perturbations	347
13.3	Parallel flow with heat dissipation	349
13.4	Freezing of water surface	352
13.5	Method of multiple scales for an oscillator	358
	13.5.1 Weakly damped harmonic oscillator	358
	13.5.2 Elastic spring with weak nonlinearity	363
13.6	Theory of homogenization	367
	13.6.1 Differential equation with periodic coefficient	367
	13.6.2 *Darcy's law in seepage flow	370
13.7	*Envelope of a propagating wave	376
13.8	Boundary-layer technique	381
13.9	Seepage flow in an aquifer with slowly varying depth	384
13.10	Water table near a cracked sheet pile	390
13.11	*Vibration of a soil layer	395
	13.11.1 Formulation	395
	13.11.2 The outer solution	397
	13.11.3 The boundary-layer correction	399
14	**Computer algebra for perturbation analysis**	**408**
14.1	Getting started	408
14.2	Algebraic and trigonometric operations	409
	14.2.1 Elementary operations	409
	14.2.2 Functions	411
	14.2.3 Algebraic reductions	412
	14.2.4 Trigonometric reductions	413
	14.2.5 Substitutions and manipulations	414
14.3	Exact and perturbation methods for algebraic equations	417
14.4	Calculus	421
	14.4.1 Differentiation	421
	14.4.2 Integration	423

14.5	Ordinary differential equations	426
14.6	Pipe flow in a vertical thermal gradient	427
14.7	Duffing problem by multiple scales	435
14.8	Evolution of wave envelope on a nonlinear string	441
Appendices		447
Bibliography		453
Index		457

Preface

This book originated from a one-semester course on introductory engineering mathematics taught at MIT over the past ten years primarily to first-year graduate students in engineering. While all students in my class have gone through standard calculus and ordinary differential equations in their undergraduate years, many still feel more awe than confidence and enthusiasm toward applied mathematics. Upon entering graduate school they need a quick and friendly exposure to the elementary techniques of partial differential equations for studying other advanced subjects and the existing literature, and for analyzing original problems. For them a popular first step is to take a course in advanced calculus, which is usually taught to large classes. To cater to a large audience with diverse backgrounds, an author or instructor tends to concentrate on mathematical principles and techniques. Applications to physics and engineering are often kept at an elementary level so that little effort is needed to set up the examples before, or interpret them after, finding the solutions. In some branches of engineering, students get further exposure to and practice in theoretical analysis in many other courses in their own fields. However, in other branches such reinforcements are less emphasized; all too often practical problems are dealt with by tentative arguments undeservingly called the *Engineering Approach*.

In engineering endeavors rooted in physical sciences, deep understanding and precise analysis cannot usually be achieved without the help of mathematics. In this book I attempt to emphasize the art of applying some of the most basic techniques of applied mathematics in the three essential phases of engineering research: formulation of the problem, solution of the problem, and analysis of the solution for its physical meaning. There are several classic books that treat all these aspects of applications in an emphatic manner. *Mathematical Methods in Engineering* by Th. v. Kármán and M.A. Biot (1944) is certainly a pioneer of this kind; it discusses a certain class of engineering problems quite thoroughly before mathematical techniques are introduced to solve them. The same spirit has been admirably extended in *Mathematics Applied to Deterministic Problems in the Natural Sciences* by C.C. Lin and L.A. Segal (1988). On a more advanced level, the celebrated two-volume treatise, *Methods for Theoretical Physics* by P.M. Morse and H. Feshbach (1953), is another; it is comprehensive in scope and depth and contains a vast number of detailed analyses of nontrivial examples, most of which are of great relevance to engineering. In the past few decades new applications as well as new analytical techniques have evolved; however, the overwhelming majority of texts on the level of this book have, in my view, been written with greater emphasis on the mathematical techniques; engineering applications do not receive a large enough share of the spotlight. In order that fewer students will repeat my own earlier frustrations in learning how to use mathematics, this book is intended to foster practical skills for examining problems quantitatively and qualitatively, and, in the long run, for carrying out numerical tasks wisely.

Guided by the philosophy stated above, I have tried in most cases to motivate first the need for mathematical topics, by introducing physical examples. The mathematics is then presented in an informal manner with a view to putting even the most reluctant student at ease. Physical examples are selected primarily from applied mechanics, a field central to many branches of engineering and applied science. While the majority of examples are designed for classroom discussions requiring no more than two lectures per example, a few lengthier ones are also included, with a view to illustrating how to juxtapose skills introduced in different parts of this book. The complexity, and the juxtaposition, are also meant to give the students a glimpse of what awaits them outside the walls of a lecture hall. These longer sections, marked by asterisks, are more suitable for assigned reading than for lectures; they can be used as reference materials or, in the language of business schools, as *case studies*. To deal with many problems that cannot be solved exactly, a quick survey of perturbation methods, which are often treated in a more advanced course, is included here. I believe that the art of making approximations should be learned as early as possible. Finally, a chapter on symbolic computation is introduced as a tool to increase the power of perturbation analysis by transferring the inevitable tedium to the computer. This chapter has occasionally been included in my own lectures but can be used for self-study. While there are still important omissions, enough material is here for a two-semester course with three hours of lectures per week. The exercises are not large in number, but many can be reasonably demanding. At MIT I typically cover two-thirds of this book in one semester with four hours of lectures per week. Except for Chapters 11 and 12, the book can also be used by undergraduate seniors in various engineering disciplines related to mechanics, and in geophysical sciences.

Finally, I hope this book will entice more theoretical engineers to engage in the teaching of applied mathematics. To them mathematics is not an end in itself, but a tool to fulfill the larger mission of solving practical problems. From them a student can learn to sort out essential ingredients in formulating a new problem, select effective mathematical ammunition, guess the outcome before solving the problem, and extract physical implications of the solution incisively. In short, how to get the most with the least – the way the *Engineering Approach* ought to be!

Acknowledgments

Since most of the mathematical substance discussed in this text is classical, I have made extensive use of existing books on applied mathematics. In particular, the texts by Kármán and Biot, Morse and Feshbach, and Koshlyakov, Smirnov and Gliner have influenced my own style and choice of subjects. I am greatly indebted to Professor Theodore Y.-T. Wu and Dr. Arthur E. Mynett for materials on Riemann–Hilbert problems. The chapter on Computer Algebra is based on the joint contribution by Dr. Mamoun Naciri, Professor Ko-Fei Liu, and Professor Tetsu Hara. I am fortunate to have received generous help from Professor Pin Tong and Professor Hung Cheng, whose critical comments improved the accuracy of many parts of this book. All the drawings were produced with the computer artistry of Dott. Paolo Sammarco and Dott. Carlo Procaccini. Mrs. Karen Blair-Joss typeset much of the first draft.

Once again, my wife, Caroline, took part in the demanding task of editing the text, in addition to helping with the typesetting. Her insistence on directness of expression helped clarify time and again what I wished to convey. I also thank the editors of Cambridge University Press and Rosenlaui Publishing Services for their meticulous attention to detail.

Part of the writing was done during my visit to the Institute of Applied Mechanics, National Taiwan University, in 1993. The hospitality of Professor Yih-Hsing Pao and his colleagues at this young and dynamic institute is as unforgettable as my student days on the same campus, long ago.

1
Formulation of physical problems

For an engineer or a physical scientist, the first necessary skill in doing theoretical analysis is to describe a problem in mathematical terms. To begin with, one must make use of the basic laws that govern the elements of the problem. In continuum mechanics, these are the conservation laws for mass, momentum (Newton's second law), energy, etc. In addition, empirical constitutive laws are often needed to relate certain unknown variables; examples are Hooke's law between stress and strain, Fourier's law between heat flux and temperature, and Darcy's law between seepage velocity and pore pressure.

To derive the conservation law one may consider an infinitesimal element (a line segment, area, or volume element), yielding a differential equation directly. Alternatively, one may consider a control volume (or area, or line segment) of arbitrary size in the medium of interest. The law is first obtained in integral form; a differential equation is then derived by using the arbitrariness of the control volume. The two approaches are completely equivalent.

Let us first demonstrate the differential approach.

1.1 Transverse vibration of a taut string

Referring to Figure 1.1, we consider a taut string stretched between two fixed points at $x = 0$ and $x = L$ and displaced laterally by a distribution of external force. Conservation of transverse momentum requires that the total lateral force on the string element be balanced by its inertia. Let the lateral displacement be $V(x, t)$ and consider a differential element between x and $x + dx$. The net transverse force due to the difference of

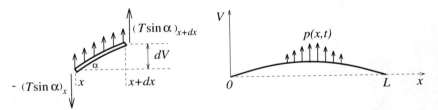

Fig. 1.1. Deformation of a taut string.

tension at both ends of the element is

$$(T \sin \alpha)_{x+dx} - (T \sin \alpha)_x,$$

where T denotes the local tension in the string and

$$\sin \alpha = \frac{dV}{\sqrt{dx^2 + dV^2}} = \frac{\frac{\partial V}{\partial x}}{\sqrt{1 + (\frac{\partial V}{\partial x})^2}}.$$

We shall assume the lateral displacement to be small everywhere so that the slope is also small: $\frac{\partial V}{\partial x} \ll 1$. The local value of $\sin \alpha$ can then be approximated by

$$\frac{\partial V}{\partial x} + O\left(\frac{\partial V}{\partial x}\right)^3,$$

where the expression $O(\delta)$ stands for *of the order of* δ. For any smooth function f, Taylor expansion gives

$$f(x + dx) - f(x) = \left(\frac{\partial f}{\partial x}\right) dx + O(dx)^2,$$

where the derivative is evaluated at x. Hence the net vertical force is

$$\frac{\partial}{\partial x}\left(T \frac{\partial V}{\partial x}\right) dx + O(dx)^2.$$

For infinitesimal stretching the string tension is proportional to the strain, according to Hooke's law. The initial strain is $\Delta L/L$ so that the initial tension is

$$T = ES\frac{\Delta L}{L},$$

where E denotes Young's modulus of elasticity and S denotes the cross-sectional area of the string. For simplicity, both E and S, hence T, will be assumed to be uniform in x. With lateral displacement, the strain must be changed. Consider the part of the string extending from 0 to x. The length $\ell(x,t)$ of this deformed part is

$$\ell(x,t) = \int_0^x dx \left[1 + \left(\frac{\partial V}{\partial x}\right)^2\right]^{1/2} = x\left[1 + O\left(\frac{\partial V}{\partial x}\right)^2\right],$$

hence the corresponding strain is

$$\frac{\ell - x}{x} = O\left(\frac{\partial V}{\partial x}\right)^2 \quad \text{for all} \quad 0 < x < L,$$

which is of second-order smallness. As long as $(\ell-x)/x = O[(\ell-L)/L] \ll \Delta L/L$, the tension is essentially unchanged, i.e., T is indistinguishable from its initial constant value. Thus the net vertical force on the string element is well represented by

$$T \frac{\partial^2 V}{\partial x^2} dx.$$

If the mass per unit length of the string is ρ, the inertia of the element is $\rho(\partial^2 V/\partial t^2)dx$. Let the applied load per unit length be $p(x,t)$. Balancing forces and inertia we get

$$\rho dx \frac{\partial^2 V}{\partial t^2} = T\frac{\partial^2 V}{\partial x^2} dx + p dx + O(dx)^2.$$

Eliminating dx and taking the limit of $dx \to 0$, we get

$$\frac{\rho}{T}\frac{\partial^2 V}{\partial t^2} - \frac{\partial^2 V}{\partial x^2} = \frac{p}{T}. \tag{1.1.1}$$

This equation, called the *wave equation*, is a partial differential equation of the second order. It is linear in the unknown V and inhomogeneous because of the forcing term on the right-hand side.

In (1.1.1) the highest derivatives with respect to x and t are of second order, hence we need two boundary conditions, one at $x = 0$ and one at $x = L$, and two initial conditions at $t = 0$. For example, if the ends of the string are fixed, then

$$V(0,t) = V(L,0) = 0. \tag{1.1.2}$$

If the displacement and velocity are known at $t = 0$, then

$$V(x,0) = f(x), \qquad \frac{\partial V}{\partial t}(x,0) = g(x). \tag{1.1.3}$$

Equations (1.1.1–1.1.3) constitute the initial-boundary-value problem for $V(x,t)$ and complete the formulation.

Is the longitudinal displacement U along the x direction important in this problem? Conservation of momentum in the x direction requires that

$$\rho dx \frac{\partial^2 U}{\partial t^2} = (T \cos \alpha)_{x+dx} - (T \cos \alpha)_x .$$

Since

$$\cos \alpha = \frac{dx}{\sqrt{(dx)^2 + (dV)^2}} = \frac{1}{\sqrt{1 + \left(\frac{\partial V}{\partial x}\right)^2}} \cong 1 + O\left(\frac{\partial V}{\partial x}\right)^2 ,$$

the acceleration is of second-order smallness

$$\frac{\rho}{T} \frac{\partial^2 U}{\partial t^2} = O\left(\frac{\partial}{\partial x}\left(\frac{\partial V}{\partial x}\right)^2\right) = O\left(\left(\frac{\partial V}{\partial x}\right) \frac{\rho}{T} \frac{\partial^2 V}{\partial t^2}\right).$$

Thus the longitudinal motion is negligible in comparison.

1.2 Longitudinal vibration of an elastic rod

Consider an elastic rod with the cross-sectional area $S(x)$ and Young's modulus E, as shown in Figure 1.2.

Let the longitudinal displacement from equilibrium be $U(x,t)$. The strain at station x is

$$\lim_{\Delta x \to 0} \frac{\Delta U}{\Delta x} = \frac{\partial U}{\partial x}.$$

By Hooke's law, the tension at x is

$$ES \frac{\partial U}{\partial x}.$$

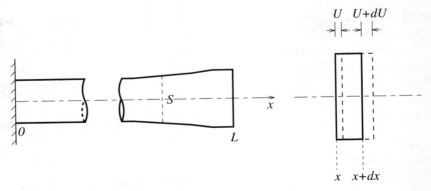

Fig. 1.2. Longitudinal deformation of an elastic rod.

1.2 Longitudinal vibration of an elastic rod

Now the net tension on a rod element from x to $x + dx$ is

$$\left(ES\frac{\partial U}{\partial x}\right)_{x+dx} - \left(ES\frac{\partial U}{\partial x}\right)_x = dx\frac{\partial}{\partial x}\left(ES\frac{\partial U}{\partial x}\right) + O\left(dx\right)^2.$$

Let the externally applied longitudinal force be $F(x,t)$ per unit length. Momentum conservation requires that

$$\rho S \frac{\partial^2 U}{\partial t^2} dx = \frac{\partial}{\partial x}\left(ES\frac{\partial U}{\partial x}\right)dx + F dx + O(dx)^2.$$

In the limit of vanishing dx, we get the differential equation

$$\rho S \frac{\partial^2 U}{\partial t^2} = \frac{\partial}{\partial x}\left(ES\frac{\partial U}{\partial x}\right) + F. \tag{1.2.1}$$

In the special case of uniform cross section, $S = $ constant, and U satisfies the inhomogeneous wave equation

$$\frac{1}{c^2}\frac{\partial^2 U}{\partial t^2} = \frac{\partial^2 U}{\partial x^2} + \frac{F}{ES}, \tag{1.2.2}$$

where $c = \sqrt{E/\rho}$ has the dimension of velocity.

The simplest boundary conditions are for fixed or free ends. If both ends are fixed, then

$$U(0,t) = 0 \quad \text{and} \quad U(L,t) = 0. \tag{1.2.3}$$

If the left end is fixed but the right end is free, then we have instead

$$U(0,t) = 0 \quad \text{and} \quad \frac{\partial U}{\partial x}(L,0) = 0, \tag{1.2.4}$$

since the stress is proportional to the strain. Again, the most natural initial conditions are

$$U(x,0) = f(x) \quad \text{and} \quad \frac{\partial U}{\partial t}(x,0) = g(x), \tag{1.2.5}$$

where f and g are prescribed functions of x for $0 < x < L$.

There are practical situations where the boundary conditions are not so simple. For example, consider a heavy mass M of small size attached to the end $x = L$, which is otherwise free, as shown in Figure 1.3. Momentum conservation of the mass M requires that

$$M\frac{\partial^2 U}{\partial t^2} + ES\frac{\partial U}{\partial x} = 0, \quad x = L. \tag{1.2.6}$$

This equation serves as a boundary condition for the rod, now involving both $\partial^2 U/\partial t^2$ and $\partial U/\partial x$ at the end.

Let us change to the integral approach in the next example.

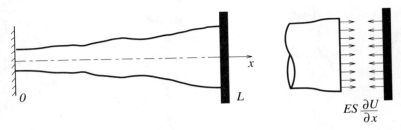

Fig. 1.3. A heavy mass at the end of a rod.

1.3 Traffic flow on a freeway

One of the mathematical models of traffic flow is the hydrodynamical theory of Lighthill and Whitham (1955). It is a simple theory capable of describing many real-life features of highway traffic with remarkable accuracy. Consider any section of a straight freeway from $x = a$ to $x = b$; see Figure 1.4. Assume for simplicity that there are no exits or entrances, and all vehicles are on the move. Let the density of cars (number of cars per unit length of highway) at x and t be $\rho(x,t)$, and the flux of cars (number of cars crossing the point x per unit time) be $q(x,t)$. By requiring that the number of cars within an arbitrary section from a to b be conserved, we have

$$-\frac{\partial}{\partial t}\int_a^b \rho(x,t)\,dx = q(b,t) - q(a,t).$$

Rewriting the right-hand side

$$q(b,t) - q(a,t) = \int_a^b \frac{\partial q}{\partial x}\,dx,$$

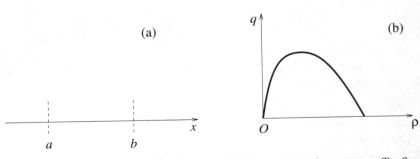

Fig. 1.4. (a) A section of the freeway. (b) The relation between traffic flux rate and traffic density.

we get

$$\int_a^b \left(\frac{\partial \rho}{\partial t} + \frac{\partial q}{\partial x} \right) dx = 0. \tag{1.3.1}$$

Since the control interval (a, b) is arbitrary, the integrand must vanish,

$$\frac{\partial \rho}{\partial t} + \frac{\partial q}{\partial x} = 0. \tag{1.3.2}$$

This result can be argued by contradiction, which is a typical reasoning needed to change an integral law to a differential law. Suppose the integrand is positive somewhere within (a, b), say in the range $(a', b') \in (a, b)$, and zero elsewhere in (a, b), then the integral in (1.3.1) must be positive. But this is a contradiction. The assumption that the integrand is positive somewhere is therefore wrong. By a similar argument, the integrand cannot be negative anywhere and hence must be zero everywhere in (a, b).

Equation (1.3.2) is the canonical differential form of all conservation laws. Having two unknowns q and ρ, a constitutive relation between ρ and q is needed and must be found by field measurements. Heuristically, q must be zero when there is no car on the road and zero again when the density attains a maximum (bumper-to-bumper traffic), hence the relation between q and ρ must be nonlinear

$$q = q(\rho), \tag{1.3.3}$$

as sketched in Figure 1.4b. With this relation, (1.3.2) becomes

$$\frac{\partial \rho}{\partial t} + \left(\frac{dq}{d\rho} \right) \frac{\partial \rho}{\partial x} = 0. \tag{1.3.4}$$

This result is a first-order nonlinear partial differential equation. As a sample initial-boundary-value problem, we may specify $\rho(x, 0) = f(x)$, where $f(x)$ vanishes outside a finite range of x, and $\rho(x, t) \to 0$ as $x \to \pm\infty$.

1.4 Seepage flow through a porous medium

Soil is a porous medium consisting of densely packed grains with fluid filling the interstitial pores. Bypassing the complicated details on the granular scale, one is usually interested only in averages over volumes or areas much larger than the size of the grains but much smaller than the scales typical of the velocity or pressure variations. Let us examine a steady seepage flow and define the seepage velocity $\mathbf{q}(x, y, z)$ to be the averaged local rate of fluid volume flowing through a unit area of the soil interior. If there is no source or sink the fluid mass in an arbitrary volume $\mathcal{V}$ must be conserved. For an incompressible fluid the net outflux through the bounding surface $\mathcal{S}$ must therefore vanish,

$$\iint_{\mathcal{S}} \mathbf{q} \cdot \mathbf{n}\, d\mathcal{S} = 0.$$

By Gauss' theorem, which is reviewed in Appendix A, the surface integral can be changed to a volume integral so that

$$\iiint_{\mathcal{V}} \nabla \cdot \mathbf{q}\, d\mathcal{V} = 0. \tag{1.4.1}$$

Now since the volume $\mathcal{V}$ is arbitrary, the integrand must be zero everywhere in the medium, thus

$$\nabla \cdot \mathbf{q} = 0. \tag{1.4.2}$$

For an isotropic and homogeneous soil there is an empirical law of momentum conservation relating $\mathbf{q}$ to the pore pressure p,

$$\mathbf{q} = -k \nabla \left(\frac{p}{\rho g} + y \right), \tag{1.4.3}$$

where ρ is the density of the pore fluid and g the gravitational acceleration. The empirical coefficient k is called the hydraulic conductivity, which is a constant if the porous medium is uniform and isotropic. Equation (1.4.3) is called *Darcy's law*, whose theoretical basis can be derived by selecting models on the granular scale, as will be described in a later chapter. It is convenient to define the velocity potential ϕ by

$$\phi = -k \left(\frac{p}{\rho g} + y \right) \tag{1.4.4}$$

such that Darcy's law becomes

$$\mathbf{q} = \nabla \phi. \tag{1.4.5}$$

Using this law in (1.4.2), we get

$$\nabla^2 \phi = \frac{\partial^2 \phi}{\partial x^2} + \frac{\partial^2 \phi}{\partial y^2} + \frac{\partial^2 \phi}{\partial z^2} = 0, \tag{1.4.6}$$

which is called the *Laplace equation*.

There can be a large variety of boundary conditions on the macro scale. For example, if the soil/water interface is immersed in water at the depth $y = -h(x,z)$ measured from the water surface, the pressure along the interface is

$$p = p_o + \rho g h(x,z), \qquad y = -h(x,z),$$

where p_o is the atmospheric pressure. In terms of the velocity potential the boundary condition is

$$\phi = -\frac{kp_o}{\rho g}, \qquad y = -h(x,z). \tag{1.4.7}$$

If the soil rests on an impervious rock, the normal velocity must vanish so that

$$\frac{\partial \phi}{\partial n} = 0. \tag{1.4.8}$$

Equation (1.4.7), which prescribes the unknown function itself, is called a *Dirichlet condition*, and (1.4.8), which prescribes the normal derivative of the unknown, is called a *Neuman condition*. In some seepage problems, however, Dirichlet condition is specified along a part of the soil boundary, while Neuman condition is specified on the remaining part. Then the boundary condition is of the mixed type, and the problem is not easy.

Still more challenging is the class of free-surface problems in which there is a water table (phreatic surface) whose location is unknown *a priori*. Two boundary conditions are needed. One is the kinematic condition that the flow must be tangential to the water table. The other is the dynamical condition that the pore pressure be prescribed (zero, if capillarity is ignored). Figure 1.5 depicts the two-dimensional cross section of a long earth dam resting on a horizontal and impermeable river bed. On the left side is a reservoir of water depth H; on the right side the river is dry. On the bottom $BCDE$ the flow must be tangential to the river bed so that

$$\frac{\partial \phi}{\partial n} = 0. \tag{1.4.9}$$

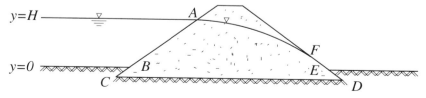

Fig. 1.5. Seepage flow through an earth dam.

On the submerged sloping surface AB, the pressure is known

$$p = p_o + \rho g(H - y), \qquad y = F(x),$$

or

$$\phi = -k\left(\frac{p}{\rho g} + y\right) = -k\left(\frac{p_o}{\rho g} + H\right) = \text{constant}, \qquad y = F(x). \tag{1.4.10}$$

On the phreatic surface $y = \eta(x)$, the kinematic condition is

$$\frac{\partial \phi}{\partial n} = 0, \qquad y = \eta(x), \tag{1.4.11}$$

while the dynamic condition is $p = p_o$, or

$$\phi + k\eta = -k\left(\frac{p_o}{\rho g}\right) = \text{constant}, \qquad y = \eta(x). \tag{1.4.12}$$

Finally, on the exposed sloping surface EF at the known position $y = G(x)$, $p = p_o$ so that

$$\phi = -k\left[\frac{p_o}{\rho g} + G(x)\right]. \tag{1.4.13}$$

The set of boundary conditions (1.4.9–1.4.13) is not only mixed but also nonlinear because the position of the water table is a part of the unknown solution. This type of problem is called a *free-boundary* problem.

1.5 Diffusion in a stationary medium

Diffusion is important in many different physical contexts. Dye can diffuse in water, dust in air, heat in solids, and pollutants in the atmosphere, lakes, rivers and oceans, etc. Whatever the cause, the process is often governed by one special type of partial differential equation.

Consider the temperature $T(x, y, z, t)$ in a homogeneous solid. By energy conservation, the rate of increase of internal energy must be equal to the sum of the rate of energy influx through the bounding surface and the rate of internal energy production.

The energy outflux through a closed geometrical surface S in the solid is

$$\iint_S \mathbf{q} \cdot \mathbf{n}\, dS.$$

Using Gauss' theorem, we can rewrite the total outflux of energy as

$$\iint_S \mathbf{q} \cdot \mathbf{n}\, dS = \iiint_V \nabla \cdot \mathbf{q}\, dV.$$

1.5 Diffusion in a stationary medium

Now the amount of heat energy per unit volume is $\rho c T$, where ρ denotes the solid density and c the specific heat. The rate of increase of heat in $\mathcal{V}$ is therefore

$$\frac{\partial}{\partial t}\iiint_{\mathcal{V}} \rho c T \, d\mathcal{V} = \iiint_{\mathcal{V}} \frac{\partial(\rho c T)}{\partial t} \, d\mathcal{V}.$$

If the internal heat source releases energy at the rate of $f(x, y, z, t)$ per unit volume per unit time, then the energy input rate is

$$\iiint_{\mathcal{V}} f \, d\mathcal{V}.$$

Now let us invoke the law of conservation of energy

$$\iiint_{\mathcal{V}} \frac{\partial(\rho c T)}{\partial t} \, d\mathcal{V} = -\iiint_{\mathcal{V}} \nabla \cdot \mathbf{q} \, d\mathcal{V} + \iiint_{\mathcal{V}} f \, d\mathcal{V}. \qquad (1.5.1)$$

Again, since $\mathcal{V}$ is arbitrary, we must have

$$\frac{\partial(\rho c T)}{\partial t} = -\nabla \cdot \mathbf{q} + f. \qquad (1.5.2)$$

This equation is the differential form of the energy conservation law. Similar to Darcy's law of seepage flow, one must add the empirical Fourier's law that the rate of energy flux through a unit area is proportional to the temperature gradient

$$\mathbf{q} = -K\nabla T, \qquad (1.5.3)$$

where K is the heat conductivity of the material. The sign is negative because heat flows from high to low temperature. Inserting (1.5.3) into (1.5.2), we get finally an inhomogeneous *diffusion equation* (or *conduction equation*)

$$\frac{\partial(\rho c T)}{\partial t} = \nabla \cdot K\nabla T + f. \qquad (1.5.4)$$

If the medium is uniform, K, ρ and c are constants; (1.5.4) reduces to

$$\frac{\partial T}{\partial t} = \kappa \nabla^2 T + \frac{f}{\rho c}, \qquad (1.5.5)$$

where $\kappa = K/\rho c$ is called the thermal diffusivity.

Now let S and V denote the physical boundary and volume of the solid. If the temperature on S is prescribed, then the boundary condition is

$$T = T_S(x, y, z, t), \qquad \mathbf{x} \in S. \qquad (1.5.6)$$

If, on the other hand, the heat flux rate is known, then

$$\mathbf{q} \cdot \mathbf{n} = -K \frac{\partial T}{\partial n} = Q(x, y, z, t), \qquad \mathbf{x} \in S. \qquad (1.5.7)$$

Since there is only one time derivative, only one initial condition is needed:

$$T(x, y, z, 0) = T_o(x, y, z), \qquad \mathbf{x} \in V. \qquad (1.5.8)$$

In most of the examples studied so far the final governing equation involves only one unknown. Now we will examine a problem with several unknowns.

1.6 Shallow water waves and linearization

1.6.1 Nonlinear governing equations

If water in a lake or along the sea coast is disturbed, waves can be created on the surface, due to the restoring force of gravity. Consider the basic laws governing the motion of long waves in shallow water of constant density and negligible viscosity. Referring to Figure 1.6, we let the z axis be directed vertically upward and the x, y plane lie in the initially calm water surface, $h(x, y)$ denote the depth below the still sea level, and $\zeta(x, y, t)$ denote the vertical displacement of the free surface. Take the differential approach again and consider the fluid flow through a vertical column with the base $dxdy$.

First, the law of mass conservation: The rate of volume increase in the column

$$\frac{\partial \zeta}{\partial t} dx dy$$

must be balanced by the net volume flux into the column from all four vertical sides. In shallow water, the horizontal length scale, characterized

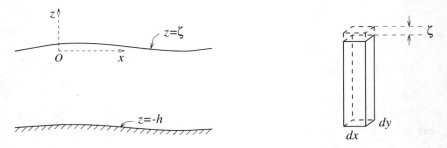

Fig. 1.6. A column element of fluid in a shallow sea.

1.6 Shallow water waves and linearization

by the wavelength λ, is much greater than the vertical length h. Water flows mainly in the horizontal planes with the velocity $\mathbf{u}(x, y, t)$, which is essentially constant in depth. Through the vertical sides normal to the x axis, the difference between influx through the left and outflux through the right is

$$-\left[u\left(\zeta+h\right)|_{x+dx} - u\left(\zeta+h\right)|_x\right] dy = -\left\{\frac{\partial}{\partial x}\left[u\left(\zeta+h\right)\right] + O(dx)\right\} dxdy.$$

Similarly, through the vertical sides normal to the y axis, the difference between influx through the front and outflux through the back is

$$-\left[v\left(\zeta+h\right)|_{y+dy} - v\left(\zeta+h\right)|_y\right] dx = -\left\{\frac{\partial}{\partial y}\left[v\left(\zeta+h\right)\right] + O(dy)\right\} dydx.$$

Omitting terms of higher order in dx, dy, we invoke mass conservation to get

$$\frac{\partial \zeta}{\partial t} dxdy = -\left\{\frac{\partial}{\partial x}\left[u\left(\zeta+h\right)\right] + \frac{\partial}{\partial y}\left[v\left(\zeta+h\right)\right] + O(dx, dy)\right\} dxdy.$$

In the limit of vanishing dx, dy, we have, in vector form,

$$\frac{\partial \zeta}{\partial t} + \nabla \cdot [\mathbf{u}(\zeta + h)] = 0, \qquad (1.6.1)$$

where $\nabla = \left(\frac{\partial}{\partial x}, \frac{\partial}{\partial y}\right)$ is the horizontal gradient operator. This equation is nonlinear because of the quadratic product of the unknowns $\mathbf{u}$ and ζ.

Now the law of conservation of momentum: In shallow water the vertical momentum balance is dominated by pressure gradient and gravity, which means that the distribution of pressure is hydrostatic

$$p = \rho g \left(\zeta - z\right), \qquad (1.6.2)$$

where the atmospheric pressure on the free surface is ignored. Consider now the momentum balance in the x direction. The time rate of momentum change in the water column is

$$\left\{\frac{\partial}{\partial t}\left[\rho u(\zeta + h)\right]\right\} dxdy.$$

On the other hand, the net pressure force on two vertical sides normal to the x direction is

$$-dxdy \frac{\partial}{\partial x}\int_{-h}^{\zeta} p\, dz = -dxdy \frac{\partial}{\partial x}\int_{-h}^{\zeta} \rho g(\zeta - z)\, dz$$

$$= -\rho g(\zeta + h)\frac{\partial(\zeta + h)}{\partial x} dxdy.$$

The hydrodynamic reaction from the sloping bottom to the fluid is

$$p\frac{\partial h}{\partial x}\,dxdy = \rho g(\zeta+h)\frac{\partial h}{\partial x}\,dxdy.$$

The net influx of momentum through four vertical sides is

$$-\frac{\partial}{\partial x}[\rho u^2(\zeta+h)]dxdy - \frac{\partial}{\partial y}[\rho uv(\zeta+h)]dydx.$$

Equating the total rate of momentum change to the sum of the net pressure force on the sides and on the bottom, and the net momentum influx, we get

$$\frac{\partial}{\partial t}[\rho u(\zeta+h)] = -\frac{\partial}{\partial x}[\rho u^2(\zeta+h)] - \frac{\partial}{\partial y}[\rho uv(\zeta+h)]$$
$$-\rho g(\zeta+h)\frac{\partial(\zeta+h)}{\partial x} + \rho g(\zeta+h)\frac{\partial h}{\partial x}.$$

The influx terms can be put on the left-hand side to give

$$\rho\left(\frac{\partial u}{\partial t} + u\frac{\partial u}{\partial x} + v\frac{\partial u}{\partial y}\right)(\zeta+h)$$
$$+\rho u\left\{\frac{\partial \zeta}{\partial t} + \frac{\partial}{\partial x}[u(\zeta+h)] + \frac{\partial}{\partial y}[v(\zeta+h)]\right\},$$

where the second line above vanishes by virtue of continuity (1.6.1). Hence the x momentum equation reduces to

$$\frac{\partial u}{\partial t} + u\frac{\partial u}{\partial x} + v\frac{\partial u}{\partial y} = -g\frac{\partial \zeta}{\partial x}. \tag{1.6.3}$$

Similarly, momentum balance in the y direction requires

$$\frac{\partial v}{\partial t} + u\frac{\partial v}{\partial x} + v\frac{\partial v}{\partial y} = -g\frac{\partial \zeta}{\partial y}. \tag{1.6.4}$$

These two equations can be summarized in the vector form

$$\frac{\partial \mathbf{u}}{\partial t} + \mathbf{u}\cdot\nabla\mathbf{u} = -g\nabla\zeta. \tag{1.6.5}$$

Equations (1.6.1) and (1.6.5) are coupled nonlinear partial differential equations for three scalar unknowns $\mathbf{u}$ and ζ.

Now the boundary and initial conditions: on a shoreline S, there can be no normal flux, therefore,

$$h\mathbf{u}\cdot\mathbf{n} = 0 \quad \text{on} \quad S, \tag{1.6.6}$$

where **n** denotes the unit normal vector pointing horizontally into the shore. This condition is applicable not only along a cliff shore where h is finite, but also on a shoreline where $h = 0$, as long as the waves are gentle enough not to break. In the latter case the whereabout of the shoreline is unknown *a priori* and must be found as a part of the solution.

At the initial instant, one may assume that the displacement $\zeta(x, y, 0)$ and the vertical velocity of the entire free surface $\frac{\partial}{\partial t}\zeta(x, y, 0)$ are known. These conditions complete the formulation of the nonlinear shallow water wave problem.

1.6.2 Linearization for small amplitude

For small amplitude waves

$$\frac{\zeta}{h} \sim \frac{A}{h} \ll 1, \tag{1.6.7}$$

where A is the characteristic amplitude, and (1.6.1) may be simplified by neglecting the quadratic term

$$\frac{\partial \zeta}{\partial t} + \nabla \cdot h\mathbf{u} = 0. \tag{1.6.8}$$

Denoting the time scale by the wave period T and the horizontal length scale by the wavelength λ, we equate the order of magnitudes of the remaining two terms above to get

$$\frac{A}{T} \sim \frac{uh}{\lambda}, \quad \text{implying} \quad \frac{A}{h} \sim \frac{uT}{\lambda} \ll 1.$$

Now let us estimate the importance of the quadratic term $\mathbf{u} \cdot \nabla \mathbf{u}$ in the momentum equation by assessing the ratio

$$\frac{\mathbf{u} \cdot \nabla \mathbf{u}}{\frac{\partial \mathbf{u}}{\partial t}} = O\left(\frac{uT}{\lambda}\right) \ll 1.$$

Clearly the quadratic term representing convective inertia can also be ignored in the first approximation, and the momentum equation becomes

$$\frac{\partial \mathbf{u}}{\partial t} = -g\nabla\zeta. \tag{1.6.9}$$

Both the continuity (1.6.8) and momentum (1.6.9) equations are now *linearized*.

In view of (1.6.9) the boundary condition on the shoreline (1.6.6) can be expressed, instead, as

$$h\frac{\partial \zeta}{\partial n} = 0 \quad \text{on} \quad S. \tag{1.6.10}$$

Consistent with the linearized approximation, the shoreline position can be prescribed *a priori*.

Equations (1.6.8) and (1.6.9) can be combined by the process of cross differentiation. First differentiate (1.6.8) with respect to t,

$$\frac{\partial}{\partial t}\left\{\frac{\partial \zeta}{\partial t} + \nabla \cdot (\mathbf{u}h)\right\} = 0,$$

then take the divergence of the product of (1.6.9) and h,

$$\nabla \cdot \left\{h\frac{\partial \mathbf{u}}{\partial t}\right\} = -\nabla(gh\nabla\zeta).$$

The difference of these two equations gives

$$\frac{\partial^2 \zeta}{\partial t^2} = \nabla \cdot (gh\nabla\zeta). \tag{1.6.11}$$

For a horizontal bottom $h = $ constant,

$$\frac{1}{c^2}\frac{\partial^2 \zeta}{\partial t^2} = \nabla^2 \zeta, \tag{1.6.12}$$

where $c = \sqrt{gh} = O(\lambda/T)$ is the characteristic velocity of infinitesimal wave motion. Equation (1.6.12) is the two-dimensional version of the wave equation. If, furthermore, all conditions are uniform in the y direction, $\partial/\partial y = 0$, (1.6.12) reduces to the familiar form

$$\frac{1}{c^2}\frac{\partial^2 \zeta}{\partial t^2} = \frac{\partial^2 \zeta}{\partial x^2}. \tag{1.6.13}$$

Other examples of formulation will be discussed or given as exercises throughout the book.

Exercises

1.1 A taut string is imbedded in an elastic medium so that any lateral displacement $u(x,t)$ of the string is resisted by a force proportional to the displacement. Derive the equation governing $u(x,t)$. Supply some physically reasonable boundary and initial conditions.

1.2 A membrane is kept taut over an area S. The membrane has a constant density ρ per unit area and is under uniform tension in all directions. Derive the equation for the lateral displacement $u(x,y,t)$ under distributed loading of $p(x,y,t)$ per unit area.

Exercises

1.3 A string stretched along the x axis is tied to a fixed wall at the end $x = 0$. At the other end $x = L$ it is tied to a massive block that can only slide in a frictionless slot perpendicular to x. Formulate the initial-boundary-value problems for the string/mass system.

1.4 The end of taut string at $x = L$ is attached to a mass M that can slide up and down in a vertical slot. The mass in turn is subject to an elastic force $-kV$ and a frictional force $-c\frac{dV}{dt}$, where V denotes the vertical displacement of the mass (hence the end of the string). Derive the boundary condition for the string at this end.

1.5 Consider the longitudinal vibration of a cylindrical rod with one end at $x = 0$ fixed and the other end at $x = L$ attached to a mass M. Before $t = 0$ the rod is compressed by the length ϵL with $\epsilon \ll 1$. At $t = 0$ the compression is released. State the governing equation and all boundary and initial conditions.

1.6 Consider a cylindrical rod of circular cross section being forced to perform torsional vibration. Let $\theta(x,t)$ = angular displacement of the cross section at x, $d\sigma$ = area element in the cross section and located at the distance r from the axis where $0 < r < a$, τ = shear stress, G = shear modulus of elasticity, and ϕ = angular displacement of a line originally parallel to the axis; see Figure 1.7.

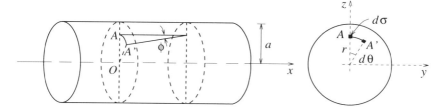

Fig. 1.7. Torsion of a circular cylinder.

Show that
$$\phi = r\frac{\partial \theta}{\partial x}.$$

Use Hooke's law $\tau = G\phi$ and show that the total torque applied to the cross section at x is
$$M = G\frac{\partial \theta}{\partial x}\iint_S r^2\, d\sigma = GJ\frac{\partial \theta}{\partial x},$$

where
$$J = \iint_S r^2 \, d\sigma$$
is the polar moment of inertia of the cross section.

Let I be the moment of inertia per unit length of the rod. Show that
$$\frac{\partial^2 \theta}{\partial t^2} = \frac{GJ}{I} \frac{\partial^2 \theta}{\partial x^2}.$$

What should be the boundary condition at a fastened end? At a free end? At an end where the total applied torque is M_o?

1.7 A heavy chain is hung at one end $x = L$. Tension is caused by its own weight. Show that for small oscillations the lateral displacement $u(x,t)$ satisfies
$$g\frac{\partial}{\partial x}\left(x\frac{\partial u}{\partial x}\right) = \frac{\partial^2 u}{\partial t^2}.$$

1.8 A cable under tension T is fixed at one end $x = 0$. The other end at $x = L$ is attached to a vertical spring of elastic constant k. What is the boundary condition at $x = L$?

1.9 To model sedimentation of suspended particles in a viscous fluid, consider the concentration $C(x,t)$ of identical particles falling at the velocity $w(x,t)$, where x points vertically downward. Deduce the law of mass conservation. For dilute concentration the falling velocity of the sediment cloud is related to the falling velocity w_o of one particle by
$$w = w_o(1 - \alpha C).$$

What is the resulting equation for C? Is it linear or nonlinear?

1.10 Consider a shallow aquifer (a porous stratum saturated with water) on an impermeable rock. Let $h(x,t)$ be the depth of the water table above the rock base, which coincides with the x axis. The horizontal seepage velocity u and the water table gradient can be described by Darcy's law
$$u = -k\frac{\partial h}{\partial x}.$$

Show first that mass conservation requires
$$\frac{\partial h}{\partial t} + \frac{\partial (uh)}{\partial x} = 0.$$

Then derive the governing equation for $h(x,t)$. If h departs from constant H only by a small amount, i.e., $h = H + h'$ with $h' \ll H$, find the linearized equation for h'.

1.11 During an earthquake, water in a reservoir exerts hydrodynamic pressure on a dam that may fail due to high internal stresses. Formulate the dam/reservoir problem under the following idealizations: The reservoir is infinitely long and is of uniform rectangular cross section. Water depth in the reservoir ($x < 0$) is constant h; there is no water on the other side of the dam ($x > 0$). Before $t = 0$, water is calm. After $t = 0$, the dam is forced to oscillate horizontally with the displacement of the vertical dam face specified as a function of time. The reservoir bottom on $z = 0$ is rigid and impermeable. The free surface is exposed to zero atmospheric pressure. Earthquake acceleration is high enough so that gravity can be neglected. Compressibility of water is important. Let the x axis coincide with the reservoir bottom and the y axis with the wetted dam surface. Let u, v be the fluid velocity components in the x, y directions, p the hydrodynamic pressure, and ρ the density departure from the mean ρ_o, assumed to be constant. The governing equations for infinitesimal motion are:

Mass conservation:
$$\frac{\partial \rho}{\partial t} + \rho_o \left(\frac{\partial u}{\partial x} + \frac{\partial v}{\partial y} \right) = 0.$$

Momentum conservation:
$$\rho_o \frac{\partial u}{\partial t} = -\frac{\partial p}{\partial x}, \qquad \rho_o \frac{\partial v}{\partial t} = -\frac{\partial p}{\partial y}.$$

Equation of state:
$$C^2 = \frac{dp}{d\rho},$$

where C is the sound speed in water, also assumed to be a constant.

Show first that p is governed by the linear wave equation
$$\frac{1}{C^2} \frac{\partial^2 p}{\partial t^2} = \nabla^2 p, \qquad x < 0, \quad 0 < z < h.$$

Write down all the initial and boundary conditions in terms of p alone.

2
Classification of equations with two independent variables

In the preceding chapter, we encountered a variety of partial differential equations that may be linear or nonlinear, and may consist of a single equation for one unknown or a system of equations for several unknowns. They may contain first-order, second-order or higher-order partial derivatives. For partial differential equations with two independent variables, an important mathematical property is the existence or nonexistence of the so-called *characteristic curves*. If these curves exist, the partial differential equations can be reduced to ordinary differential equations. The existence of these curves suggests not only the type of boundary or initial conditions that are appropriate, but also the principles of solution. We start from the simplest example.

2.1 A first-order equation

For one unknown with two independent variables, the following first-order partial differential equation

$$A\frac{\partial \rho}{\partial x} + B\frac{\partial \rho}{\partial y} = C \qquad (2.1.1)$$

frequently appears in practical problems. If A and B depend on x, y only and C is at most linear in ρ, this equation is linear; if A and B depend on ρ also, it is nonlinear.† The traffic flow equation of §1.3 is a special case of (2.1.1) if x here represents time and y represents space. Consider the x, y plane. At each point a unit vector $\mathbf{s}$ can be defined by

$$\mathbf{s} \equiv \left(\frac{A}{\sqrt{A^2 + B^2}}, \frac{B}{\sqrt{A^2 + B^2}}\right)$$

---

† It is customary to call a nonlinear partial differential equation quasilinear if it is nonlinear in the unknowns but linear in the derivatives of the unknowns.

so that (2.1.1) is written as

$$\mathbf{s} \cdot \nabla \rho = \frac{C}{\sqrt{A^2 + B^2}} \equiv D. \qquad (2.1.2)$$

Starting from any point on an initial curve I, one can draw a trajectory in the x, y plane so that every point along it is tangent to the local vector $\mathbf{s}$. The slope of the local tangent is

$$\frac{dy}{dx} = \frac{B}{A}, \qquad (2.1.3)$$

which is just the differential equation of the trajectory. As long as the coefficients A, B are real, the trajectories can always be found and are called the *characteristics*. Let σ denote the arc length along a characteristic. Since

$$\mathbf{s} = \left(\frac{dx}{d\sigma}, \frac{dy}{d\sigma} \right)$$

(2.1.2) may be written as an ordinary differential equation

$$\frac{d\rho}{d\sigma} = D. \qquad (2.1.4)$$

When $A(x, y), B(x, y)$ are known functions of the independent variables, one can integrate (2.1.3) first to find the characteristic curves and then (2.1.4) to get ρ. However, if A and B are also functions of $\rho(x, y)$ but not of the derivatives of ρ, the characteristics are not known *a priori*. Both (2.1.3) and (2.1.4) must be solved together to find the characteristics and the unknown ρ. In general this task can be carried out by a discrete numerical method. Let the initial value of ρ be prescribed along any curve I that is not one of the characteristics; see Figure 2.1. From (2.1.3) and the initial value of ρ at some chosen point on I, the next incremental point along the characteristic is found by a finite difference approximation. With two end points of a segment known, (2.1.4) or, equivalently, (2.1.2) can be integrated for ρ at the end of the segment. In this manner the solution can be continued along one characteristic. This process can be repeated by starting from all other points along I; all the characteristics and the solution along them are found.

It is important that I cannot coincide with any characteristic curve; otherwise one cannot integrate the differential equation to find ρ away from I, hence cannot solve for $\rho(x, y)$ in the rest of the x, y plane.

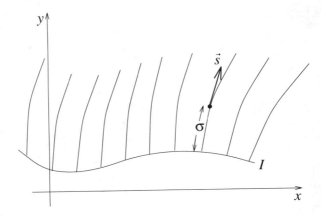

Fig. 2.1. Characteristics from an initial curve in the x, y plane.

2.2 System of first-order equations

Long waves in shallow water and gas dynamics (see, e.g., Liepman and Roshko, 1957) are two prominent examples governed by a system of first-order equations involving two or more unknowns. In the one-dimensional case the system can be cast in the following general form:

$$\mathcal{L}_{ij} u^j = a_{ij}\frac{\partial u^j}{\partial t} + b_{ij}\frac{\partial u^j}{\partial x} + c_i = 0 \qquad (2.2.1)$$

for n unknowns u^j with $j = 1, 2, \ldots, n$, where we have employed the Einstein convention that summation over repeated indices is implied without the summation symbol. The system is *linear* if the coefficients a_{ij}, b_{ij} and c_i depend only on t and x, and *quasilinear* if they also depend on u^j but not on the derivatives of u^j. Let us look for a linear combination of (2.2.1) so that it can be reduced to an ordinary differential equation with derivatives taken along a characteristic curve in the x, t plane. If the characteristic is parametrically represented by $t = t(\sigma)$, $x = x(\sigma)$, where σ denotes the distance along the characteristic, the linear combination should yield

$$\lambda_i \mathcal{L}_{ij} u^j = P_j \frac{du^j}{d\sigma} + R = 0. \qquad (2.2.2)$$

Let us multiply (2.2.1) by λ_i and sum over i

$$\lambda_i a_{ij}\frac{\partial u^j}{\partial t} + \lambda_i b_{ij}\frac{\partial u^j}{\partial x} + \lambda_i c_i = 0, \qquad (2.2.3)$$

2.2 System of first-order equations

which can be written in the form of (2.2.2) only if

$$\lambda_i a_{ij} = P_j t_\sigma \quad \text{and} \quad \lambda_i b_{ij} = P_j x_\sigma$$

for all j, with $x_\sigma = dx/d\sigma$ and $t_\sigma = dt/d\sigma$. Equivalently, we have

$$\lambda_i a_{ij} \frac{dx}{dt} = \lambda_i b_{ij}, \quad j = 1, 2, \ldots, n, \tag{2.2.4}$$

where

$$\frac{dx}{dt} = \frac{x_\sigma}{t_\sigma}.$$

Equation (2.2.4) is a homogeneous algebraic system for the coefficients λ_i. Nontrivial solutions can be found only if the following determinant vanishes

$$\left\| a_{ij} \frac{dx}{dt} - b_{ij} \right\| = 0, \tag{2.2.5}$$

which leads to an nth order polynomial equation for the characteristic slope dx/dt. If n real solutions ζ_m can be found, say

$$\frac{dx}{dt} = \frac{x_\sigma}{t_\sigma} = \zeta_m, \quad m = 1, 2, \ldots, n, \tag{2.2.6}$$

then there are n real characteristic directions, hence n characteristic curves passing through each point x, t. The quasilinear system is then called *hyperbolic*. Along each characteristic the unknowns are governed by the ordinary differential equation (2.2.2), which can be integrated together with (2.2.6), usually by numerical means.

If all ζ_m are real and distinct, the system is called *totally hyperbolic*. If some ζ_m are complex, the system is *elliptic*.

Let us apply the general criterion to one-dimensional shallow water waves on a horizontal bottom. Let $H = \zeta + h$ be the total depth; the governing equations are recalled from (1.6.1) and (1.6.5)

$$\frac{\partial H}{\partial t} + u \frac{\partial H}{\partial x} + H \frac{\partial u}{\partial x} = 0 \tag{2.2.7}$$

and

$$\frac{\partial u}{\partial t} + u \frac{\partial u}{\partial x} + g \frac{\partial H}{\partial x} = 0. \tag{2.2.8}$$

Let $u^1 = H$, $u^2 = u$, then

$$(a_{ij}) = \begin{pmatrix} 1 & 0 \\ 0 & 1 \end{pmatrix} \quad (b_{ij}) = \begin{pmatrix} u & H \\ g & u \end{pmatrix}.$$

Equation (2.2.5) gives

$$\begin{vmatrix} \frac{dx}{dt} - u & H \\ g & \frac{dx}{dt} - u \end{vmatrix} = 0,$$

hence

$$\frac{dx}{dt} = u \pm \sqrt{gH}. \tag{2.2.9}$$

Once the characteristic directions are formally found, the factors λ_i can be solved from (2.2.4), which in the present example reads

$$\pm\sqrt{gH}\lambda_1 - g\lambda_2 = 0$$
$$H\lambda_1 \pm \sqrt{gH}\lambda_2 = 0.$$

The solutions are

$$\lambda_1 = g$$

and

$$\lambda_2 = \pm\sqrt{gH}.$$

Equation (2.2.3) becomes

$$g\frac{\partial H}{\partial t} \pm \sqrt{gH}\frac{\partial u}{\partial t} + g(u \pm \sqrt{gH})\frac{\partial H}{\partial x} + (gH \pm u\sqrt{gH})\frac{\partial u}{\partial x} = 0 \tag{2.2.10}$$

along

$$\frac{dx}{dt} = \pm\sqrt{gH}.$$

With the change of symbols $C = \sqrt{gH}$, (2.2.10) can be written as

$$\left[\frac{\partial}{\partial t} + (u \pm C)\frac{\partial}{\partial x}\right](u \pm 2C) = 0.$$

Thus

$$\frac{d}{d\sigma}(u \pm 2C) = 0 \quad \text{along} \quad \frac{dx}{dt} = u \pm C. \tag{2.2.11}$$

Now the ordinary differential equations along the characteristics can be integrated trivially to yield the so-called Riemann invariants

$$u \pm 2C = \text{constant} \quad \text{along} \quad \frac{dx}{dt} = u \pm C. \tag{2.2.12}$$

Many interesting physical aspects of shallow water waves can be deduced from these Riemann invariants. For further information interested readers may consult Stoker (1957) and Mei (1989). Applications of the theory of characteristics to gas dynamics are fully described in Courant and Friedrichs (1948) and to several other fields in Whitham (1976).

2.3 Linear second-order equations
2.3.1 Constant coefficients

Let us recall the three linear equations deduced in the last chapter and restrict to two independent variables:
the wave equation

$$\frac{\partial^2 u}{\partial x^2} - \frac{1}{c^2}\frac{\partial^2 u}{\partial t^2} = 0, \qquad (2.3.1)$$

the diffusion equation

$$\frac{\partial^2 u}{\partial x^2} - \frac{1}{k}\frac{\partial u}{\partial t} = 0, \qquad (2.3.2)$$

and the Laplace equation

$$\frac{\partial^2 u}{\partial x^2} + \frac{\partial^2 u}{\partial y^2} = 0. \qquad (2.3.3)$$

Can characteristics be found so that these partial differential equations can be reduced to ordinary differential equations?

For the wave equation this is a simple matter. It is straightforward to show by the transformation

$$\xi = x - ct \quad \text{and} \quad \eta = x + ct \qquad (2.3.4)$$

and the chain rule of differentiation that (2.3.1) can be reduced to

$$u_{\xi\eta} = 0. \qquad (2.3.5)$$

The lines $\xi =$ constant and $\eta =$ constant are the characteristics. Equation (2.3.5) may be integrated with respect to ξ and η successively to give

$$u(\xi, \eta) = f(\xi) + g(\eta),$$

where f and g are yet to be determined by information along an initial curve, as will be discussed further in §3.1.

Let us examine the following general linear partial differential equation of second order

$$\mathcal{L}u = Au_{xx} + 2Bu_{xy} + Cu_{yy} + Du_x + Eu_y + Fu + G = 0, \qquad (2.3.6)$$

where partial derivatives are abbreviated by subscripts $u_x = \partial u/\partial x$, $u_{xx} = \partial^2 u/\partial x^2$, etc., and A, B, C, D, E are constants. Clearly, all three equations (2.3.1–2.3.3) are special cases of (2.3.6). Consider the transformation from x, y to ξ, η, where

$$\xi = y - \alpha_1 x \quad \text{and} \quad \eta = y - \alpha_2 y, \qquad (2.3.7)$$

which are straight lines with slopes α_1 and α_2. According to the chain rule of differentiation, the derivatives are transformed as follows:

$$u_x = u_\xi \xi_x + u_\eta \eta_x = -(\alpha_1 u_\xi + \alpha_2 u_\eta)$$

$$u_y = u_\xi \xi_y + u_\eta \eta_y = u_\xi + u_\eta$$

$$u_{xx} = \left(-\alpha_1 \frac{\partial}{\partial \xi} - \alpha_2 \frac{\partial}{\partial \eta}\right)^2 u = \alpha_1^2 u_{\xi\xi} + 2\alpha_1\alpha_2 u_{\xi\eta} + \alpha_2^2 u_{\eta\eta}$$

$$u_{yy} = \left(\frac{\partial}{\partial \xi} + \frac{\partial}{\partial \eta}\right)^2 u = u_{\xi\xi} + 2u_{\xi\eta} + u_{\eta\eta}$$

$$u_{xy} = -\left(\alpha_1 \frac{\partial}{\partial \xi} + \alpha_2 \frac{\partial}{\partial \eta}\right)(u_\xi + u_\eta)$$

$$= -\alpha_1 u_{\xi\xi} - \alpha_2 u_{\eta\eta} - (\alpha_1 + \alpha_2) u_{\xi\eta}.$$

Substituting these derivatives into (2.3.6), we get

$$\left(A\alpha_1^2 - 2B\alpha_1 + C\right) u_{\xi\xi} + \left(A\alpha_2^2 - 2B\alpha_2 + C\right) u_{\eta\eta}$$
$$+ 2\left[\alpha_1\alpha_2 A - (\alpha_1 + \alpha_2)B + C\right] u_{\xi\eta}$$
$$+ (-\alpha_1 D + E) u_\xi + (-\alpha_2 D + E) u_\eta + Fu + G = 0. \quad (2.3.8)$$

Clearly, if α_1 and α_2 are the roots of the quadratic equation

$$A\alpha^2 - 2B\alpha + C = 0,$$

i.e.,

$$\begin{pmatrix} \alpha_1 \\ \alpha_2 \end{pmatrix} = \frac{B}{A} \pm \frac{\sqrt{B^2 - AC}}{A}, \quad (2.3.9)$$

then the terms $u_{\xi\xi}$ and $u_{\eta\eta}$ will disappear. Three cases can be distinguished:

Case (i): $B^2 - AC > 0$.

Both α_1 and α_2 are real and distinct. The coefficient of $u_{\xi\eta}$ is

$$2\left[\alpha_1\alpha_2 A - (\alpha_1 + \alpha_2)B + C\right] = -\frac{4}{A}(B^2 - AC) \neq 0$$

and (2.3.8) may be written

$$u_{\xi\eta} + D'u_\xi + E'u_\eta + F'u + G' = 0, \quad (2.3.10)$$

where D', E', F' and G' are some constants. Equation (2.3.6) is called *hyperbolic*. The real coordinate curves (ξ, η) given by (2.3.7) are called *characteristics*, and (2.3.10) is the canonical form of a hyperbolic equation.

By further transformation

$$\sigma = \frac{1}{2}(\xi + \eta), \qquad \tau = \frac{1}{2}(\xi - \eta) \qquad (2.3.11)$$

(2.3.10) can be changed to

$$u_{\sigma\sigma} - u_{\tau\tau} + D''u_\sigma + E''u_\tau + F''u + G'' = 0, \qquad (2.3.12)$$

where the second derivative terms are clearly the same, except for a scaling factor, as those of the standard wave equation.

Case (ii): $B^2 - AC = 0$.

There is only one family of real characteristics with the slope $\alpha = \alpha_1 = \alpha_2 = B/A$. Equation (2.3.6) is called *parabolic*. Let us change to the new coordinates

$$\xi = y - \frac{B}{A}x \quad \text{and} \quad \eta = y - \beta x,$$

where $\beta \neq \alpha$; then the coefficient of $u_{\xi\eta}$ in (2.3.8) also vanishes since

$$2\left[\alpha\beta A - (\alpha + \beta)B + C\right] = 2B\beta - 2\left(\frac{B}{A} + \beta\right)B + 2C = 0.$$

Furthermore, since $A\beta^2 - 2B\beta + C \neq 0$, (2.3.8) may be written as

$$u_{\eta\eta} + d'u_\xi + e'u_\eta + f'u + g' = 0, \qquad (2.3.13)$$

where d', e', f' and g' are some constants. In this equation the highest derivative in η is of the second order, while the highest derivative in ξ is of the first order; this feature is shared by the standard diffusion equation.

Case (iii): $B^2 - AC < 0$.

The roots α_1 and α_2 are complex and no real characteristics exist. Equation (2.3.4) is called *elliptic*. Formally, we can still employ the complex coordinates ξ, η defined by (2.3.7) and obtain (2.3.8) with complex coefficients. It is, however, preferable to modify (2.3.11) to

$$\sigma = \frac{1}{2}(\xi + \eta) \quad \text{and} \quad \tau = \frac{1}{2i}(\xi - \eta) \qquad (2.3.14)$$

in order to transform the second-order derivatives to a Laplacian. Denoting

$$\begin{pmatrix} \alpha_1 \\ \alpha_2 \end{pmatrix} = a \pm ib$$

with a and b being real

$$a = \frac{B}{A}, \qquad b = \frac{\sqrt{AC - B^2}}{A},$$

we may rewrite (2.3.14) as a real transformation

$$\sigma = \frac{1}{2}(y - \alpha_1 x + y - \alpha_2 x) = y - ax$$

$$\tau = \frac{1}{2i}(y - \alpha_1 x - y + \alpha_2 x) = -bx,$$

which may be substituted in (2.3.6) directly to obtain

$$u_{\sigma\sigma} + u_{\tau\tau} + d'' u_\sigma + e'' u_\tau + f'' u + g'' = 0. \qquad (2.3.15)$$

The highest-order derivatives are now Laplacian, and the coefficients d'', e'', f'' and g'' are real constants. Details are left as an exercise.

The terms hyperbolic, elliptic and parabolic are borrowed from analytical geometry, where a quadratic curve

$$Ax^2 + 2Bxy + Cy^2 + Dx + Ey + F = 0$$

is classified into three types according to the sign of $B^2 - AC$.

2.3.2 Variable coefficients

We now sketch an extension of the preceding section by allowing the coefficients $A, B, \ldots, F$ in (2.3.6) to be known functions of x and y. Characteristic curves are first sought in the general form $\xi = \xi(x, y), \eta = \eta(x, y)$. For $u = u(\xi, \eta)$, the derivatives are transformed by the chain rule

$$u_x = u_\xi \xi_x + u_\eta \eta_x, \qquad u_y = u_\xi \xi_y + u_\eta \eta_y.$$

Higher-order derivatives are easily evaluated so that (2.3.6) may be rewritten as

$$A' u_{\xi\xi} + 2B' u_{\xi\eta} + C' u_{\eta\eta} + D' u_\xi + E' u_\eta + F' = 0, \qquad (2.3.16)$$

where

$$A' = Q(\xi, \xi), \quad B' = Q(\xi, \eta), \quad C' = Q(\eta, \eta) \qquad (2.3.17)$$

with $Q(\xi, \eta)$ being the quadratic form

$$Q(\xi, \eta) \equiv A\xi_x \eta_x + B(\xi_x \eta_y + \xi_y \eta_x) + C\xi_y \eta_x. \qquad (2.3.18)$$

By straightforward algebra, it may be shown that

$$B'^2 - A'C' = (B^2 - AC)(\xi_x \eta_y - \xi_y \eta_x)^2$$

2.3 Linear second-order equations

so that the signs of $B^2 - AC$ and $B'^2 - A'C'$ are the same. We now seek $\xi(x,y)$ and $\eta(x,y)$ so that (2.3.16) takes the simplest (canonical) form. Similar to §2.3.1, there are three such forms:

(i) $A' = C' = 0$, $B' \neq 0$, (hyperbolic);
(ii) $B' = 0$, $A' = C' \neq 0$, (elliptic); and
(iii) $C' \neq 0$, $A' = B' = 0$, (parabolic).

To have the canonical form (i), we need to find two real and distinct solutions to the first-order partial differential equation

$$Q(f,f) = Af_x^2 + 2Bf_xf_y + Cf_y^2 = 0. \tag{2.3.19}$$

If found, the solutions are just $f = \xi$ and η. Let us consider the following cases.

Case (i): $B^2 - AC > 0$.

If $A = C = 0$, (2.3.6) is in the canonical form; no further transformation is needed. Assume therefore $A \neq 0$. Equation (2.3.19) implies that there are two solutions for

$$\frac{f_x}{f_y} = -\frac{B}{A} \pm \frac{\sqrt{B^2 - AC}}{A},$$

which are real and distinct because of the assumption $B^2 - AC > 0$. The above equation may be written as

$$Af_x + (B^2 \mp \sqrt{B^2 - AC})f_y = 0. \tag{2.3.20}$$

Clearly, along the two directions,

$$\frac{dy}{dx} = \frac{f_x}{f_y} = -\frac{B}{A} \pm \frac{\sqrt{B^2 - AC}}{A}, \tag{2.3.21}$$

$df = f_x dx + f_y dy = 0$ so that $f = $ constant, which corresponds to two real and distinct curves $\xi(x,y) = $ constant and $\eta(x,y) = $ constant. In principle, these sets of curves can be found by integrating (2.3.21). Note also that $B'^2 - A'C' = B'^2 > 0$. Thus under the condition $B^2 - AC > 0$ we have found a set of real curves ξ, η such that (2.3.16) takes the first canonical form of a hyperbolic equation

$$u_{\xi\eta} + D'u_\xi + E'u_\eta + F' = 0. \tag{2.3.22}$$

The families of curves ξ, η are called the characteristics; their existence makes the preceding form possible.

To get the second canonical form (2.3.12) we need only introduce the transformation (2.3.11) into (2.3.22) as before.

Case (ii): $B^2 - AC = 0$.

Now there is only one family of characteristics $\xi(x,y) = f(x,y) = $ constant, which makes one of the coefficients, say A', vanish. Let us choose any other family of curves $\eta(x,y) = $ constant that is not parallel to $\xi = $ constant, i.e.,

$$\frac{\xi_x}{\xi_y} \neq \frac{\eta_x}{\eta_y} \quad \text{or} \quad \xi_x \eta_y - \eta_x \xi_y \neq 0. \tag{2.3.23}$$

The coefficient B' is

$$B' = Q(\xi, \eta) = (A\xi_x + B\xi_y)\eta_x + (B\xi_x + C\xi_y)\eta_y.$$

Recall from (2.3.20) that $B^2 - AC = 0$ implies

$$A\xi_x + B\xi_y = 0.$$

Multiplying the above equation by C gives

$$AC\xi_x + BC\xi_y = 0 \quad \text{or} \quad B\xi_x + C\xi_y = 0,$$

hence $B' = 0$. On the other hand

$$C' = Q(\eta, \eta) = (A\eta_x + B\eta_y)\eta_x + (B\eta_x + C\eta_y)\eta_y$$
$$= \left(\eta_x + \frac{C}{B}\eta_y\right)(A\eta_x + B\eta_y) \neq 0,$$

otherwise $A\eta_x + B\eta_y = 0$, which would violate (2.3.23). Thus (2.3.16) reduces to the canonical form of the parabolic equation

$$u_{\eta\eta} + D'u_\xi + E'u_\eta + F' = 0.$$

Case (iii): $B^2 - AC < 0$.

The two solutions (ξ, η) of (2.3.21) are now complex conjugates of each other. Let us denote

$$\sigma + i\tau = \xi, \qquad \sigma - i\tau = \eta,$$

where σ, τ are real. Substituting the former into (2.3.18), we get

$$Q(\sigma, \sigma) - Q(\tau, \tau) + 2iQ(\sigma, \tau) = 0.$$

Equating the real and imaginary parts to zero separately, we obtain

$$Q(\sigma, \sigma) = Q(\tau, \tau), \qquad Q(\sigma, \tau) = 0.$$

Thus if we change from (x, y) to (σ, τ), (2.3.16) will have the coefficients $A' = C', B' = 0$ and takes the canonical form of the elliptic equation

$$u_{\sigma\sigma} + u_{\tau\tau} + D'u_\sigma + E'u_\tau + F' = 0.$$

The task of classifying (2.3.6) with variable coefficients is now complete.

For variable coefficients there are situations where real characteristics exist only over part of the entire domain of interest. The equation may then change its type from one region to another. This change occurs in transonic gas dynamics.

For different types of equations the boundary or initial conditions must also be different, in general, to make the problem meaningful. We shall not discuss these matters in general terms, but will bring them up from physical considerations later.

Extensions to second-order partial differential equations with more than two independent variables can be found in Koshlyakov et al. (1964).

Exercises

2.1 Let us rewrite (2.3.16) as a system of three first-order partial differential equations for three unknowns u, p and q

$$p = u_x, \quad q = u_y$$

$$Ap_x + Bp_y + Cq_y + Du_x + Eu + y + F + G = 0.$$

Find the conditions where real characteristics exist.

2.2 In high-speed aerodynamics, the variation of air density due to motion is important. For one-dimensional isentropic flow of a perfect gas the equations governing the density ρ, pressure p and velocity u are:

Mass:
$$\frac{\partial \rho}{\partial t} + \frac{\partial (\rho u)}{\partial x} = 0$$

Momentum:
$$\rho \left(\frac{\partial u}{\partial t} + u \frac{\partial u}{\partial x} \right) = -\frac{\partial p}{\partial x}$$

State:
$$\frac{p}{p_o} = \left(\frac{\rho}{\rho_o} \right)^\gamma,$$

where p_o, ρ_o are reference values of p, ρ, respectively, and γ is the ratio of specific heats. Determine the differential equations for the characteristics and find the Riemann invariants.

32   Classification of equations with two independent variables

2.3 Consider a one-dimensional flood wave in a wide river of bed inclination θ. The depth h and velocity u are governed by the following equations:
$$\frac{\partial h}{\partial t} + \frac{\partial (uh)}{\partial x} = 0$$
and
$$\frac{\partial u}{\partial t} + u\frac{\partial u}{\partial x} = g\sin\theta - \frac{fu^2}{h},$$
where f is the empirical coefficient of bed friction. Determine the characteristics.

2.4 Reduce each of the following equations to its canonical form:
(i)
$$\frac{\partial^2 u}{\partial x^2} + 4\frac{\partial^2 u}{\partial x \partial y} + 5\frac{\partial^2 u}{\partial y^2} + 7u = 5$$
(ii)
$$\frac{\partial^2 u}{\partial x^2} - 3\frac{\partial^2 u}{\partial x \partial y} + \frac{\partial^2 u}{\partial y^2} - \frac{\partial u}{\partial x} + 2\frac{\partial u}{\partial y} = 0$$
(iii)
$$\frac{\partial^2 u}{\partial x^2} - 2\frac{\partial^2 u}{\partial x \partial y} + \frac{\partial^2 u}{\partial y^2} + \frac{\partial u}{\partial y} + 7u = 0.$$

2.5 For each equation below find the region of hyperbolicity and then reduce the equation to canonical form:
(i)
$$x^2\frac{\partial^2 u}{\partial x^2} + 2xy\frac{\partial^2 u}{\partial x \partial y} + y^2\frac{\partial^2 u}{\partial y^2} = 0$$
(ii)
$$y\frac{\partial^2 u}{\partial x^2} + 5x\frac{\partial^2 u}{\partial y^2} = 0$$
(iii)
$$\cos y\frac{\partial^2 u}{\partial x^2} + 2\sin y\frac{\partial^2 u}{\partial x \partial y} - \frac{\partial^2 u}{\partial y^2} = 0.$$

3
One-dimensional waves

The effort needed in solving a boundary-value problem depends in part on the complexity of the boundary geometry. Other things being equal, an infinitely large domain is often simpler than a finite domain. Hence, we begin by examining one-dimensional wave propagation in an infinite domain.

First we study the linear wave equation. The method of characteristics is used to obtain the solution to an initial-value problem. The concepts of domain of dependence and range of influence are then introduced. The relation between the characteristics and the physics of wave propagation is explained through several examples. Finally, some effects of nonlinearity are illustrated by a simple model of traffic flow.

3.1 Waves due to initial disturbances

Recall the governing equation for one-dimensional waves in a taut string

$$\frac{\partial^2 u}{\partial t^2} - c^2 \frac{\partial^2 u}{\partial x^2} = 0, \qquad -\infty < x < \infty. \tag{3.1.1}$$

Let the initial transverse displacement and velocity be given along the entire string

$$u(x, 0) = f(x) \tag{3.1.2}$$

$$\frac{\partial u}{\partial t}(x, t) = g(x), \tag{3.1.3}$$

where $f(x)$ and $g(x)$ are nonzero only in the finite domain of x. At infinities $x \to \pm\infty$, u and $\partial u/\partial t$ are zero for any finite t. These conditions are best displayed in the space-time diagram, as shown in Figure 3.1. In (3.1.1) the highest time derivative is of the second order and initial

33

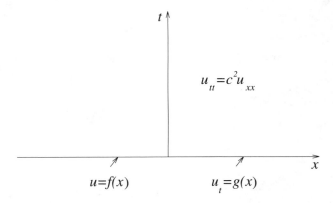

Fig. 3.1. Summary of the initial-boundary-value problem.

data are prescribed for u and $\partial u/\partial t$. Initial conditions that specify all derivatives of all orders less than the highest in the differential equation are called the *Cauchy initial conditions*.

Recall that in terms of the characteristic coordinates $\xi = x - ct$ and $\eta = x + ct$, the wave equation can be reduced to the canonical form

$$\frac{\partial^2 u}{\partial \xi \partial \eta} = 0. \qquad (3.1.4)$$

The general solution is

$$u = \phi(\xi) + \psi(\eta)$$
$$= \phi(x - ct) + \psi(x + ct), \qquad (3.1.5)$$

where ϕ and ψ are arbitrary functions to be determined. As time proceeds ϕ remains constant along $\xi = x - ct = $ constant. Thus to an observer traveling at the speed c to the right, the spatial distribution of ϕ remains always the same as the initial distribution $\phi(x)$ at $t = 0$. In other words, the initial profile of ϕ is propagated at the speed c to the right without changing its form. Similarly, the initial distribution of ψ is propagated to the left at the speed c without change of form. This feature is clearly a distinguishing characteristic of waves, hence c is the speed of wave propagation.

From the initial conditions we get

$$u(x,0) = \phi(x) + \psi(x) = f(x) \qquad (3.1.6)$$

$$\frac{\partial u}{\partial t}(x,0) = -c\frac{d\phi(x)}{dx} + c\frac{d\psi(x)}{dx} = g(x). \qquad (3.1.7)$$

3.1 Waves due to initial disturbances

The last equation may be integrated with respect to x

$$-\phi + \psi = \frac{1}{c}\int_{x_o}^{x} g(x')dx' - K, \tag{3.1.8}$$

where x_o and K are some arbitrary constants. Now ϕ and ψ can be solved from (3.1.6) and (3.1.8) as functions of x,

$$\phi(x) = \frac{1}{2}[f(x) + K] - \frac{1}{2c}\int_{x_o}^{x} g(x')\,dx'$$

$$\psi(x) = \frac{1}{2}[f(x) - K] + \frac{1}{2c}\int_{x_o}^{x} g(x')\,dx'.$$

Replacing the arguments of ϕ by $x-ct$ and of ψ by $x+ct$ and substituting the results in u, we get

$$\begin{aligned}u(x,t) &= \frac{1}{2}f(x-ct) - \frac{1}{2c}\int_{x_o}^{x-ct} g\,dx' \\ &\quad + \frac{1}{2}f(x+ct) + \frac{1}{2c}\int_{x_o}^{x+ct} g\,dx' \\ &= \frac{1}{2}[f(x-ct) + f(x+ct)] + \frac{1}{2c}\int_{x-ct}^{x+ct} g(x')\,dx',\end{aligned} \tag{3.1.9}$$

which is d'Alembert's solution to the homogeneous wave equation subject to general Cauchy initial conditions.

To see the physical meaning, let us draw in the space-time diagram a triangle formed by two characteristic lines passing through the observer at x,t, as shown in Figure 3.2. The base of the triangle along the initial axis $t = 0$ begins at $x - ct$ and ends at $x + ct$. The solution (3.1.9) depends on the initial displacement at just the two corners $x - ct$ and $x + ct$, and on the initial velocity only along the segment from $x - ct$ to $x + ct$. Nothing outside the triangle matters. Therefore, to the observer at x,t, the *domain of dependence* is the base of the characteristic triangle formed by two characteristics passing through x,t. On the other hand, the data at any point x on the initial line $t = 0$ must influence all observers in the wedge formed by two characteristics drawn from $x, 0$ into the region of $t > 0$; this characteristic wedge is called the *range of influence*.

Let us illustrate the physical effects of initial displacement and velocity separately.

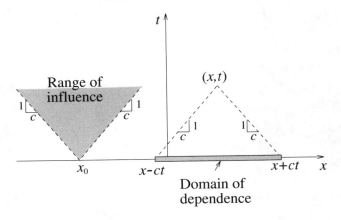

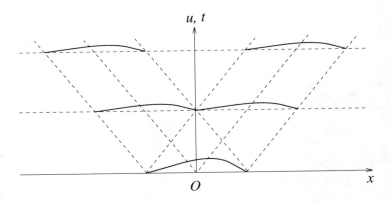

Fig. 3.2. Domain of dependence and range of influence.

Fig. 3.3. Waves due to initial displacement.

Case (i): Initial displacement only: $f(x) \neq 0$ and $g(x) = 0$. The solution is

$$u(x,t) = \frac{1}{2}f(x-ct) + \frac{1}{2}f(x+ct)$$

and is shown in Figure 3.3 for a simple $f(x)$ at successive time steps. Clearly, the initial disturbance is split into two equal waves propagating in opposite directions at the speed c. The outgoing waves preserve the initial profile, although their amplitudes are reduced by half.

Case (ii): Initial velocity only: $f(x) = 0$ and $g(x) \neq 0$. Consider

3.1 Waves due to initial disturbances

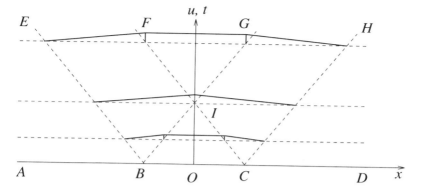

Fig. 3.4. Waves due to initial velocity.

the simple example where

$$g(x) = g_o \quad \text{when} \quad |x| < b, \quad \text{and}$$
$$= 0 \quad \text{when} \quad |x| > b.$$

Referring to Figure 3.4, we divide the $x \sim t$ diagram into six regions by the characteristics with B and C lying on the t axis at $x = -b$ and $+b$, respectively. The solution in various regions is

$$u = 0$$

in the wedge ABE;

$$u = \frac{1}{2c}\int_{-b}^{x+ct} g_o \, dx' = \frac{g_o}{2c}(x + ct + b)$$

in the strip $EBIF$;

$$u = \frac{1}{2c}\int_{x-ct}^{x+ct} g_o \, dx' = g_o t$$

in the triangle BCI;

$$u = \frac{1}{2c}\int_{-b}^{b} g_o \, dx' = \frac{g_o b}{c}$$

in the wedge FIG;

$$u = \frac{1}{2c}\int_{x-ct}^{b} g_o \, dx' = \frac{g_o}{2c}(b - x + ct)$$

in the strip $GICH$; and
$$u = 0$$
in the wedge HCD. The spatial variation of u is plotted for several instants in Figure 3.4. Note that the wave fronts in both directions advance at the speed c. In contrast to Case (i), the disturbance persists for all time in the region between the two fronts.

3.2 Reflection from the fixed end of a string

Let us use the d'Alembert solution to a problem in a half infinite domain $x > 0$. Consider a long and taut string stretched from $x = 0$ to infinity. How do disturbances generated near the left end propagate as the result of initial displacement and velocity?

At the left boundary $x = 0$, we must now add the condition
$$u = 0, \quad x = 0, \quad t > 0. \qquad (3.2.1)$$

In the space-time diagram let us draw two characteristics passing through x, t. For an observer in the region $x > ct$, the characteristic triangle does not intersect the time axis because t is still too small. The observer does not feel the presence of the fixed end at $x = 0$, hence the solution (3.1.9) for an infinitely long string applies,
$$u = \frac{1}{2}[f(x+ct) + f(x-ct)] + \frac{1}{2c}\int_{x-ct}^{x+ct} g(\tau)\,d\tau, \quad x > ct. \qquad (3.2.2)$$

But for $x < ct$, this result is no longer valid. To ensure that the boundary condition is satisfied we employ the idea of mirror reflection. Consider a fictitious extension of the string to $-\infty < x \leq 0$. If on the side $x < 0$ the initial data are imposed such that $f(x) = -f(-x)$, $g(x) = -g(-x)$, then $u(0,t) = 0$ is assured by symmetry. We now have initial conditions stated over the entire x axis
$$u(x,0) = F(x) \quad \text{and} \quad u_t(x,0) = G(x), \quad -\infty < x < \infty,$$
where
$$F(x) = \begin{cases} f(x) & \text{if } x > 0 \\ -f(-x) & \text{if } x < 0 \end{cases}$$

$$G(x) = \begin{cases} g(x) & \text{if } x > 0 \\ -g(-x) & \text{if } x < 0. \end{cases}$$

3.2 Reflection from the fixed end of a string

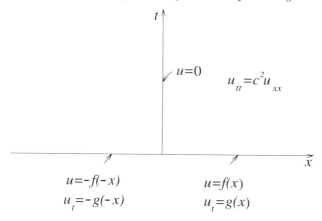

Fig. 3.5. Initial-boundary-value problem and the mirror reflection.

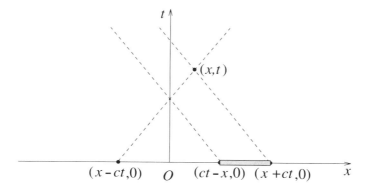

Fig. 3.6. Reflection from a fixed end.

These conditions are summarized in Figure 3.5 and can be substituted into (3.1.9) to get the solution for $0 < x < ct$

$$
\begin{aligned}
u &= \frac{1}{2}[F(x+ct) + F(x-ct)] + \frac{1}{2c}\left(\int_{x-ct}^{0} + \int_{0}^{x+ct}\right) G(x')\,dx' \\
&= \frac{1}{2}[f(x+ct) - f(ct-x)] + \frac{1}{2c}\left(\int_{ct-x}^{0} + \int_{0}^{x+ct}\right) g(x')\,dx' \\
&= \frac{1}{2}[f(x+ct) - f(ct-x)] + \frac{1}{2c}\int_{ct-x}^{ct+x} g(x')\,dx'. \quad (3.2.3)
\end{aligned}
$$

The domain of dependence is shown by the hatched segment on the x axis in Figure 3.6.

3.3 Specification of initial and boundary data

While the prescription of auxiliary (initial and boundary) conditions can be easily guided by physics when the initial curve is either the t or x axis, more care is needed if the initial curve is anywhere else in the space-time diagram. Referring to Figure 3.7, a convenient rule is that the number of conditions along the curve C is equal to the number of characteristics pointing into the region of interest in the direction of increasing t. There are two kinds of initial curves. If the slope of C_1 is $|dx/dt| > c$, where c is the characteristic wave speed, two characteristics point into the region of increasing t; therefore two auxiliary conditions are needed, which are called the initial conditions, and the curve C_1 is called *space-like*. On the other hand, if along C_2 the slope is $|dx/dt| < c$, only one characteristic points into the region; only one auxiliary condition, now called the boundary condition, is needed. The curve C_2 is called *time-like*. The reason is as follows. The integral of (3.1.4) has two integration constants, hence requires two conditions at the intersections of the characteristics $\xi = $ constant, $\eta = $ constant passing through the observer, and the initial curve. If the observer is near a space-like initial curve C, there are two intersection points; hence two conditions are needed on C. If, however, the observer is near a time-like curve C, only one of the characteristics, say $\xi = $ constant, intersects C and only one condition is needed. Mathematically, the governing equation and the auxiliary conditions that define the initial-boundary-value problem must be checked to ensure that the solution exists, is unique and stable. If these criteria, due to Hadamard, are satisfied, the initial-boundary-value problem is said to be well-posed. The examination of such criteria can be abstract and is an important part in mathematics texts. We shall only touch upon this point later.

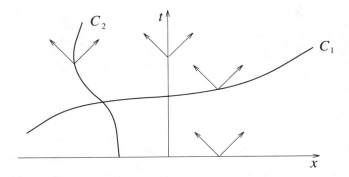

Fig. 3.7. Time- and space-like initial curves.

3.4 Forced waves in a long string

We now derive the complete d'Alembert solution to the initial-value problem for an infinite string with forcing. The wave equation is now inhomogeneous

$$\frac{\partial^2 u}{\partial t^2} - c^2 \frac{\partial^2 u}{\partial x^2} = h(x,t), \quad -\infty < x < \infty \qquad (3.4.1)$$

and so are the initial conditions

$$u(x,0) = f(x) \qquad (3.4.2)$$

and

$$\frac{\partial u}{\partial t}(x,0) = g(x), \qquad (3.4.3)$$

where h, f and g are prescribed. Let us make a change of variables

$$y = ct$$

to remove the coefficient c^2. Then (3.4.1–3.4.3) become

$$\frac{\partial^2 u}{\partial x^2} - \frac{\partial^2 u}{\partial y^2} = H(x,y), \quad \text{where} \quad H = -\frac{h}{c^2} \qquad (3.4.4)$$

$$u(x,0) = f(x) \qquad (3.4.5)$$

$$\frac{\partial u}{\partial y}(x,0) = G(x) \quad \text{and} \quad G = \frac{g}{c}. \qquad (3.4.6)$$

Referring to the $x \sim y$ plane in Figure 3.8, consider a characteristic triangle formed by two characteristic lines passing through the observer at (x_o, y_o). The slopes of the lines are ± 1 so that the two corners on the x axis are at $Q(x_o, -y_o, 0)$ and $R(x_o, y_o, 0)$. We now integrate (3.4.4) over the triangle

$$\iint_\Delta \left(\frac{\partial^2 u}{\partial x^2} - \frac{\partial^2 u}{\partial y^2} \right) dx\,dy = \iint_\Delta H\,dx\,dy. \qquad (3.4.7)$$

According to Stokes' theorem, proven in Appendix B, for any vector $\mathbf{q}$ the following is true:

$$\iint_S \nabla \times \mathbf{q}\,dx\,dy = \oint_C \mathbf{q} \cdot d\mathbf{s},$$

where C is the boundary of the area S. If we put

$$\mathbf{q} = \left(\frac{\partial u}{\partial y}, \frac{\partial u}{\partial x} \right)$$

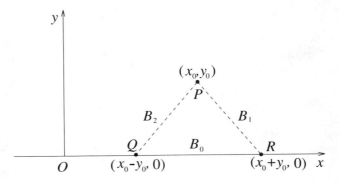

Fig. 3.8. Characteristic triangle passing through (x_o, y_o).

in Stokes' theorem, take $\mathcal{S}$ to be the characteristic triangle and let $\mathcal{C}$ be the three sides B_0, B_1 and B_2, as shown in Figure 3.8, then

$$\iint_\Delta \left(\frac{\partial^2 u}{\partial x^2} - \frac{\partial^2 u}{\partial y^2}\right) dx dy = \oint_\mathcal{C} \left(\frac{\partial u}{\partial x} dy + \frac{\partial u}{\partial y} dx\right). \qquad (3.4.8)$$

Along B_0, $y = 0$ and $dy = 0$ so that

$$\int_{B_0} \left(\frac{\partial u}{\partial x} dy + \frac{\partial u}{\partial y} dx\right) = \int_{x_o-y_o}^{x_o+y_o} \frac{\partial u}{\partial y}(x, 0) dx = \int_{x_o-y_o}^{x_o+y_o} G(\tau)\, d\tau,$$

where the boundary condition (3.4.6) has been used. Along B_1, $dy = -dx$ so that

$$\int_{B_1} \left(\frac{\partial u}{\partial x} dy + \frac{\partial u}{\partial y} dx\right) = -\int_{B_1} \left(\frac{\partial u}{\partial x} dx + \frac{\partial u}{\partial y} dy\right),$$

which is a total derivative and can be integrated to give

$$-\int_R^P du = -u(P) + u(R) = -\left[u(x_o, y_o) - u(x_o + y_o, 0)\right]$$
$$= -u(x_o, y_o) + f(x_o + y_o)$$

after using (3.4.5). Lastly, along B_2, $dy = dx$, the line integral is

$$\int_{B_2} \left(\frac{\partial u}{\partial x} dy + \frac{\partial u}{\partial y} dx\right) = \int_{B_2} \left(\frac{\partial u}{\partial x} dx + \frac{\partial u}{\partial y} dy\right),$$

which is again a total derivative from P to Q and is equal to

$$\int_P^Q du = u(Q) - u(P) = u(x_o - y_o, 0) - u(x_o, y_o)$$
$$= f(x_o - y_o) - u(x_o, y_o).$$

Combining these results, we get the right-hand side of (3.4.8)

$$\oint \left(\frac{\partial u}{\partial x}dy + \frac{\partial u}{\partial y}dx\right) = \int_{x_o-y_o}^{x_o+y_o} G(\tau)\,d\tau$$
$$-2u(x_o,y_o) + f(x_o+y_o) + f(x_o-y_o).$$

On the other hand, the left-hand side of (3.4.8) is given by (3.4.7), therefore,

$$\iint_\Delta H\,dxdy = \int_{x_o-y_o}^{x_1+y_o} G(\tau)\,d\tau - 2u(x_o,y_o) + f(x_o+y_o) + f(x_o-y_o),$$

or

$$u(x_o,y_o) = \frac{1}{2}\left[f(x_o+y_o) + f(x_o-y_o)\right] + \int_{x_o-y_o}^{x_o+y_o} G(\tau)\,d\tau$$
$$-\frac{1}{2}\iint_\Delta H\,dxdy.$$

Interchanging the variables (x_o, y_o) with (x, y), we get

$$u(x,y) = \frac{1}{2}\left[f(x+y) + f(x-y)\right]$$
$$+\frac{1}{2}\int_{x-y}^{x+y} G(\tau)\,d\tau + \frac{1}{2}\iint_\Delta H(x_o,y_o)\,dx_ody_o. \quad (3.4.9)$$

Finally, let us return to the original notations

$$y \to ct, \qquad G \to \frac{g}{c}, \qquad H \to -\frac{h}{C^2},$$

then

$$u(x,t) = \frac{1}{2}\left[f(x+ct) + f(x-ct)\right]$$
$$+\frac{1}{2c}\int_{x-ct}^{x+ct} g(\tau)\,d\tau + \frac{1}{2c}\iint_\Delta h(x_o,t_o)\,dx_odt_o, \quad (3.4.10)$$

which is the complete solution of d'Alembert.

Compared to (3.1.9), the domain of dependence now includes the entire interior of the characteristic triangle as the consequence of the transient forcing $h(x,t)$.

3.5 Uniqueness of the Cauchy problem

Consider the forced waves in a taut string of infinite length

$$T\frac{\partial^2 V}{\partial x^2} - \rho\frac{\partial^2 V}{\partial t^2} = F(x,t), \quad -\infty < x < \infty, \quad t > 0 \tag{3.5.1}$$

with the boundary condition

$$V(\pm\infty, t) = 0 \tag{3.5.2}$$

and the initial conditions

$$V(x,0) = f(x), \quad \frac{\partial V}{\partial t}(x,0) = g(x). \tag{3.5.3}$$

Are there two different solutions V_1 and V_2 that satisfy the same conditions? If so, the difference $V' = V_1 - V_2$ must be a nontrivial homogeneous solution, i.e.,

$$T\frac{\partial^2 V'}{\partial x^2} - \rho\frac{\partial^2 V'}{\partial t^2} = 0, \quad -\infty < x < \infty, \quad t > 0 \tag{3.5.4}$$

with

$$V'(\pm\infty, t) = 0 \tag{3.5.5}$$

and

$$V'(x,0) = \frac{\partial V'}{\partial t}(x,0) = 0. \tag{3.5.6}$$

Let us examine how energy associated with V' varies in time. The total kinetic energy in the string is

$$\frac{1}{2}\int_{-\infty}^{\infty} \rho\left(\frac{\partial V'}{\partial t}\right)^2 dx,$$

while the total potential energy is the work done by the tension T when the string is stretched from its original length,

$$T\int_{-\infty}^{\infty} dx \left[\left(1 + \frac{\partial V'^2}{\partial x}\right)^{1/2} - 1\right] \cong \frac{T}{2}\int_{-\infty}^{\infty} dx \left(\frac{\partial V'}{\partial x}\right)^2.$$

Therefore, the total energy in the string is

$$E(t) = \frac{1}{2}\int_{-\infty}^{\infty} dx \left[\rho\left(\frac{\partial V'}{\partial t}\right)^2 + T\left(\frac{\partial V'}{\partial x}\right)^2\right], \tag{3.5.7}$$

whose time rate of change is

$$\frac{dE}{dt} = \int_{-\infty}^{\infty} \left[\frac{\partial V'}{\partial t} \left(\rho \frac{\partial^2 V'}{\partial t^2} - T \frac{\partial^2 V}{\partial x^2} \right) + T \frac{\partial}{\partial x} \left(\frac{\partial V'}{\partial x} \frac{\partial V'}{\partial t} \right) \right] dx$$

$$= \int_{-\infty}^{\infty} \left[V'_t \left(\rho \frac{\partial^2 V'}{\partial t^2} - T \frac{\partial^2 V'}{\partial x^2} \right) \right] + [TV'_t V'_x]_{-\infty}^{\infty} = 0,$$

which vanishes on account of (3.5.4) to (3.5.6). Since $E(0) = 0$, it follows that $E(t) \equiv 0$ for all t. But this is possible if and only if each term in the energy integral (3.5.7) vanishes identically, i.e.,

$$\frac{\partial V'}{\partial t} = \frac{\partial V'}{\partial x} = 0, \quad -\infty < x < \infty,$$

which means $V' \equiv 0$ everywhere and for all time, or that $V_1 = V_2 = V$ is unique.

The reader should prove that the same result holds for a string of finite length.

3.6 Traffic flow – a taste of nonlinearity

In this section we illustrate the use of characteristics and study the nonlinear problem of one-dimensional traffic flow. As derived in Chapter 1, the density of cars on a one-way road is related to the flux q by

$$\frac{\partial \rho}{\partial t} + \frac{\partial q}{\partial x} = 0, \tag{3.6.1}$$

where $q(\rho)$ is an empirical function of ρ, as sketched in Figure 1.4b. For explicitness let us assume that $q(\rho)$ is quadratic in the region $0 < \rho < \rho_m$ and zero otherwise,

$$\frac{q}{q_m} = \frac{\rho}{\rho_m} \left(1 - \frac{\rho}{\rho_m} \right), \quad 0 < \rho < \rho_m, \tag{3.6.2}$$

where ρ_m is the maximum density. Note that the maximum flux rate is $q_m/4$, which occurs at $\rho = \rho_m/2$. The implied velocity of traffic flow is $u = q/\rho$ or

$$\frac{u}{u_m} = \left(1 - \frac{\rho}{\rho_m} \right), \quad 0 < \rho < \rho_m, \tag{3.6.3}$$

where $u_m = q_m/\rho_m$ is the maximum speed when $\rho = 0$. The flux q can now be eliminated from (3.6.1) to get an equation for ρ alone

$$\frac{\partial \rho}{\partial t} + \frac{dq}{d\rho} \frac{\partial \rho}{\partial x} = \frac{\partial \rho}{\partial t} + u_m \left(1 - \frac{2\rho}{\rho_m} \right) \frac{\partial \rho}{\partial x} = 0, \tag{3.6.4}$$

which may be written as

$$\frac{\partial \rho}{\partial t}dt + \frac{\partial \rho}{\partial x}dx = d\rho = 0 \tag{3.6.5}$$

if

$$\frac{dx}{dt} = u_m\left(1 - \frac{2\rho}{\rho_m}\right). \tag{3.6.6}$$

Thus ρ remains constant along a characteristic curve described by (3.6.5). Clearly, all characteristics are straight lines with the slope $V(\rho) = V(\rho_o)$, where ρ_o is the initial density. In the following sections we shall examine the implications of these seemingly simple statements under various initial conditions.

3.7 Green light at the head of traffic

Consider a long road with cars stopped by a red light at $x = 0$. The road is packed initially behind the light and there is no traffic ahead of the light. What happens after the light turns green? The initial condition is

$$\rho(x,0) = \begin{cases} \rho_m, & x < 0, \quad t = 0 \\ 0, & x > 0, \quad t = 0, \end{cases} \tag{3.7.1}$$

as shown in Figure 3.9. Since ρ is constant along a characteristic, the slope of a characteristic curve is constant. Any characteristic intersecting the positive x axis at $x_o > 0$ has the slope

$$\frac{dx}{dt} = u_m\left(1 - \frac{2\rho}{\rho_m}\right) = u_m\left(1 - \frac{2\rho(x,0)}{\rho_m}\right) = u_m \tag{3.7.2}$$

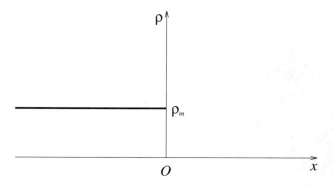

Fig. 3.9. An initial density distribution.

3.7 Green light at the head of traffic

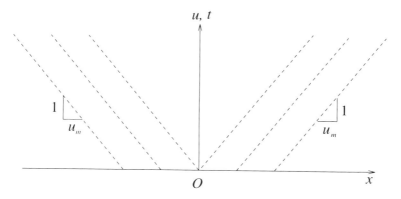

Fig. 3.10. Characteristics due to a discontinuous traffic density.

since $\rho(x,0) = 0$. The characteristics are the right-leaning straight lines

$$x = u_m t + x_o. \qquad (3.7.3)$$

Any characteristic intersecting the negative x axis at $x_o < 0$ has the slope

$$\frac{dx}{dt} = u_m \left(1 - \frac{2\rho(x,0)}{\rho_m}\right) = u_m \left(1 - \frac{2\rho_m}{\rho_m}\right) = -u_m \qquad (3.7.4)$$

since $\rho = \rho(x,0) = \rho_m$. Hence, the characteristics are left-leaning straight lines

$$x = -u_m t + x_o. \qquad (3.7.5)$$

These characteristic lines are depicted in Figure 3.10. In the wedge $-u_m t < x < u_m t$, all characteristics must begin from the origin and must be straight, hence $dx/dt = x/t$. It follows from (3.6.6) that

$$\frac{x}{t} = u_m \left(1 - \frac{2\rho}{\rho_m}\right)$$

or

$$\frac{\rho}{\rho_m} = \frac{1}{2}\left(1 - \frac{x}{u_m t}\right), \qquad -u_m t < x < u_m t. \qquad (3.7.6)$$

At any instant the density varies linearly in x from ρ_m at $x = -u_m t$ to zero at $x = u_m t$. The final solution for ρ for all x at any t is summarized in Figure 3.11.

While $\rho(x,t)$ is useful information to the highway designer, a driver is more concerned with the movement of the car. In fluid mechanics the

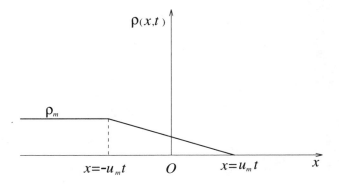

Fig. 3.11. Traffic density after light turns green.

motion can be described either by the velocity at fixed points in space (the Eulerian description) or by the instantaneous positions of marked material particles (the Lagrangian description). To derive particle movement from the Eulerian solution is called the Euler-Lagrange problem. Let us study such a problem for traffic flow.

Let $X(t, x_o)$ denote the Lagrangian position of a car whose initial position is x_o, i.e., $X(0, x_o) = x_o$. Consider a car initially at some distance behind the traffic light $x_o < 0$. Since the car begins to move only at $t = t_o = -x_o/u_m$, its initial condition is

$$X\left(-\frac{x_o}{u_m}, x_o\right) = x_o. \qquad (3.7.7)$$

Now for any $t > t_o$

$$\frac{dX}{dt} = u(X, t) = u_m \left(1 - \frac{\rho(X, t)}{\rho_m}\right)$$

$$= u_m \left[1 - \frac{1}{2}\left(1 - \frac{X}{u_m t}\right)\right] = \frac{u_m}{2} + \frac{X}{2t}.$$

Note that we have replaced x in ρ by X, since at time t the geometric point x is occupied at the instant by the car in question. This differential equation for X can be rearranged

$$t\frac{dX}{dt} - \frac{X}{2} = \frac{u_m t}{2}$$

whose general solution

$$X = Bt^{1/2} + u_m t$$

3.7 Green light at the head of traffic

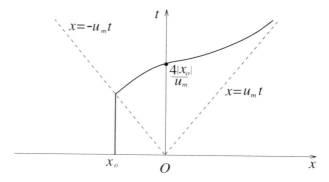

Fig. 3.12. Trajectory of a car crossing the green light.

is found by adding homogeneous and inhomogeneous solutions. Applying the initial condition (3.7.7), we find

$$B = -2\sqrt{-x_o u_m} = -2\sqrt{|x_o| u_m}$$

and the instantaneous position of the car that started at $-|x_o|$ at $t = t_o$ is

$$X = u_m t - 2\sqrt{|x_o| u_m t}. \qquad (3.7.8)$$

A typical path trajectory in the $x \sim t$ plane is shown in Figure 3.12. The speed of the car is

$$\frac{dX}{dt} = u_m - \sqrt{\frac{|x_o| u_m}{t}}. \qquad (3.7.9)$$

As a check, we see that $dX/dt = 0$ at $t = t_o = |x_o|/m$ and $dX/dt = u_m$ at $t \to \infty$.

How long does it take for the car to cross the traffic light? We set $X = 0$ in (3.7.8) to get

$$t = \frac{4|x_o|}{u_m}. \qquad (3.7.10)$$

This result can be used to estimate the number of cars passing the green light in a given period, say $0 < t < T$. During this period, all cars located within $0 > x_o > -u_m T/4$ will cross the light; the corresponding number of cars crossing the light is

$$\rho_m |x_o| = \frac{\rho_m u_m T}{4}. \qquad (3.7.11)$$

Let the typical car length be 6 m, the bumper-to-bumper density is $\rho_m = 1000/6 \cong 167$ cars per km. Let $u_m = 30$ km/hour and $T = 1$ minute. The number of cars passing through the light during the first minute after the green light is 21.

3.8 Traffic congestion and jam

Let the traffic be heavier ahead and lighter behind, as sketched in Figure 3.13. Recall that along any characteristic the density remains constant; the characteristics are all straight with the slope given by (3.6.6). Those beginning from the zone of denser traffic are steeper than those from the zone of lighter traffic. Hence a characteristic from the back tends to catch up with a characteristic in the front. As a consequence the distance between two points associated with the fixed densities ρ_1 and ρ_2 tend to shorten, and the instantaneous density profile becomes steeper. At a critical instant t_c, the profile becomes vertical. If the mathematical solution is continued, the profile overturns and becomes triple valued. This process represents the onset of a traffic jam. Since a triple-valued solution is meaningless, we trace the further development for $t > t_c$ by allowing the density to be discontinuous, i.e., by allowing *shocks*.

The relation between ρ and u on both sides of the shock is called the shock condition and must be deduced from the conservation law (3.6.1). Let $X_1(t)$ and $X_2(t)$ be two arbitrary moving stations. By integrating the conservation equation between them, we get

$$\int_{X_1}^{X_2} \left(\frac{\partial \rho}{\partial t} + \frac{\partial q}{\partial x} \right) dx = 0.$$

By partial integration and Leibniz's rule, it follows that

$$\frac{\partial}{\partial t} \int_{X_1}^{X_2} \rho\, dt - \frac{dX_2}{dt} [\rho]_2 - \frac{dX_1}{dt} [\rho]_1 + [q]_2 - [q]_1 = 0,$$

where the subscripts $[\cdot]_{1,2}$ denote quantities at stations $(1,2)$. Now let the two stations collapse onto the two sides of the shock, i.e., $X_1 = X_s - 0, X_2 = X_s + 0$, then the integral vanishes and

$$\frac{dX_s}{dt} = \frac{[q]_+ - [q]_-}{[\rho]_+ - [\rho]_-}, \qquad (3.8.1)$$

where the subcripts $(+,-)$ represent the (right, left) sides of the shock.

Let us apply this result to examine the effect of a red light.

3.8 Traffic congestion and jam   51

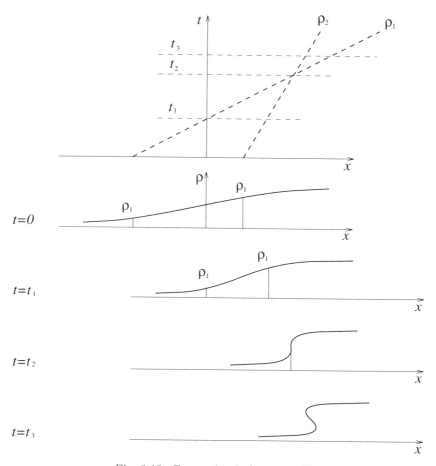

Fig. 3.13. Congestion in heavy traffic.

In Figure 1.4b, the part where $dq/d\rho > 0$ corresponds to light traffic, the characteristics lean forward with increasing time. On the other hand, the part $\rho > \rho_m/2$ where $dq/d\rho < 0$ corresponds to heavy traffic; the characteristics lean backward with increasing time. Now consider an initial density distribution of $\rho = \rho_o = $ constant with $\rho_o < \rho_m$. At $t = 0$ the light at $x = 0$ turns red so that the density at the light becomes ρ_m instantly. As time passes the zone of $\rho = \rho_m$, where the traffic is stopped, extends backward. On the other hand, the traffic far behind

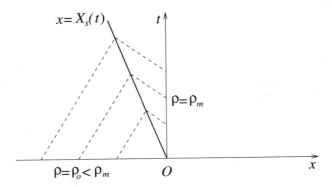

Fig. 3.14. The shock path.

still moves at the speed

$$u = u_m \left(1 - \frac{\rho_o}{\rho_m}\right)$$

before feeling the congestion ahead. The characteristics lean forward and must cross the backward characteristics at some later time. At the moment of intersection, there are two densities at the same location where a shock must occur. Let us examine how the shock propagates backward.

Since on two sides of the shock the densities are constants, the path of the shock must be a straight line, described by its differential equation

$$\frac{dX_s}{dt} = \frac{\rho_m u(\rho_m) - \rho_o u(\rho_o)}{\rho_m - \rho_o}.$$

Since $u(\rho_m) = 0$, we get

$$\frac{dX_s}{dt} = -\frac{\rho_o u(\rho_o)}{\rho_m - \rho_o} = \text{constant} < 0.$$

The shock path is therefore a straight line, as shown in Figure 3.14,

$$X_s = -\frac{\rho_o u(\rho_o)}{\rho_m - \rho_o} t. \qquad (3.8.2)$$

A snapshot of the density profile is sketched in Figure 3.15.

The description of a traffic jam by discontinuity is obviously a mathematical simplification. In reality, upon sensing the congestion ahead,

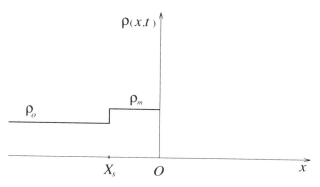

Fig. 3.15. Traffic density behind a shock.

a driver must step on the brakes. This human response smoothens the traffic flow and has been modeled by adding $-\nu \partial \rho/\partial x$ to the flux rate q. The revised conservation equation is then

$$\frac{\partial \rho}{\partial t} + \frac{dq}{d\rho}\frac{\partial \rho}{\partial x} = \nu \frac{\partial^2 \rho}{\partial x^2}. \tag{3.8.3}$$

Thus the driver reaction has a diffusive effect. For the quadratic model of q, we have

$$\frac{\partial \rho}{\partial t} + u_m \left(1 - \frac{2\rho}{\rho_m}\right)\frac{\partial \rho}{\partial x} = \nu \frac{\partial \rho^2}{\partial x^2}. \tag{3.8.4}$$

By the transformation

$$s = u_m \left(1 - \frac{2\rho}{\rho_m}\right),$$

the preceding equation becomes the celebrated Burger's equation

$$\frac{\partial s}{\partial t} + s\frac{\partial s}{\partial x} = \nu \frac{\partial^2 s}{\partial x^2}, \tag{3.8.5}$$

which has been studied extensively in the literature of fluid turbulence.

For other interesting aspects of traffic-flow theory, the book by Haberman (1977) is delightful to read.

Exercises

3.1 Consider a very long rope under tension T. At time $t = 0$ an acrobat of mass M jumps on the rope at $x = 0$ with the initial velocity V. Let u_- and u_+ denote the lateral displacement of the rope for $x < 0$ and $x > 0$, respectively, and U the lateral displacement of the acrobat. Show that if gravity is negligible, the boundary conditions at $x = 0$ are

$$M\frac{d^2U}{dt^2} = -T\frac{\partial u_-}{\partial x} + T\frac{\partial u_+}{\partial x}$$

$$U = u_- = u_+$$

for all $t > 0$. Complete the formulation by giving the governing equations for $u_-(x,t)$ and $u_+(x,t)$ as well as the initial conditions for U.

Show by the method of characteristics that the displacement of the acrobat is

$$U(t) = \frac{McV}{2T}\left[1 - \exp\left(-\frac{2Tt}{Mc}\right)\right]$$

with $c = \sqrt{T/\rho}$. Show also that

$$u_+(x,t) = 0, \quad x > ct$$

$$u_+(x,t) = \frac{McV}{2T}\left[1 - \exp\left(\frac{2T}{Mc^2}(x - ct)\right)\right], \quad 0 < x < ct.$$

Describe the physical phenomenon (Koshlyakov et al., 1964).

3.2 In a shallow channel the depth changes abruptly at $x = 0$. On the left side $x < 0$ the depth is h_-, on the right side $x > 0$, h_+. A wave pulse approaches from $x \sim -\infty$ whose free-surface displacement is given by $\zeta = f(x - c_- t)$, where f is non-zero only within a finite wavelength. Upon arrival at the depth discontinuity, part of the wave is transmitted to the right of the step and part is reflected to the left. Formulate the governing equations on each side of the step and the initial and boundary conditions. Note that at $x = 0$ the free-surface displacement and the rate of volume flux must be continuous at all times. Far from the step the transmitted wave must propagate to the right and the reflected wave to the left.

Use the method of characteristics to find the transmitted and reflected waves.

3.3 A long string $x \leq L$ is imbedded in an elastic medium and fixed at the end $x = L$. A pulse-like disturbance is incident toward the end and is reflected. Find by the method of characteristics the string deformation during and after the reflection process

(Graf, 1975). Consider a specific initial pulse that is centered at 0 and has the following shape

$$v(x,0) = \begin{cases} v_o \left(1 - \frac{|x|}{a}\right), & |x| < a, \\ 0, & |x| > a, \end{cases}$$

where $a < L$.

3.4 Two semi-infinite rods A and B of identical cross sections and material lie on the x axis. Before $t = 0$ rod B on the right is at rest. Rod A on the left moves at velocity V toward $x = 0$ and collides with the left end of B at $t = 0$. Afterward the two rods adhere to each other, and stress waves are generated in both. Find the longitudinal displacement in the rods at all $t > 0$.

3.5 Consider one-dimensional flow of traffic described by (3.6.1) and (3.6.2). Before any disturbance occurs the density of vehicles is uniform $\bar{\rho} = $ constant. At $t = 0$ infinitesimal disturbances begin in a finite region of x. How does the disturbance propagate if $\bar{\rho} < \rho_m/2$ (light traffic) and if $\bar{\rho} > \rho_m/2$ (heavy traffic)?

3.6 Let a one-way highway pass through a small town over the small stretch $0 < x < a$, where cars are allowed to enter the highway at the steady rate of q_o per kilometer per hour. Show that the governing equation for the car density $\rho(x,t)$ is

$$\frac{\partial \rho}{\partial t} + u_m \left(1 - \frac{2\rho}{\rho_m}\right) \frac{\partial \rho}{\partial x} = q(x,t),$$

where

$$q = \begin{cases} 0 & \text{if } x < 0 \\ q_o & \text{if } 0 < x < a \\ 0 & \text{if } x > a. \end{cases}$$

Find the density $\rho(x,t)$ for all $t > 0$. Compare the results for $\rho_o > \rho_m/2$ and $\rho_o < \rho_m/2$.

3.7 From Exercise 1.9 the concentration of a cloud of uniform sedimenting particles is found to satisfy

$$\frac{\partial C}{\partial t} + w_o \frac{\partial}{\partial x}[C(1 - \alpha C)] = 0$$

(Kynch, 1952). If the bottom at $x = 0$ is impermeable and the initial concentration is uniform C_o for all $x < 0$, describe the change of $C(x,t)$ for all $x < 0, t > 0$.

3.8 Photosensitive molecules are suspended uniformly in a liquid to form an emulsion in $x > 0$. A light beam entering the emulsion at $x = 0$ is absorbed by the molecules, which in turn decompose to make the emulsion more transparent to light. The problem is to find the light intensity and the concentration of emulsion as a function of x and t (Cheng, 1984).

Let I and N denote the light intensity and the concentration of molecules. The absorption rate of light $\partial I/\partial x$ is proportional to I and N according to

$$\frac{\partial I}{\partial x} = -aNI, \tag{E3.1}$$

where a is the absorption constant. The rate of decay of molecules is proportional to the rate of light absorption

$$\frac{\partial N}{\partial t} = b\frac{\partial I}{\partial x}, \tag{E3.2}$$

where b is the decay coefficient.

Is the system of partial differential equations for I and N hyperbolic, elliptic, or parabolic?

Let the initial condition be

$$N(x, 0) = N_o \tag{E3.3}$$

and the boundary condition be

$$I(0, t) = I_o. \tag{E3.4}$$

Obtain the exact solution by first rewriting (E3.1) as

$$\frac{\partial \ln I}{\partial x} = -aN$$

and then use (E3.2) to eliminate N to get

$$\frac{\partial \ln I}{\partial t} = -abI + F(t).$$

Use (E3.3), (E3.4) and (E3.1) to show that

$$I(x,t) = \frac{I_o}{1 + e^{-abI_o t}\left(e^{aN_o x} - 1\right)}$$

$$N(x,t) = \frac{N_o}{1 + e^{-aN_o x}\left(e^{abI_o t} - 1\right)}.$$

Sketch this result.

4
Finite domains and separation of variables

Since linear ordinary differential equations are simpler than partial differential equations, it is natural to reduce a problem involving several independent variables to several problems each of which involves just one variable. This is the idea behind the so-called method of *separation of variables*. Typically, one first expresses the original unknown as a product of new unknowns in the form of $u(x,y) = X(x)Y(y)$, in case there are only two independent variables. From the governing partial differential equation for $u(x,y)$, two ordinary differential equations for $X(x)$ and $Y(y)$ are deduced. If the boundary conditions for $u(x,y)$ are given on curves describable by a single coordinate, it may be possible to break the whole problem down to the solution of several boundary-value problems involving just ordinary differential equations.

The procedure described above leads to the important concepts of eigenfunctions, eigenvalues and orthogonality, all of which are very general and powerful for dealing with linear problems. These concepts are first introduced through examples involving rectangular coordinates and ordinary differential equations with constant coefficients, then generalized for the so-called Sturm–Liouville problems. Some examples involving cylindrical polar coordinates are also given near the end of the chapter.

4.1 Separation of variables

Let us discuss one-dimensional water waves of infinitesimal amplitude in a shallow lake of depth h and length L. In the absence of atmospheric forcing, the linearized governing equation for the surface displacement is

$$\frac{\partial^2 \zeta}{\partial t^2} = c^2 \frac{\partial^2 \zeta}{\partial x^2}, \qquad 0 < x < L, \qquad t > 0, \qquad (4.1.1)$$

where $c = \sqrt{gh}$. Let the initial surface displacement and velocity be given by
$$\zeta(x,0) = f(x) \tag{4.1.2}$$
and
$$\frac{\partial \zeta}{\partial t}(x,0) = g(x). \tag{4.1.3}$$

Along the banks the horizontal velocity u must vanish. Consequently, one must require
$$\frac{\partial \zeta}{\partial x}(0,t) = 0 \quad \text{and} \quad \frac{\partial \zeta}{\partial x}(L,t) = 0 \tag{4.1.4}$$

(cf. (1.6.10)). In the space-time diagram the problem is defined in a semi-infinite rectangular strip. The boundary conditions are specified along the lines $x = $ constants, while the initial conditions are specified on the line $t = 0$.

Let us try a solution in the form of a product of two single-valued functions
$$\zeta(x,t) = X(x)T(t), \tag{4.1.5}$$
where X is a function only of x and T is a function only of t, then (4.1.1) requires
$$T''X = c^2 X''T$$
or, after division by XT,
$$\frac{X''}{X} = \frac{1}{c^2} \frac{T''}{T}. \tag{4.1.6}$$

Since the left-hand side is purely a function of x while the right-hand side is purely a function of t, they must both be equal to the same constant, say λ. This leads to two ordinary differential equations
$$X'' - \lambda X = 0 \tag{4.1.7}$$
and
$$T'' - c^2 \lambda T = 0. \tag{4.1.8}$$

From the boundary conditions we get
$$\frac{\partial \zeta}{\partial x}(0,t) = X'(0)T(t) = 0; \quad \frac{\partial \zeta}{\partial x}(L,t) = X'(L)T(t) = 0.$$

Since $T(t)$ cannot be zero for all t, we must have
$$X'(0) = 0 \quad \text{and} \quad X'(L) = 0. \tag{4.1.9}$$

4.1 Separation of variables

Thus X is governed by a homogeneous boundary-value problem defined by (4.1.7) and (4.1.9). The general solution is

$$X(x) = Ae^{-\sqrt{\lambda}x} + Be^{\sqrt{\lambda}x}. \tag{4.1.10}$$

The corresponding solution for T is

$$T(t) = a\cosh\left(\sqrt{|\lambda|}ct\right) + b\sinh\left(\sqrt{|\lambda|}ct\right).$$

Applying the boundary conditions on (4.1.10), we get

$$\sqrt{\lambda}(-A + B) = 0$$

$$\sqrt{\lambda}(-Ae^{-\sqrt{\lambda}L} + Be^{\sqrt{\lambda}L}) = 0.$$

In order that the coefficients A and B are not identically zero, i.e., the solution of X is nontrivial, the determinant of coefficients must vanish,

$$\sqrt{\lambda}\begin{vmatrix} -1 & 1 \\ -e^{-\sqrt{\lambda}L} & e^{\sqrt{\lambda}L} \end{vmatrix} = \sqrt{\lambda}\left(-e^{\sqrt{\lambda}L} + e^{-\sqrt{\lambda}L}\right)$$

$$= -2\sqrt{\lambda}\sinh\sqrt{\lambda}L = 0, \tag{4.1.11}$$

implying that λ must be a root of this transcendental equation. Clearly, λ cannot be real and positive. Let us examine all other possibilities.

(i) $\lambda = 0$: The solution is

$$X = A + Bx.$$

To satisfy the boundary conditions, $X = A$ = constant, which is trivial. This option is of limited interest.

(ii) $\lambda < 0$: For this last choice (4.1.11) becomes

$$\sinh i\sqrt{|\lambda|}L = i\sin\sqrt{|\lambda|}L = 0. \tag{4.1.12}$$

Possible roots are

$$\sqrt{|\lambda|}L = n\pi, \quad \text{i.e.,} \quad -\lambda = \lambda_n \equiv \left(\frac{n\pi}{L}\right)^2, \quad n = 0, 1, 2, 3, \ldots. \tag{4.1.13}$$

These special values λ_n, which render the solution of the homogeneous problem nontrivial, are called the *eigenvalues*. Equation (4.1.11) is called the *eigenvalue condition*. The governing equations (4.1.7) and (4.1.9.a,b) constitute an eigenvalue problem. The corresponding nontrivial solutions

$$X_n = \frac{1}{2}\left(e^{-i\sqrt{\lambda_n}x} + e^{i\sqrt{\lambda_n}x}\right) = \cos\frac{n\pi x}{L} \tag{4.1.14}$$

are called *eigenfunctions*.

Using (4.1.13), we get the corresponding solution for T,

$$T_n(t) = a_n \cos \frac{n\pi ct}{L} + b_n \sin \frac{n\pi ct}{L}.$$

Because the problem is linear, the most general solution is the linear superposition of all the products $X_n T_n$, $n = 0, 1, 2, \ldots$, i.e.,

$$u = \sum_{n=0}^{\infty} \cos \frac{n\pi x}{L} \left(a_n \cos \frac{n\pi ct}{L} + b_n \sin \frac{n\pi ct}{L} \right), \qquad (4.1.15)$$

where the expansion coefficients a_n and b_n are yet unknown.

To determine the expansion coefficients let us apply the initial conditions

$$\zeta(x,0) = f(x) = \sum_{n=0}^{\infty} a_n \cos \frac{n\pi x}{L} \qquad (4.1.16)$$

and

$$\frac{\partial \zeta}{\partial t}(x,0) = g(x) = \sum_{n=0}^{\infty} \frac{n\pi c}{L} b_n \cos \frac{n\pi x}{L}. \qquad (4.1.17)$$

Note the following identities:

$$\int_0^\pi \cos my \cos ny \, dy = \left[\frac{\sin(m-n)y}{2(m-n)} + \frac{\sin(m+n)y}{2(m+n)} \right]_0^\pi = 0$$

if $m \neq n$,

$$\int_0^\pi \cos^2 my \, dy = \pi/2$$

if $m = n \neq 0$, and

$$\int_0^\pi \cos^2 my \, dy = \pi$$

if $m = n = 0$. By the change of variables, $y = \pi x/L$, these identities become

$$\int_0^L dx \cos \frac{m\pi x}{L} \cos \frac{n\pi x}{L} = \begin{cases} 0, & m \neq n \\ L/2, & m = n \neq 0 \\ L, & m = n = 0. \end{cases} \qquad (4.1.18)$$

Multiplying (4.1.16) by $\cos n\pi x/L$ and integrating with respect to x from 0 to L, we get

$$a_0 = \frac{1}{L} \int_0^L f(x) dx \equiv \overline{f}$$

4.1 Separation of variables

$$a_n = \frac{2}{L} \int_0^L f(x) \cos \frac{n\pi x}{L} dx.$$

For compactness, these two equations can be combined by introducing the Jacobi symbol ϵ_n, where

$$\epsilon_0 = 1, \quad \epsilon_n = 2, \quad n = 1, 2, 3, \ldots,$$

then

$$a_n = \frac{\epsilon_n}{L} \int_0^L f(x) \cos \frac{n\pi x}{L} dx, \quad n = 0, 1, 2, \ldots. \tag{4.1.19}$$

Similarly, we multiply (4.1.17) by $\cos n\pi x/L$ and integrate to get

$$b_n = \frac{2}{n\pi c} \int_0^L g(x) \cos \frac{n\pi x}{L} dx. \tag{4.1.20}$$

Thus all the coefficients are determined, and the final solution is

$$\zeta(x,t) = \sum_{n=0}^{\infty} \cos \frac{n\pi x}{L} \left\{ \left[\frac{\epsilon_n}{L} \int_0^L f(x') \cos \frac{n\pi x'}{L} dx' \right] \cos \frac{n\pi ct}{L} \right.$$
$$\left. \left[\frac{2}{n\pi c} \int_0^L g(x') \cos \frac{n\pi x'}{L} dx' \right] \sin \frac{n\pi ct}{L} \right\}. \tag{4.1.21}$$

Note that the success of the method of separation of variables depends not only on reducing the partial differential equation to two ordinary differential equations, but also on expressing the boundary and initial conditions in terms of a single variable only. The latter is possible only if the boundary or data are specified on lines that are described by a single coordinate.

Let us examine the physical meaning of a typical term in the series solution. The quantity $\omega \equiv \pi c/L$ may be called the fundamental frequency of the basin oscillation. The nth term of the series

$$\cos \frac{n\pi x}{L} \left\{ a_n \cos \frac{n\pi ct}{L} + b_n \sin \frac{n\pi ct}{L} \right\} \tag{4.1.22}$$

may then be called the nth mode of oscillation. The factor $\cos \frac{n\pi x}{L}$ describes the spatial structure of the nth mode. For $n = 1$ the water surface has one node at $x = \frac{L}{2}$, as shown in Figure 4.1. The water surface rises on one side but falls on the other. This mode is the sloshing mode, or the fundamental mode, because it has the fundamental frequency ω.

For $n = 2$ the surface has two nodes at

$$\frac{2\pi x}{L} = \frac{\pi}{2}, \frac{3\pi}{2}, \quad \text{i.e.,} \quad x = \frac{L}{4}, \frac{3L}{4},$$

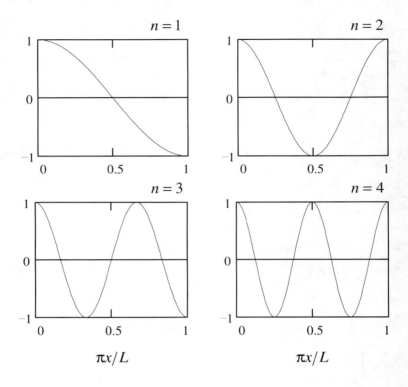

Fig. 4.1. The first few eigenmodes in a narrow lake: $\cos n\pi x/L$ with $n = 1, 2, 3$ and 4.

as shown in Figure 4.1. Similarly, for the nth mode, there are n nodes at

$$\frac{n\pi x}{L} = \frac{\pi}{2}, \frac{3\pi}{2}, \ldots, \left(n - \frac{1}{2}\right)\pi.$$

The number of nodes of an eigenfunction can be predicted for more general eigenvalue problems by the so-called oscillation theorems (Ince, 1956).

Because of the trigonometric identities

$$\cos\frac{n\pi x}{L}\cos\frac{n\pi ct}{L} = \frac{1}{2}\left[\cos\frac{n\pi}{L}(x+ct) + \cos\frac{n\pi}{L}(x-ct)\right]$$

and

$$\cos\frac{n\pi x}{L}\sin\frac{n\pi ct}{L} = \frac{1}{2}\left[\sin\frac{n\pi}{L}(x+ct) - \sin\frac{n\pi}{L}(x-ct)\right],$$

(4.1.21) can also be viewed as the superposition of progressive waves propagating in opposite directions, as the result of reflection from both banks.

The term with $n = 0$ corresponds to a mode uniform in space and stationary in time. The modal amplitude is

$$a_0 = \frac{1}{L}\int_0^L f(x)dx = \overline{f},$$

which is the spatial average of the initial free-surface displacement. If no fluid is added or taken away initially, a_0 must vanish; this mode cannot be present.

4.2 One-dimensional diffusion

As a second illustration for the use of separation of variables, we consider the diffusion of heat in a slab of thickness L. The governing equation for the temperature, denoted here by u, is

$$\frac{\partial u}{\partial t} = k\frac{\partial^2 u}{\partial x^2}, \quad t > 0, \quad 0 < x < L. \tag{4.2.1}$$

If we let the boundaries be kept at the same temperature, say zero, then the boundary conditions are homogeneous:

$$u(0, t) = u(L, t) = 0, \quad t > 0. \tag{4.2.2}$$

Initially, there is a nonuniform distribution of temperature in the slab

$$u(x, 0) = f(x), \quad 0 < x < L. \tag{4.2.3}$$

Let us try separation of variables again

$$u = X(x)T(t).$$

Omitting the trivial cases in which the separation constant is either positive or zero, we write

$$\frac{X''}{X} = \frac{T'}{kT} = -\lambda^2,$$

where λ is real. This equality leads to two ordinary differential equations

$$T' + \lambda^2 kT = 0$$

and

$$X'' + \lambda^2 X = 0.$$

The general solutions are
$$T = e^{-k\lambda^2 t}$$
and
$$X = A \sin \lambda x + B \cos \lambda x.$$
The boundary conditions for X are
$$X(0) = X(L) = 0.$$
Consequently, $B = 0$ and the eigenvalue condition is
$$\sin \lambda L = 0, \tag{4.2.4}$$
which gives the eigenvalues
$$\lambda_n = \frac{n\pi}{L}, \quad n = 1, 2, 3, \ldots. \tag{4.2.5}$$
The eigenfunctions are
$$X_n = \sin \frac{n\pi x}{L}, \quad n = 1, 2, 3, \ldots,$$
and the corresponding time factor is
$$T_n = \exp\left(-\left(\frac{n\pi}{L}\right)^2 kt\right), \quad n = 1, 2, 3, \ldots.$$
Hence the total solution is of the form
$$u = \sum_{n=0}^{\infty} a_n \exp\left(-\left(\frac{n\pi}{L}\right)^2 kt\right) \sin \frac{n\pi x}{L}. \tag{4.2.6}$$
Now the initial condition (4.2.3) implies
$$f(x) = \sum_{n=1}^{\infty} a_n \sin \frac{n\pi x}{L}. \tag{4.2.7}$$
For a pair of sines, the following identities hold:
$$\int_0^L \sin \frac{m\pi x}{L} \sin \frac{n\pi x}{L} \, dx = \begin{cases} 0, & m \neq n \\ \frac{L}{2}, & m = n. \end{cases} \tag{4.2.8}$$
Multiplying both sides of (4.2.7) by $\sin \frac{n\pi x}{L}$ and integrating from 0 to L, we get
$$a_n = \frac{2}{L} \int_0^L f(x) \sin \frac{n\pi x}{L} dx, \quad n = 1, 2, 3, \ldots. \tag{4.2.9}$$

Thus the solution is completely determined

$$U(x,t) = \sum_{n=1}^{\infty} \left(\frac{2}{L} \int_0^L f(x') \sin \frac{n\pi x'}{L} dx' \right) \sin \frac{n\pi x}{L} \exp\left(-\left(\frac{n\pi}{L}\right)^2 kt \right).$$
(4.2.10)

Unlike the wave equation whose solution oscillates in time, the diffusion process is distinguished by attenuation in time. In particular the decay time scale for the nth mode is $(L^2/k)(1/n^2\pi^2)$, which is proportional to the square of the thickness but inversely proportional to the thermal diffusivity. It is also inversely proportional to n. The characteristic decay time for the slab is given by the term $n = 1$, i.e., $L^2/k\pi^2$. This result can be independently confirmed by balancing the order of magnitudes of the two terms in the diffusion equation without carrying out the solution. As the famed chef Julia Child can testify, if it takes three minutes per side to grill a steak two centimeters thick, nine minutes per side are needed for a four-centimeter steak.

4.3 Eigenfunctions and base vectors

The property represented by (4.1.18) and (4.2.8) is called *orthogonality*, and is really a generalization of a concept from three dimensions to infinite dimensions. To see the parallelism let us first list the well-known features of the vector space in three dimensions.

i) In Cartesian coordinates the base vectors of unit length are directed along the three coordinate axes: $\mathbf{e}_1, \mathbf{e}_2, \mathbf{e}_3$.

ii) The base vectors are mutually orthogonal:

$$\langle \mathbf{e}_i, \mathbf{e}_j \rangle \equiv \mathbf{e}_i \cdot \mathbf{e}_j = \delta_{ij},$$

where $\langle \mathbf{a}, \mathbf{b} \rangle$ is the general symbol for the scalar product of two vectors $\mathbf{a}$ and $\mathbf{b}$, and δ_{ij} denotes the Kronecker delta.

iii) Any vector $\mathbf{A}$ in the three-dimensional vector space can be expanded as a linear sum of the base vectors

$$\mathbf{A} = \sum_{j=1}^{3} a_j \mathbf{e}_j.$$

iv) The expansion coefficient a_j is just the projection of $\mathbf{A}$ in the direction of the base vector $\mathbf{e}_j$. Therefore, it is found by taking the scalar product of $\mathbf{A}$ and $\mathbf{e}_i$,

$$a_j = \mathbf{A} \cdot \mathbf{e}_j.$$

Now let us list the parallel features for the infinite dimensional space of eigenfunctions. We first extend the definition of the scalar product of any two functions defined in the interval $(0, L)$:

$$\langle f, g \rangle = \int_0^L f(x)g(x)\, dx.$$

i) For the Fourier sine series the infinite sequence of eigenfunctions forms the base vectors

$$\sin\frac{\pi x}{L}, \sin\frac{2\pi x}{L}, \ldots, \sin\frac{n\pi x}{L}, \ldots$$

ii) The base vectors are orthogonal in the sense that

$$\left\langle \sin\frac{n\pi x}{L}, \sin\frac{m\pi x}{L} \right\rangle \equiv \int_0^L \sin\frac{n\pi x}{L} \sin\frac{m\pi x}{L}\, dx = \frac{L}{2}\delta_{nm}.$$

iii) Any $f(x)$ can be expanded as an infinite sum of the base vectors

$$f(x) = \sum_{n=1}^{\infty} f_n \sin\frac{n\pi x}{L}.$$

iv) The nth expansion coefficient is obtained by taking the scalar product of $f(x)$ and $\sin n\pi x/L$

$$\left\langle f, \sin\frac{n\pi x}{L} \right\rangle = \sum_{m=1}^{\infty} f_m \left\langle \sin\frac{n\pi x}{L}, \sin\frac{m\pi x}{L} \right\rangle.$$

The expansion coefficient

$$f_n = \frac{2}{L} \int_0^L f(x) \sin\frac{n\pi x}{L}\, dx$$

is the projection of f along the base vector $\sin\frac{n\pi x}{L}$.

4.4 Partially insulated slab

A slab of thickness L and diffusivity k is shielded by two insulating layers of thicknesses L_1, L_2 and diffusivities k_1, k_2, as shown in Figure 4.2. The insulating layers are surrounded by heat reservoirs of temperature $T_1(t)$ on the left and $T_2(t)$ on the right. Assume that k_1, k_2 and k are of the same order but $L_1, L_2 \ll L$. What is the temperature in the main slab if it is $f(x)$ initially?

In principle, one can formulate the entire problem by writing down a diffusion equation for each layer and invoking continuity of temperature

4.4 Partially insulated slab

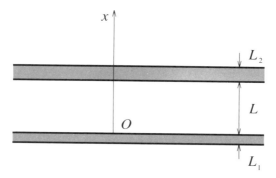

Fig. 4.2. An insulated slab.

and heat fluxes at the interfaces $x = 0$ and $x = L$, as well as the fixed-temperature conditions at $x = -L_1$ and $x = L + L_2$. This laborious procedure is necessary if all three layers are comparable in thickness. For the present case an approximation suffices.

Because of its dominant thickness, the center slab controls the time scale of the diffusion process

$$T = \frac{L^2}{k}.$$

Consider the diffusion equation in the left insulator, $-L_1 < x < 0$,

$$\frac{\partial u_1}{\partial t} = k_1 \frac{\partial^2 u_1}{\partial x^2}.$$

The two terms are of the order

$$\frac{\partial u_1}{\partial t} \sim \frac{U}{L^2/k}, \qquad k_1 \frac{\partial^2 u_1}{\partial x^2} \sim \frac{k_1}{L_1^2} U,$$

where U is the typical scale of the temperature in the system. Hence their ratio is

$$\frac{\partial u_1}{\partial t} \bigg/ k_1 \frac{\partial^2 u_1}{\partial x^2} \sim \frac{k}{k_1} \frac{L_1^2}{L^2} \ll 1.$$

In other words,

$$\frac{\partial^2 u_1}{\partial x^2} \cong 0$$

with a relative error of $O(L_1/L)^2$. Physically, the diffusion process appears so slow to the insulators as to be essentially steady. Obviously, the temperature in the left insulator is approximately given by

$$u_1(x,t) = T_1(t) + \left(\frac{\partial u_1}{\partial x}\right)(x + L_1), \qquad (4.4.1)$$

which satisfies the boundary condition that

$$u_1 = T_1(t) \quad \text{on} \quad x = -L_1,$$

and

$$\frac{\partial u_1}{\partial x} = \frac{u_1(0,t) - T_1(t)}{L_1} \qquad (4.4.2)$$

is constant in x. Since

$$u = u_1 \quad \text{and} \quad K\frac{\partial u}{\partial x} = K_1\frac{\partial u_1}{\partial x}, \quad x = 0, \qquad (4.4.3)$$

with $k = K/c\rho$ and $k_1 = K_1/c_1\rho$, it follows that

$$K_1\frac{\partial u_1}{\partial x} = \frac{K_1}{L_1}[u(0,t) - T_1] = K\frac{\partial u}{\partial x}, \quad x = 0,$$

i.e.,

$$\frac{\partial u}{\partial x} - \frac{K_1}{KL_1}u = -\frac{K_1}{KL_1}T_1(t), \quad x = 0, \qquad (4.4.4)$$

which serves as a boundary condition for u at $x = 0$.

Similarly, the temperature u_2 in the right insulator is also quasi-steady

$$u_2(x,t) = T_2(t) + \left(\frac{\partial u_2}{\partial x}\right)(x - L - L_2),$$

where

$$\frac{\partial u_2}{\partial x} = \frac{T_2(t) - u_1(L,t)}{L_2}$$

is constant in x. Hence

$$k_2\frac{\partial u_2}{\partial x} = \frac{K}{L_2}[T_2 - u_1(L,t)] = K\frac{\partial u}{\partial x}, \quad x = L,$$

or

$$\frac{\partial u}{\partial x} + \frac{K_2}{KL_2}u = \frac{K_2}{KL_2}T_2(t), \quad x = L, \qquad (4.4.5)$$

which is another boundary condition for u at $x = L$. Together with the initial condition

$$u(x,0) = f(x), \quad 0 < x < L, \qquad (4.4.6)$$

we now have a problem just for the center slab. The two boundary

4.4 Partially insulated slab

conditions (4.4.4) and (4.4.5) involve the unknown and its derivative and is of the mixed type.

Now let us solve this problem for $u(x,t)$ with zero ambient temperature

$$T_1 = T_2 = 0. \tag{4.4.7}$$

By separation of variables

$$u = X(x)T(t),$$

we get

$$T' + k\lambda^2 T = 0, \qquad X'' + \lambda^2 X = 0.$$

Let the insulators be of the same thickness and made of the same materials so that

$$k_1 = k_2, \qquad L_1 = L_2, \qquad h = \frac{K_1}{KL_1} = \frac{K_2}{KL_2}.$$

Then X must obey the boundary conditions

$$X' - hX = 0, \qquad x = 0; \qquad X' + hX = 0, \qquad x = L.$$

The general solution to the homogeneous equation for X is

$$X = C_1 \cos \lambda x + C_2 \sin \lambda x.$$

From the two boundary conditions for X, we get

$$hC_1 - \lambda C_2 = 0$$

and

$$(h \cos \lambda L - \lambda \sin \lambda L)C_1 + (h \sin \lambda L + \lambda \cos \lambda L)C_2 = 0.$$

For C_1 and C_2 to be nontrivial, the coefficient determinant must vanish, giving

$$2 \cot \mu = \frac{\mu}{p} - \frac{p}{\mu}, \tag{4.4.8}$$

where $\mu = \lambda L$ and $p = hL$. Equation (4.4.8) is the eigenvalue condition for μ, with p being a parameter. The solution of this transcendental equation can be obtained numerically, but will be examined graphically here. Let us plot the two sides of the equation as functions of μ, i.e.,

$$y = \cot \mu \qquad \text{and} \qquad y = \frac{1}{2}\left(\frac{\mu}{p} - \frac{p}{\mu}\right),$$

as shown in Figure 4.3. Intersections of the two sets of curves give the eigenvalues μ_n

$$\mu_1 < \mu_2 < \mu_3 < \cdots.$$

Correspondingly, we have

$$\lambda_n = \frac{\mu_n}{L}$$

with

$$\lambda_1 < \lambda_2 < \lambda_3 \cdots.$$

The eigenfunctions are

$$X_n = \cos\frac{\mu_n x}{L} + \frac{p}{\mu_n}\sin\frac{\mu_n x}{L}$$

and

$$T_n = \exp\left(-k\frac{\mu_n^2}{L^2}t\right).$$

The general solution is

$$u = \sum_{n=1}^{\infty} a_n \exp\left(-k\frac{\mu_n^2}{L^2}t\right)\left[\cos\frac{\mu_n x}{L} + \frac{p}{\mu_n}\sin\frac{\mu_n x}{L}\right]. \qquad (4.4.9)$$

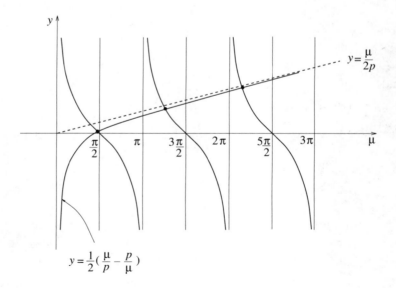

Fig. 4.3. Graphical solution of (4.4.8).

By a lengthy algebra and the use of eigenvalue condition (4.4.8), it can be shown that $X_n, n = 0, 1, 2, 3, \ldots$ form an orthogonal sequence

$$\int_0^L X_n X_m \, dx = 0, \quad m \neq n \tag{4.4.10}$$

and that

$$\int_0^L X_n^2 \, dx = \frac{L}{2} \frac{p(p+2) + \mu_n^2}{\mu_n^2}. \tag{4.4.11}$$

Hence from the initial condition the expansion coefficients can be determined individually

$$a_n = \frac{2}{L} \frac{\mu_n^2}{p(p+2) + \mu_n^2} \int_0^L f(x) \left(\cos \frac{\mu_n x}{L} + \frac{p}{\mu_n} \sin \frac{\mu_n x}{L} \right) dx. \tag{4.4.12}$$

The solution is now complete.

Detailed proof of (4.4.10) is omitted here since there is a more general theory assuring its validity, as discussed below.

4.5 Sturm–Liouville problems

Let us generalize the method of separation of variables and the associated eigenvalue problems by considering the wave equation with variable coefficients

$$\frac{\partial}{\partial x}\left(p\frac{\partial u}{\partial x}\right) - qu = \rho\frac{\partial^2 u}{\partial t^2} \tag{4.5.1}$$

subject to the boundary conditions

$$\alpha u - \beta \frac{\partial u}{\partial x} = 0, \quad x = 0; \quad \gamma u + \delta \frac{\partial u}{\partial x} = 0, \quad x = L \tag{4.5.2}$$

and the initial conditions

$$u(x, 0) = f(x), \quad \frac{\partial u(x, 0)}{\partial t} = g(x). \tag{4.5.3}$$

For reasons to be given shortly, it is assumed that p, q and ρ are continuous functions of x and p, q, ρ are all nonnegative, while α, β, γ and δ are real and positive constants such that

$$\alpha^2 + \beta^2 \neq 0, \quad \gamma^2 + \delta^2 \neq 0. \tag{4.5.4}$$

Using separation of variables again

$$u = X(t)T(t), \tag{4.5.5}$$

we get
$$T'' + \lambda T = 0 \tag{4.5.6}$$
and
$$(pX')' + (\lambda\rho - q)X = 0. \tag{4.5.7}$$
The boundary conditions for X are
$$\alpha X - \beta X' = 0, \quad x = 0; \quad \gamma X + \delta X' = 0, \quad x = L. \tag{4.5.8}$$
What are the eigenvalues of λ that will lead to nontrivial solutions for X?

Equation (4.5.7) is of the Sturm–Liouville type and the eigenvalue problem defined by (4.5.7) and (4.5.8) is called a Sturm–Liouville problem. It can be easily shown that the eigenvalue problems discussed in earlier sections are special cases of the Sturm–Liouville problem, with constant coefficients. Accordingly, infinitely many eigenvalues $\lambda_1 < \lambda_2 < \lambda_3 \cdots$ are expected. To each λ_k corresponds an eigenfunction X_k defined up to a constant multiplier. For convenience let us normalize X such that
$$\int_0^L X_n^2 \rho \, dx = 1. \tag{4.5.9}$$
Under the conditions stated before and in (4.5.4), a number of general theorems can be derived.

Let the Sturm–Liouville operator be defined by
$$\mathcal{L} \stackrel{\text{def}}{=} \frac{d}{dx}\left(p\frac{d}{dx}\right) + (\lambda\rho - q).$$
We first show that the Sturm–Liouville operator is *self-adjoint*, i.e., for any twice differentiable functions Y and Z, $\mathcal{L}$ satisfies the following Green's identity:
$$\int_a^b dx (Z\mathcal{L}Y - Y\mathcal{L}Z) = \left(pZ\frac{dY}{dx} - pY\frac{dZ}{dx}\right)_a^b. \tag{4.5.10}$$
This assertion is proven below by partial integration of the left-hand side
$$LHS = \int_a^b dx \left\{Z(pY')' - Y(pZ')' + Z(\lambda\rho - q)Y - Y(\lambda\rho - q)Z\right\}$$
$$= \int_a^b dx \left\{(pZY' - pYZ')' - pZ'Y' + pY'Z'\right\}$$
$$= \left[pZY' - pYZ'\right]_a^b = RHS.$$

4.5 Sturm–Liouville problems

We now prove three more theorems:

Theorem 1 *All eigenfunctions are orthogonal with respect to the weighting function ρ,*

$$\int_0^L \rho(x) X_k(x) X_n(x)\, dx = 0, \quad k \neq n. \qquad (4.5.11)$$

By definition X_k and X_n satisfy the following differential equations

$$(pX_k')' + (\lambda_k \rho - q) X_k = 0,$$

$$(pX_n')' + (\lambda_n \rho - q) X_n = 0,$$

and (4.5.8). By following the steps leading to (4.5.10), we get

$$(\lambda_k - \lambda_n) \int_0^L \rho X_k X_n\, dx = [pX_k' X_n - pX_n' X_k]_0^L.$$

If $\beta \neq 0$ and $\delta \neq 0$, the boundary terms on the right become

$$\left[p\left(-\frac{\gamma}{\delta}\right) X_k X_n - pX_k \left(-\frac{\gamma}{\delta}\right) X_n\right]_0$$

$$- \left[p\left(\frac{\alpha}{\beta}\right) X_k X_n - pX_n \left(\frac{\alpha}{\beta}\right) X_k\right]_L = 0$$

after using (4.5.8). It follows that

$$(\lambda_k - \lambda_n) \int_0^L \rho X_k X_n\, dx = 0.$$

Since the two eigenvalues are different, Theorem 1 is established.

Theorem 2 *All eigenvalues of the Sturm–Liouville problem are real.*

Let us assume the contrary, i.e., that λ_k is complex, hence X_k is complex in general. Since

$$(pX_k')' + (\lambda_k \rho - q) X_k = 0, \qquad (4.5.12)$$

the complex conjugate of (4.5.12) is also true,

$$(pX_k^{*'})' + (\lambda_k^* \rho - q) X_k^* = 0, \qquad (4.5.13)$$

where the superscript ( )* denotes the complex conjugate. Again, by the steps leading to (4.5.10), we get

$$[pX_k' X_k^* - pX_k^{*'} X_k]_0^L = 0 = (\lambda_n - \lambda_k^*) \int_0^L \rho\, |X_k|^2\, dx.$$

The integrated terms vanish because of the boundary conditions, hence

$$(\lambda_k - \lambda_k^*) \int_0^L \rho |X_k|^2 \, dx = 0 = 2i \text{Im} \, \lambda_k \int_0^L \rho |X|^2 \, dx,$$

where $\text{Im} \, \lambda_k =$ Imaginary part of λ_k. Since $\rho > 0$ by assumption, the integral is positive. We must have $\text{Im} \, \lambda_k = 0$, which means that λ_k is real for all k.

Theorem 3 *The eigenvalues of the Sturm–Liouville problem are all non-negative.*

Rewriting the governing equation as

$$(pX_k')' - qX_k = -\lambda_k \rho X_k ,$$

multiplying both sides by X_k, and integrating the result from 0 to L, we have

$$\lambda_k \int_0^L \rho X_k^2 \, dx = \lambda_k = -\int_0^L X_k \left[(pX_k')' - qX_k \right] dx$$

$$= \int_0^L \left[pX_k' X_k' + qX_k^2 \right] dx - (pX_k X_k')_0^L ,$$

where the normalization condition has been used. Since $p > 0$ and $q > 0$, the last integral is non-negative. In view of the homogeneous boundary conditions and the restrictions on their coefficients, the boundary terms (including the sign) above are also non-negative, hence $\lambda_k \geq 0$.

With these nice properties, we may now summarize the typical procedure for solving the initial-boundary-value problem for a partial differential equation that can be reduced to a Sturm–Liouville problem with homogeneous boundary conditions.

i) Let $u = X(x)T(t)$.

ii) Solve the eigenvalue problem for X to get X_n and λ_n.

iii) Express the general solution as the superposition of all eigenfunctions

$$u = \sum_{n=0}^{\infty} X_n T_n .$$

For (4.5.1) the ordinary differential equation governing T is of the second order, therefore T_n must involve two unknown coefficients, say a_n and b_n. In general if the highest order of time derivatives is m, T_n involves m coefficients for each n.

iv) The initial conditions require

$$u(x,0) = f(x) = \sum_{n=0}^{\infty} X_n(x) T_n(0)$$

$$\frac{\partial u}{\partial t}(x,0) = g(x) = \sum_{n=0}^{\infty} X_n T_n'(0).$$

v) Use orthogonality of X_n to get

$$T_n(0) = \int_0^L \rho f X_n \, dx$$

and

$$T_n'(0) = \int_0^L \rho g X_n \, dx$$

for $n = 0, 1, 2, \ldots$. These equations can be explicitly solved for $a_n, b_n, \ldots$.

4.6 Steady forcing

Let us continue to use the diffusion equation to demonstrate other complications of initial-boundary-value problems. Here we allow steady forcing at the boundary and in the interior. The governing conditions are

$$\frac{\partial u}{\partial t} = k \frac{\partial^2 u}{\partial x^2} + q(x), \quad 0 < x < L, \quad t > 0. \tag{4.6.1}$$

$$\frac{\partial u}{\partial x}(0,t) = A, \quad u(L,t) = B; \quad t > 0 \tag{4.6.2}$$

$$u(x,0) = f(x), \tag{4.6.3}$$

where A and B are constants. Since all forcings are independent of time, a steady state can be expected ultimately. Taking advantage of linearity of the problem, we split the solution into two parts: the steady state $u_o(x)$ and the transient $v(x,t)$, i.e.,

$$u(x,t) = u_o(x) + v(x,t). \tag{4.6.4}$$

The steady state must satisfy

$$k \frac{\partial^2 u_o}{\partial x^2} = -q(x), \quad 0 < x < L$$

$$\frac{\partial u_o}{\partial x}(0) = A, \quad u_o(L) = B$$

and is solvable by inspection

$$u_o = A(x-L) + B + \int_x^L dx' \int_0^{x'} \frac{q(x'')}{k} dx''. \qquad (4.6.5)$$

The transient part must then satisfy the homogeneous equation

$$\frac{\partial v}{\partial t} = k\frac{\partial^2 v}{\partial x^2}, \quad 0 < x < L,$$

the homogeneous boundary conditions

$$\frac{\partial v}{\partial x}(0) = 0, \quad v(L) = 0,$$

and the inhomogeneous initial condition

$$v(x,0) = f(x) - u_o(x) = F(x).$$

The solution for v can be obtained as in §4.2, except that the eigenfunctions must be

$$\cos\left[\left(n + \frac{1}{2}\right)\frac{\pi x}{L}\right].$$

4.7 Transient forcing

Consider the inhomogeneous initial-boundary-value problem defined by

$$\frac{\partial u}{\partial t} - k\frac{\partial^2 u}{\partial x^2} = q(x,t), \quad 0 < x < L, \quad t > 0 \qquad (4.7.1)$$

$$\frac{\partial u}{\partial x}(0,t) = A(t), \quad t > 0 \qquad (4.7.2)$$

$$u(L,t) = B(t), \quad t > 0 \qquad (4.7.3)$$

and

$$u(x,0) = f(x), \quad 0 < x < L. \qquad (4.7.4)$$

Let

$$u = W(x,t) + K(x,t), \qquad (4.7.5)$$

then

$$\frac{\partial W}{\partial t} - k\frac{\partial^2 W}{\partial x^2} + \frac{\partial K}{\partial t} - k\frac{\partial^2 K}{\partial x^2} = q(x,t)$$

$$\frac{\partial W}{\partial x} + \frac{\partial K}{\partial x} = A(t), \quad x = 0$$

4.7 Transient forcing

$$W + K = B(t), \quad x = L$$
$$W + K = f(x), \quad t = 0.$$

Now let us choose K just to satisfy the boundary conditions

$$\frac{\partial K(0,t)}{\partial x} = A(t) \quad \text{and} \quad K(L,t) = B(t),$$

hence

$$K = A(x - L) + B. \qquad (4.7.6)$$

The remaining part W must satisfy

$$\frac{\partial W}{\partial t} - k\frac{\partial^2 W}{\partial x^2} = q(x,t) - \dot{A}(t)(x - L) - \dot{B}(t) \equiv Q(x,t) \qquad (4.7.7)$$

with homogeneous boundary conditions

$$\frac{\partial W}{\partial x}(0,t) = 0, \quad W(L,t) = 0 \qquad (4.7.8)$$

and a modified initial condition

$$W(x,0) = f(x) - A(0)(x - L) - B(0) \equiv F(x). \qquad (4.7.9)$$

To solve for W, we utilize the eigenfunctions that satisfy the homogeneous boundary conditions

$$W(x,t) = \sum_{n=0}^{\infty} T_n(t) \cos\left[\left(n + \frac{1}{2}\right)\frac{\pi x}{L}\right]. \qquad (4.7.10)$$

The expansion coefficients $T_n(t)$ are now unknown functions of t. To determine them we substitute the above equation into (4.7.7)

$$\sum_{n=0}^{\infty} \left\{T_n' + \left[\left(n + \frac{1}{2}\right)\frac{\pi}{L}\right]^2 k T_n\right\} \cos\left[\left(n + \frac{1}{2}\right)\frac{\pi x}{L}\right] = Q(x,t).$$

It follows from orthogonality that

$$T_n' + \left[\left(n + \frac{1}{2}\right)\frac{\pi}{L}\right]^2 k T_n$$
$$= \frac{2}{L}\int_0^L Q(x,t) \cos\left[\left(n + \frac{1}{2}\right)\frac{\pi x}{L}\right] dx \equiv Q_n(t), \qquad (4.7.11)$$

which is an inhomogeneous initial-value problem for $T_n(t)$. The initial condition for $T_n(0)$ is found from (4.7.9)

$$W(x,0) = F(x) = \sum_{n=0}^{\infty} T_n(0) \cos\left[\left(n+\frac{1}{2}\right)\frac{\pi x}{L}\right]$$

and by orthogonality

$$T_n(0) = \frac{2}{L}\int_0^L F(x)\cos\left[\left(n+\frac{1}{2}\right)\frac{\pi x}{L}\right] dx. \qquad (4.7.12)$$

Finally, (4.7.11) can be rewritten as

$$e^{-\lambda_n k t}\left[e^{\lambda_n k t} T_n\right]' = Q_n,$$

where

$$\lambda_n = \left[\left(n+\frac{1}{2}\right)\frac{\pi}{L}\right]^2.$$

Upon integration we get

$$T_n(t) = T_n(0)\, e^{-\lambda_n k t} + \int_0^t Q_n(\tau)\, e^{-\lambda_n k(t-\tau)}\, d\tau. \qquad (4.7.13)$$

The mathematical solution is now complete with W given by (4.7.10) and K given by (4.7.6).

4.8 Two-dimensional diffusion

The method of separation of variables can be extended to higher dimensions as long as the boundaries are parallel to the coordinate axes. For illustration we consider diffusion in a rectangular domain with heat baths on two opposite sides and insulation on the other two. The governing equation now reads

$$\frac{\partial u}{\partial t} = k\left(\frac{\partial^2 u}{\partial x^2} + \frac{\partial^2 u}{\partial y^2}\right), \quad 0 < x < a,\, 0 < y < b. \qquad (4.8.1)$$

We assume the initial condition

$$u(x,y,0) = f(x,y) \qquad (4.8.2)$$

and the boundary conditions

$$\frac{\partial u}{\partial x}(0,y,t) = \frac{\partial u}{\partial x}(a,y,t) = 0, \quad 0 < y < b \qquad (4.8.3)$$

and

$$u(x,0,t) = u(x,b,t) = 0, \quad 0 < x < a. \qquad (4.8.4)$$

4.8 Two-dimensional diffusion

Let
$$u = X(x)Y(y)T(t),$$
then
$$\frac{X''}{X} + \frac{Y''}{Y} = \frac{T'}{kT}.$$
This equality implies that
$$T' + \lambda kT = 0,$$
$$X'' - \mu X = 0$$
and
$$Y'' + (\lambda + \mu)Y = 0,$$
where λ and μ are separation constants. From the original boundary conditions on u we find
$$X'(0) = X'(a) = 0$$
$$Y(0) = Y(b) = 0.$$
With $\mu = -\alpha^2$ where α is real, the eigenvalue and eigensolutions for $X(x)$ are
$$\alpha = \alpha_m = \frac{m\pi}{a}, \quad m = 1, 2, 3, \ldots,$$
$$X_m(x) = A_m \cos\frac{m\pi x}{a}, \quad m = 0, 1, 2, \ldots.$$
Now to seek the eigenfunctions for Y, we first write
$$\beta^2 = \lambda + \mu = \lambda - \alpha^2. \tag{4.8.5}$$
The equation and the boundary conditions for Y then require that
$$Y = \sin \beta u$$
and
$$\beta = \beta_n \equiv \frac{n\pi}{b}, \quad n = 1, 2, 3, \ldots.$$
Thus the eigenfunctions are
$$Y_n = \sin\frac{n\pi y}{b}, \quad n = 1, 2, 3, \ldots.$$
Equation (4.8.5) implies that
$$\lambda_n = \alpha_n^2 + \beta_n^2 = \left(\frac{m\pi}{a}\right)^2 + \left(\frac{n\pi}{b}\right)^2.$$

The corresponding time factor is

$$T_{mn}(t) = a_{mn} \exp\left\{-\left[\left(\frac{m\pi}{a}\right)^2 + \left(\frac{n\pi}{b}\right)^2\right]kt\right\}.$$

The general solution is then found by superposition

$$u = \sum_{m=0}^{\infty}\sum_{n=1}^{\infty} a_{mn} \exp\left\{-\left[\left(\frac{m\pi}{a}\right)^2 + \left(\frac{n\pi}{b}\right)^2\right]kt\right\} \cos\frac{m\pi x}{a} \sin\frac{n\pi y}{b}.$$

Let us now apply the initial conditions

$$u(x,y,0) = f(x,y) = \sum_{m=0}^{\infty}\sum_{n=1}^{\infty} a_{mn} \cos\frac{m\pi x}{a} \sin\frac{n\pi y}{b}.$$

By using orthogonality in both directions of x and y, we find

$$a_{0n} = \frac{2}{ab}\int_0^a\int_0^b f(x,y) \sin\frac{n\pi y}{b}\,dx dy$$

$$a_{mn} = \frac{4}{ab}\int_0^a\int_0^b dx dy\, f(x,y) \cos\frac{m\pi x}{a}\sin\frac{n\pi y}{b}\,dx dy.$$

The solution is mathematically complete.

4.9 Cylindrical polar coordinates

So far we have discussed only rectangular domains. There are also many other geometries that can be described by the so-called orthogonal curvilinear coordinates.† In particular for domains with circular boundaries the cylindrical polar coordinates are best suited for analytical purposes. Again, if the boundary conditions are specified by one coordinate, separation of variables can be tried.

Let us first change all space derivatives from rectangular to cylindrical polar coordinates.

As shown in Figure 4.4, any given point in space is equally defined by either (x,y,z) or (r,θ,z), which are related by

$$x = r\cos\theta, \qquad y = r\sin\theta, \qquad z = z. \qquad (4.9.1)$$

We remark that the cylindrical polar coordinate system is but a special case of the curvilinear orthogonal coordinate system for which a general theory of transformation of derivatives is available (see, e.g., Hildebrand,

---

† An encyclopedic collection of such coordinate systems is given in Morse and Feshbach (1953).

4.9 Cylindrical polar coordinates

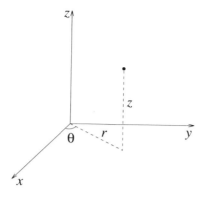

Fig. 4.4. Cylindrical polar coordinates.

1964). This general theory is, however, lengthy, hence we shall discuss only the specific case on hand. First, by the chain rule of differentiation, we have for any function $f(x(r,\theta), y(r,\theta), z)$

$$\frac{\partial f}{\partial r} = \frac{\partial x}{\partial r}\frac{\partial f}{\partial x} + \frac{\partial y}{\partial r}\frac{\partial f}{\partial y} = \cos\theta \frac{\partial f}{\partial x} + \sin\theta \frac{\partial f}{\partial y}$$

$$\frac{1}{r}\frac{\partial f}{\partial \theta} = \frac{1}{r}\left(\frac{\partial x}{\partial \theta}\frac{\partial f}{\partial x} + \frac{\partial y}{\partial \theta}\frac{\partial f}{\partial y}\right) = -\sin\theta \frac{\partial f}{\partial x} + \cos\theta \frac{\partial f}{\partial y}$$

$$\frac{\partial f}{\partial z} = \frac{\partial f}{\partial z}.$$

The first two equations above can be solved for $\partial f/\partial x$ and $\partial f/\partial y$

$$\frac{\partial f}{\partial x} = \cos\theta \frac{\partial f}{\partial r} - \frac{\sin\theta}{r}\frac{\partial f}{\partial \theta} \qquad (4.9.2)$$

$$\frac{\partial f}{\partial y} = \frac{\cos\theta}{r}\frac{\partial f}{\partial \theta} + \sin\theta \frac{\partial f}{\partial r}. \qquad (4.9.3)$$

The Laplace operator is transformed by repeated use of the preceding rules

$$\frac{\partial^2 f}{\partial x^2} + \frac{\partial^2 f}{\partial y^2} + \frac{\partial^2 f}{\partial z^2} = \left(\cos\theta \frac{\partial}{\partial r} - \frac{\sin\theta}{r}\frac{\partial}{\partial \theta}\right)^2 f$$

$$+ \left(\sin\theta \frac{\partial}{\partial r} + \frac{\cos\theta}{r}\frac{\partial}{\partial \theta}\right)^2 f + \frac{\partial^2 f}{\partial z^2}.$$

After expanding the right-hand side and using trigonometric identities, we get

$$\frac{\partial^2 f}{\partial x^2} + \frac{\partial^2 f}{\partial y^2} + \frac{\partial^2 f}{\partial z^2} = \frac{\partial^2 f}{\partial r^2} + \frac{1}{r}\frac{\partial f}{\partial r} + \frac{1}{r^2}\frac{\partial^2 f}{\partial \theta^2} + \frac{\partial^2 f}{\partial z^2}$$

$$= \frac{1}{r}\frac{\partial}{\partial r}\left(r\frac{\partial f}{\partial r}\right) + \frac{1}{r^2}\frac{\partial^2 f}{\partial \theta^2} + \frac{\partial^2 f}{\partial z^2}. \quad (4.9.4)$$

4.10 Steady heat conduction in a circle

In steady equilibrium, the temperature field $T(r,\theta)$ in the circle $0 < r < a$ is governed by Laplace's equation

$$\nabla^2 T(r,\theta) = \frac{\partial^2 T}{\partial r^2} + \frac{1}{r}\frac{\partial T}{\partial r} + \frac{1}{r^2}\frac{\partial^2 T}{\partial \theta^2} = 0, \quad 0 \le r \le a. \quad (4.10.1)$$

Let the temperature on the circumference be specified

$$T(a,\theta) = f(\theta), \quad 0 \le \theta \le 2\pi. \quad (4.10.2)$$

The temperature should be finite everywhere inside the circle.

Assuming a solution of the separable form

$$T(r,\theta) = R(r)\Theta(\theta), \quad (4.10.3)$$

we get from (4.9.1)

$$\left(R'' + \frac{1}{r}R'\right)\Theta + \frac{1}{r^2}\Theta'' = 0$$

or

$$\frac{r^2 R'' + rR'}{R} = -\frac{\Theta''}{\Theta} = \lambda,$$

where λ is a separation constant. It follows that

$$r^2 R'' + rR' - \lambda R = 0, \quad 0 \le r \le a \quad (4.10.4)$$

$$\Theta'' + \lambda\Theta = 0, \quad -\pi \le \theta \le \pi. \quad (4.10.5)$$

The solution for Θ is

$$\Theta = A\cos\sqrt{\lambda}\theta + B\cos\sqrt{\lambda}\theta.$$

Since the range of θ covers the entire circle, Θ and Θ' must be periodic in θ with the period 2π, i.e., $\Theta(\theta) = \Theta(\theta + 2\pi)$ and $\Theta'(\theta) = \Theta'(\theta + 2\pi)$, which are possible if and only if $\sqrt{\lambda}$ is an integer, or

$$\lambda = n^2, \quad n = 0, 1, 2, \ldots. \quad (4.10.6)$$

4.10 Steady heat conduction in a circle

Now (4.10.4) becomes

$$r^2 R'' + rR' - n^2 R = 0,$$

which admits the general solution

$$R_0 = C_0 + D_0 \ln r, \qquad R_n = C_n r^n + \frac{D_n}{r^n}, \qquad n = 1, 2, 3, \ldots.$$

Since T should be bounded everywhere in the circle, it is necessary to set $D_n = 0$ for all n. Therefore the most general solution must be

$$T = \sum_{n=0}^{\infty} r^n (A_n \cos n\theta + B_n \sin n\theta). \tag{4.10.7}$$

The boundary condition on $r = a$ demands that

$$f(\theta) = \sum_{n=0}^{\infty} a^n (A_n \cos n\theta + B_n \sin n\theta).$$

As will be explained in the next chapter, the above equation amounts to expanding $f(\theta)$ as a trigonometric Fourier series. The expansion coefficients can be easily found by using the following orthogonality properties of $\{\cos n\theta\}$ and $\{\sin n\theta\}$ over the interval $(-\pi, \pi)$:

$$\int_{-\pi}^{\pi} \cos m\theta \cos n\theta \, d\theta = \left[\frac{\sin(m-n)\theta}{2(m-n)} + \frac{\sin(m+n)\theta}{2(m+n)} \right]_{-\pi}^{\pi}$$

$$= \begin{cases} 0 & \text{if } m \neq n \\ \pi & \text{if } m = n \neq 0 \\ 2\pi & \text{if } m = n = 0 \end{cases} \tag{4.10.8}$$

$$\int_{-\pi}^{\pi} \sin m\theta \sin n\theta \, d\theta = \left[\frac{\sin(m-n)\theta}{2(m-n)} - \frac{\sin(m+n)\theta}{2(m+n)} \right]_{-\pi}^{\pi}$$

$$= \begin{cases} 0 & \text{if } m \neq n \\ \pi & \text{if } m = n \neq 0 \end{cases} \tag{4.10.9}$$

$$\int_{-\pi}^{\pi} \sin m\theta \cos n\theta \, d\theta = 0 \quad \text{all} \quad m, n. \tag{4.10.10}$$

Therefore,

$$A_n = \frac{\epsilon_n}{2\pi a^n} \int_{-\pi}^{\pi} F(\theta) \cos n\theta \, d\theta, \tag{4.10.11}$$

where ϵ_n is the Jacobi symbol, and

$$B_n = \frac{1}{2\pi a^n} \int_{-\pi}^{\pi} f(\theta) \sin n\theta \, d\theta. \tag{4.10.12}$$

The solution is complete.

Note at the center, $r = 0$,

$$T(0, \theta) = A_0 = \frac{1}{2\pi} \int_{-\pi}^{\pi} f(\theta) \, d\theta.$$

Thus the temperature at the center of the circle is equal to the circumferential average of the boundary value. The preceding formula is called the *mean-value theorem* for harmonic functions.

We postpone the discussion of wave and diffusion problems involving circular boundaries until a later chapter.

Lest the reader should think that the routine of separation of variables is infallible, we point out that the success in the present example can be easily spoiled by a seemingly small change of conditions. The success here is associated with the property of orthogonality because the domain in this problem is the full circle. In an incomplete circle, e.g., a fan-like domain, one may lose the advantage of orthogonality.

To demonstrate this point, we modify the steady heat-conduction problem by assuming that there is a thin insulator along the radius $\theta = \pi$, $0 \leq r < a$ so that there is no heat flux across the radius

$$\frac{\partial T}{\partial \theta} = 0, \quad 0 \leq r < a, \quad \theta = \pm \pi. \tag{4.10.13}$$

On the circular boundary we still prescribe the temperature

$$T(a, \theta) = f(\theta), \quad -\pi < \theta < \pi. \tag{4.10.14}$$

The physical domain is no longer a full circle.†

By separation of variables we get

$$T = r^\lambda (A \cos \lambda \theta + B \lambda \sin \lambda \theta).$$

For T to be finite at the origin λ cannot be negative. The condition on the insulated surfaces requires that

$$-\lambda(A \sin \lambda \pi - B \cos \lambda \pi) = 0$$

$$\lambda(A \sin \lambda \pi + B \cos \lambda \pi) = 0.$$

† This example is also relevant to the displacement field of a plane crack in an elastic solid.

In order that A and B are not zeros simultaneously, λ must satisfy

$$\sin 2\lambda\pi = 0 \quad \text{or} \quad \lambda = \frac{n}{2}, \quad n = 0, 1, 2, \ldots.$$

The general solution is, therefore,

$$w(r, \theta) = \sum_{n=0}^{\infty} r^{n/2} \left(A_{\frac{n}{2}} \cos\frac{n}{2}\theta + B_{\frac{n}{2}} \sin\frac{n}{2}\theta \right). \tag{4.10.15}$$

Now the boundary condition on the circle $r = a$ requires that

$$f(\theta) = \sum_{n=0}^{\infty} a^{n/2} \left(A_{\frac{n}{2}} \cos\frac{n}{2}\theta + B_{\frac{n}{2}} \sin\frac{n}{2}\theta \right).$$

The sequences $\{\cos\frac{n}{2}\theta\}$ and $\{\sin\frac{n}{2}\theta\}$ are, unfortunately, not orthogonal in the range $(-\pi, \pi)$, hence the coefficients $A_{\frac{n}{2}}, B_{\frac{n}{2}}$ cannot be determined one at a time, but must be calculated by numerical means.

The reader should try a potential problem in a fan-like domain with a Dirichlet or Neuman condition on the three sides: ($\theta = 0, 0 < r < a$); ($\theta = \alpha, 0 < r < a$); and ($r = a, 0 < \theta < \alpha$), where α is smaller than 2π.

Exercises

4.1 The spreading of a toxic waste that decays in time is often modeled by the diffusion equation

$$\frac{\partial C}{\partial t} - k\frac{\partial^2 C}{\partial x^2} + aC = f(x, t),$$

where a denotes the decay rate and $f(x, t)$ the source strength per unit distance and unit time. Let the domain be $0 < x < L$ and no flux be permitted at both ends. Initially the concentration is zero everywhere. Find $C(x, t)$ for $t > 0$ for general f. Discuss the result for the following special source function f:

$$f(x, t) = \begin{cases} f_o \sin\left(\frac{\pi t}{T}\right) & \text{for } 0 \leq t \leq T \\ 0 & \text{for } t > T, \end{cases}$$

where f_o is a constant. Comment on the effect of the decay rate.

4.2 A long cylindrical pipe of circular cross section has the inner radius a and outer radius b. The pipe is heated nonuniformly from within and insulated from without so that

$$T(a, \theta) = f(\theta), \quad \frac{\partial T(b, \theta)}{\partial r} = 0.$$

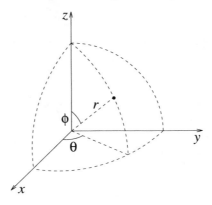

Fig. 4.5. Spherical polar coordinates.

Find the steady-state temperature $T(r,\theta)$ in the pipe wall for $a < r \leq b$.

4.3 As shown in Figure 4.5, a point in space can be represented by the spherical polar coordinates (r, θ, ϕ), which are related to rectilinear coordinates by

$$x = r \cos\theta \sin\phi, \quad y = r \sin\theta \sin\phi, \quad z = r \cos\phi. \qquad (E4.1)$$

Use the chain rule to show that the Laplacian operator in spherical coordinates takes the form

$$\nabla^2 T = \frac{1}{r^2}\frac{\partial}{\partial r}\left(r^2 \frac{\partial T}{\partial r}\right) + \frac{1}{r^2 \sin\phi}\frac{\partial}{\partial \phi}\left(\sin\phi \frac{\partial T}{\partial \phi}\right)$$
$$+ \frac{1}{r^2 \sin^2\phi}\frac{\partial^2 T}{\partial \theta^2}. \qquad (E4.2)$$

Solve the initial-boundary-value problem for a sphere of radius a heated from within by a uniform source. Assume $T(a,t) = 0$ and $T(r,0) = 0$. (*Hint:* Let $T(r,t) = V(r,t)/r$.)

4.4 Referring to Exercise 1.10, let us consider a shallow aquifer with a constant mean depth, flanked by two rivers along $x = 0$ and $x = L$. Starting from $t = 0$, flooding in the west river raises the water level suddenly by a small amount h_1. How does $h(x,t)$ in the aquifer change?

4.5 A cable is tautly stretched between $x = 0$ and $x = L$ under tension T. It is also surrounded by a viscous medium, which gives a resistance force proportional to the lateral velocity of the cable. With both ends fixed, the cable is given the following initial displacement:

$$u(x,0) = \begin{cases} 2Ax/L, & 0 < x < L/2 \\ 2A(1 - x/L), & -L/2 < x < L, \end{cases}$$

while the initial velocity is zero. Find the transient solution and discuss the effect of damping.

4.6 An elastic rod of uniform cross section extends from the fixed end at $x = 0$ to the free end at $x = L$. Before $t = 0$ the rod is compressed at the right end by a force so that the length is reduced slightly by ℓ, where $\ell \ll L$. At $t = 0$ the force is suddenly removed. Find the stress wave in the rod for all $t > 0$.

4.7 An elastic rod of length L and uniform cross section is fixed at the left end $x = 0$ and attached to a mass M at the right end. Show that the boundary condition at $x = L$ is

$$\frac{\partial u}{\partial x} + \frac{M}{EA}\frac{\partial^2 u}{\partial t^2} = 0.$$

Let the initial data be

$$u(x,0) = u_o x/L, \qquad \frac{\partial u(x,0)}{\partial t} = 0$$

for all $0 < x < L$. Find $u(x,t)$ by separation of variables. Verify that the eigenfunctions $\{X_n\}$ are orthogonal.

4.8 In a river of constant depth h, there is a steady uniform flow in the x direction. Sediments are suspended in the flow by turbulent diffusion. Because of their weight, particles also tend to settle downward. Let the settling velocity of a particle be w and the coefficient of turbulent diffusion be k (assumed to be constant for simplicity). Show that the sediment concentration $C(z,t)$ satisfies the following equation:

$$\frac{\partial C}{\partial t} - w\frac{\partial C}{\partial z} = k\frac{\partial^2 C}{\partial z^2}, \qquad 0 < z < h.$$

Assume that the vertical flux rate $-wC - k\,\partial C/\partial z$ at the top is zero, and the bottom is an infinite reservoir of sediment supply so that $C(0,t) = C_o$. Find $C(z,t)$ if $C(z,0) = 0$. What is the ultimate concentration for large t? Check the result by solving the steady-state problem directly.

4.9 Consider the hydrodynamic pressure on a dam in an earthquake, as formulated in Exercise 1.11. Let the earthquake-induced horizontal velocity of the dam surface be

$$u(0, z, t) = \operatorname{Re} U e^{-i\omega t},$$

where U is a constant. Consider only a simple harmonic response in pressure

$$p = \operatorname{Re} P(x, z) e^{-i\omega t}$$

and find the boundary-value problem for P in the semi-infinite strip $0 < x < \infty, 0 < z < h$. Impose the radiation condition that all propagating waves must travel away from the dam. Obtain a formula for the pressure distribution on the dam surface. Discuss the number of propagating modes as a function of the dam frequency ω.

4.10 Pollutant is introduced from the left end of a long lake $0 < x < L$ at the steady rate Q over a finite duration $0 < t < T$ and is diffused toward the right according to the one-dimensional heat equation. At the right end vigilance is maintained so that $C(L, t) = 0$ at all times. Find $C(x, t)$ and the rate of cleanup at the right end.

4.11 Solve the following forced-vibration problem:

$$\frac{\partial^2 u}{\partial t^2} - c^2 \frac{\partial^2 u}{\partial x^2} = q(x, t)$$

$$u = (0, t), \quad u(L, t) = 0;$$

$$u(x, 0) = f(x), \quad \frac{\partial u}{\partial t}(x, 0) = g(x).$$

4.12 The method of separation of variables can be extended to three-dimensional problems. Consider the vibration of a compressible fluid in a box $0 < x < a, 0 < y < b, 0 < z < c$, within which the fluid pressure is governed by

$$\frac{1}{c^2} \frac{\partial^2 p}{\partial t^2} = \frac{\partial^2 p}{\partial x^2} + \frac{\partial^2 p}{\partial y^2} + \frac{\partial^2 p}{\partial z^2}.$$

Let the normal derivatives of p vanish on all six sides of the box and the initial values be

$$p = f(x, y, z), \quad \frac{\partial p}{\partial t} = g(x, y, z), \quad t = 0$$

inside the box. Find $p(x, y, z, t)$.

4.13 Consider a problem that must be described by several independent variables. Moist soil in a cold region such as Alaska expands or consolidates as the seasons change. Consider a uniform soil layer of thickness h, $0 < z < h$. Under certain conditions the soil temperature is governed simply by

$$\frac{\partial T}{\partial t} = k\frac{\partial^2 T}{\partial z^2}, \quad 0 < z < h.$$

The motions of pore fluid and soil matrix are coupled. As will be explained in Chapter 13, conservation of mass of fluid and solid phases leads to

$$(1-n)\frac{\partial v}{\partial z} + n\frac{\partial u}{\partial z} = 0, \quad 0 < z < h,$$

where u and v are the fluid and solid velocities, respectively, and n denotes the original static porosity. Momentum conservation in the pore fluid is expressed by Darcy's law

$$\frac{\partial p}{\partial z} = -\frac{n}{K}(u-v),$$

and K is the soil permeability. Quasi-static equilibrium in the solid matrix requires

$$0 = -(1-n)\frac{\partial p}{\partial z} + \frac{\partial \sigma}{\partial z} + \frac{n^2}{K}(u-v).$$

The time derivative of Hooke's law, including thermal effect, is

$$\frac{\partial \sigma}{\partial t} = (\lambda + 2\mu)\frac{\partial v}{\partial z} + \alpha(\lambda + \frac{2\mu}{3})\frac{\partial T}{\partial t},$$

where λ, μ are elastic constants and α is the coefficient of thermal expansion.

Show first from mass conservation that

$$\frac{\partial p}{\partial z} = \frac{v}{K}, \quad 0 < x < h$$

and then from Hooke's law that

$$(\lambda + 2\mu)\frac{\partial^2 \sigma}{\partial z^2} - \frac{1}{K}\frac{\partial \sigma}{\partial t} = \frac{\alpha}{K}\left(\lambda + \frac{2\mu}{3}\right)\frac{\partial T}{\partial t}, \quad 0 < z < h.$$

Let the soil layer rest on an impermeable bedrock

$$u = v = 0, \quad z = 0.$$

Show that

$$\frac{\partial p}{\partial z} = \frac{\partial \sigma}{\partial z} = 0, \quad z = 0.$$

Let $T(0,t) = 0$ for $t > 0$. On the ground surface $z = h$ the solid stress is zero but the temperature is raised from zero to a finite value T_o for all $t > 0$. For the initial conditions
$$T(z,0) = \sigma(z,0) = 0,$$
solve for the solid stress $\sigma(z,t)$ and the ground deformation
$$V(z,t) = \int_0^t v(z,t)dt$$
for $0 < z < h$.

5
Elements of Fourier series

In the last chapter we were led from separation of variables to the concept of eigenfunctions. When the boundary value is expressed as a series of these eigenfunctions the coefficients are conveniently found by using the property of orthogonality of eigenfunctions. In this chaper we examine, without reference to physics, a special but most useful type of orthogonal series, the trigonometric Fourier series.

5.1 General Fourier series

A motivation of Fourier series is to approximate a given function $f(x)$ by a finite sum of base functions $\{\phi_n(x)\}$

$$f(x) \cong S_N = \sum_{n=0}^{N} c_n \phi_n(x), \tag{5.1.1}$$

where f and ϕ_n's are defined in the interval $(0, L)$, and $\{\phi_n(x)\}$ are orthogonal with respect to the positive weighting function ρ:

$$\int_a^b \rho \phi_m \phi_n \, dx = 0, \quad m \neq n. \tag{5.1.2}$$

The expansion coefficients c_n can be obtained by multiplying both sides of the sum by $\rho \phi_m$ and integrating from 0 to L,

$$\int_0^L \rho \phi_m f \, dx = c_m \int_0^L \rho \phi_m^2 \, dx. \tag{5.1.3}$$

The limit of S_N as $N \to \infty$ is called the Fourier series.

We shall now show that S_N gives the best approximation of f by having the smallest mean-square error, defined by

$$\Delta_N = \int_a^b \rho \left[f - \sum_{n=0}^N c_n \phi_n \right]^2 dx$$

$$= \int_a^b \rho f^2 \, dx - 2 \sum_{n=0}^N c_n \int_a^b \rho f \phi_n \, dx + \int_a^b \rho \left(\sum_{n=0}^N c_n \phi_n \right)^2 dx$$

$$= \int_a^b \rho f^2 \, dx - 2 \sum_{n=0}^N c_n \int_a^b \rho f \phi_n \, dx + \sum_{n=0}^N c_n^2 \int_a^b \rho \phi_n^2 \, dx, \quad (5.1.4)$$

where orthogonality of ϕ_n has been used. Let us choose the coefficients c_n so as to minimize Δ_N

$$\frac{\partial \Delta_N}{\partial c_n} = 0, \quad n = 0, \ldots, N,$$

hence

$$-2 \int_a^b \rho \phi_n f \, dx + 2 c_n \int_a^b \rho \phi_n^2 \, dx = 0$$

for every $n = 0, \ldots, N$. Clearly,

$$c_n = \int_a^b \rho f \phi_n \, dx \bigg/ \int_a^b \rho \phi_n^2 \, dx, \quad (5.1.5)$$

which is precisely the nth Fourier coefficient. Note that the second derivatives

$$\frac{\partial^2 \Delta_N}{\partial c_n^2} = \int_a^b \rho \phi_n^2 \, dx, \quad n = 0, \ldots, N$$

are all positive. Hence, the error is indeed a minimum. Thus, the best approximation of a function f by a finite sum of orthogonal sequence is a truncated Fourier series.

Use of (5.1.5) in the last of (5.1.4) gives

$$\Delta_N = \int_a^b \rho f^2 \, dx - 2 \sum_{n=0}^N c_n^2 \int \rho \phi_n^2 \, dx + \sum_{n=0}^N c_n^2 \int \rho \phi_n^2 \, dx$$

$$= \int_a^b \rho f^2 \, dx - \sum_{n=0}^N c_n^2 \int_a^b \rho \phi_n^2 \, dx.$$

Thus, as N increases, the mean-square error decreases; S_N gives a better approximation to $f(x)$. By definition, $\Delta_N \geq 0$, hence,

$$\int_a^b \rho f^2\, dx \geq \sum_{n=0}^{N} c_n^2 \int_a^b \rho \phi_n^2\, dx. \tag{5.1.6}$$

As N goes to infinity, the sum on the right of (5.1.6) must be bounded by the integral on the left. Thus, if this integral is bounded,

$$\int_a^b \rho f^2\, dx < \infty, \tag{5.1.7}$$

the following must be true,

$$\int_a^b \rho f^2\, dx \geq \sum_{n=0}^{\infty} c_n^2 \int_a^b \rho \phi_n^2\, dx. \tag{5.1.8}$$

This result is called *Bessel's inequality*. The series on its right must converge, hence

$$\lim_{N\to\infty} c_N \int_a^b \rho \phi_N^2\, dx = 0,$$

implying, in turn, that $c_N \to 0$ as $N \to \infty$. We say that S_N converges to f in the mean-square sense if

$$\lim_{N\to\infty} \int_a^b [f(x) - S_N(x)]^2\, \rho\, dx \to 0, \tag{5.1.9}$$

which implies

$$\int_a^b \rho f^2 dx = \sum_{n=0}^{\infty} c_n^2 \int_a^b \rho \phi_n^2\, dx. \tag{5.1.10}$$

If, furthermore, S_N converges to f in the mean-square sense for *every* f that satisfies (5.1.7), then the set of base functions $\{\phi_n\}$ is said to be complete, which means that every f can be expanded as

$$f = \sum_{n=0}^{\infty} c_n \phi_n(x). \tag{5.1.11}$$

Other implications of the completeness condition may be found in Tolstov (1962).

5.2 Trigonometric Fourier series

5.2.1 Full Fourier series

The best-known Fourier series is the trigonometric series, where the base functions are sines and cosines. Recall from (4.10.8–4.10.10) that the set of base functions $\{\sin mx, \cos mx\}$ are orthogonal as follows:

$$\int_{-\pi}^{\pi} \cos mx \cos nx \, dx = \begin{cases} 0 & \text{if } m \neq n \\ \pi & \text{if } m = n \neq 0 \\ 2\pi & \text{if } m = n = 0 \end{cases}$$

$$\int_{-\pi}^{\pi} \sin mx \sin nx \, dx = \begin{cases} 0 & \text{if } m \neq n \\ \pi & \text{if } m = n \neq 0 \end{cases}$$

$$\int_{-\pi}^{\pi} \sin mx \cos nx \, dx = 0 \quad \text{all} \quad m, n.$$

If $f(x), x \in (-\pi, \pi)$ is expanded as a trigonometric Fourier series

$$f(x) = \frac{a_o}{2} + \sum_{n=0}^{N} (a_n \cos nx + b_n \sin nx), \tag{5.2.1}$$

the expansion coefficients are, by orthogonality,

$$\left.\begin{array}{l} a_o = \frac{1}{\pi} \int_{-\pi}^{\pi} f(x) \, dx \\ a_n = \frac{1}{\pi} \int_{-\pi}^{\pi} f(x) \cos nx \, dx \\ b_n = \frac{1}{\pi} \int_{-\pi}^{\pi} f(x) \sin nx \, dx. \end{array}\right\} \tag{5.2.2}$$

If these coefficients are substituted in the series, we get

$$f(x) = \frac{1}{2\pi} \int_{-\pi}^{\pi} f(\xi) \, d\xi + \sum_{n=1}^{\infty} \frac{1}{\pi} \int_{-\pi}^{\pi} f(\xi) \cos n(x - \xi) \, d\xi, \tag{5.2.3}$$

which is called the Fourier theorem.

For the trigonometric Fourier series, (5.1.10) takes the form

$$\frac{1}{\pi} \int_{-\pi}^{\pi} f^2 dx = \frac{a_o^2}{2} + \sum_{n=1}^{\infty} (a_n^2 + b_n^2), \tag{5.2.4}$$

which is called the Parseval theorem. If $f(x)$ is a time series, where x represents time and a_n and b_n represent the amplitude of the nth harmonic, the Parseval theorem states that the total energy is the sum of the energy of all the harmonics.

Now we introduce the trigonometric series consisting of either cosine or sine terms but not both.

Let us first define even and odd functions. A function $f(x)$ is even in the interval $-L < x < L$ if $f(-x) = f(x)$ and odd if $f(-x) = -f(x)$.

5.2 Trigonometric Fourier series

Note that any function can be considered as the sum of an even and odd function, since

$$f(x) = \frac{1}{2}[f(x) + f(-x)] + \frac{1}{2}[f(x) - f(-x)]. \quad (5.2.5)$$

Clearly, the first bracket is even and the second is odd.

5.2.2 Fourier cosine and sine series

Consider a function f that is even in the interval $(-\pi, \pi)$, then

$$a_n = \frac{1}{\pi} \int_{-\pi}^{\pi} f(x) \cos nx \, dx = \frac{2}{\pi} \int_{0}^{\pi} f(x) \cos nx \, dx \quad (5.2.6)$$

$$b_n = \frac{1}{\pi} \int_{-\pi}^{\pi} f(x) \sin nx \, dx = 0 \quad (5.2.7)$$

for all $n = 0, 1, 2, \ldots$. Hence,

$$f(x) = \frac{a_o}{2} + \sum_{n=1}^{\infty} a_n \cos nx \quad (5.2.8)$$

consists only of cosine terms and is called a Fourier cosine series.

On the other hand, if f is odd in x over $(-\pi, \pi)$, then

$$a_n = \frac{1}{\pi} \int_{-\pi}^{\pi} f(x) \cos nx \, dx = 0 \quad (5.2.9)$$

$$b_n = \frac{1}{\pi} \int_{-\pi}^{\pi} f(x) \sin nx \, dx = \frac{2}{\pi} \int_{0}^{\pi} f(x) \sin nx \, dx \quad (5.2.10)$$

and

$$f = \sum_{n=1}^{\infty} b_n \sin nx, \quad (5.2.11)$$

which consists only of sine series and is called a Fourier sine series.

Any function defined in $(0, \pi)$ can be regarded as one-half of an even function in $(-\pi, \pi)$, hence it can be expressed as a cosine series. On the other hand, the same f can also be regarded as one-half of an odd function in $(-\pi, \pi)$. Hence it can also be represented alternatively by a sine series.

Since the trigonometric terms are periodic in x with the period 2π, the Fourier cosine series represents not only $f(x)$ in $(0, \pi)$ and its even image in $(-\pi, 0)$, but also their periodic extensions in all other 2π periods, as shown in Figure 5.1. In other words, the Fourier cosine series represents a periodic function that is even in the basic period $(-\pi, \pi)$.

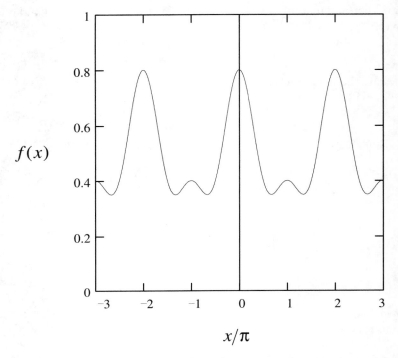

Fig. 5.1. $f(x)$, its even image, and their periodic extensions.

Similarly, the Fourier sine series represents not only $f(x)$ in $(0, \pi)$ and its odd image in $(-\pi, 0)$, but also their periodic extensions in all other 2π periods, as shown in Figure 5.2. In other words, the Fourier sine series represents a periodic function that is odd in the basic period $(-\pi, \pi)$.

So far the Fourier series and the function it represents are defined inside the interval $(-\pi, \pi)$ only. Extension to any other finite interval involves only a change of variables, as shown below.

5.2.3 Other intervals

If a function $F(y)$ is defined in the interval from $-L$ to L instead of $(-\pi, \pi)$, we can rescale the coordinates by letting

$$y = Lx/\pi,$$

5.2 Trigonometric Fourier series

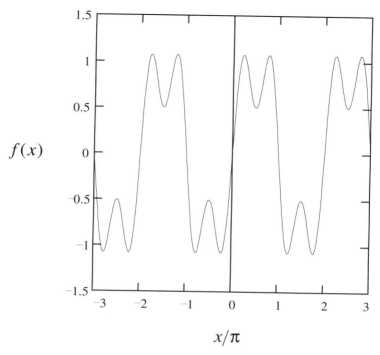

Fig. 5.2. $f(x)$, its odd image, and their periodic extensions.

then $F(y)$ can be expanded as

$$F(y) \stackrel{\text{def}}{=} f\left(\frac{\pi y}{L}\right) = \frac{a_o}{2} + \sum_{n=1}^{\infty}\left(a_n \cos\frac{n\pi y}{L} + b_n \sin\frac{n\pi y}{L}\right) \quad (5.2.12)$$

with the coefficients given by

$$a_n = \frac{1}{L}\int_{-L}^{L} F(y)\cos\frac{n\pi y}{L}\,dy$$

$$b_n = \frac{1}{L}\int_{-L}^{L} F(y)\sin\frac{n\pi y}{L}\,dy. \quad (5.2.13)$$

The Fourier theorem now takes the form

$$F(y) = \frac{1}{2L}\int_{-L}^{L} F(\xi)\,d\xi + \sum_{n=1}^{\infty}\frac{1}{L}\int_{-L}^{L} F(\xi)\cos\frac{n\pi}{L}(x-\xi)\,d\xi. \quad (5.2.14)$$

More generally, if a function $F(y)$ is defined in the interval (a,b), then we let

$$y = \frac{1}{2}(b+a) + \frac{b-a}{2\pi}x$$

so that $y = a$ when $x = -\pi$ and $y = b$ when $x = \pi$. By this transformation the standard Fourier series defined for the interval $-\pi < x < \pi$

$$f(x) = \frac{a_o}{2} + \sum_{n=1}^{\infty}(a_n \cos nx + b_n \sin nx)$$

can be rewritten as

$$F(y) \stackrel{\text{def}}{=} f\left[\frac{2\pi}{b-a}\left(x - \frac{a+b}{2}\right)\right]$$

$$= \frac{a_o}{2} + \sum_{n=1}^{\infty}\left(a_n \cos\frac{n\pi(2x-b-a)}{b-a} + b_n \sin\frac{n\pi(2x-b-a)}{b-a}\right) \tag{5.2.15}$$

with

$$a_n = \frac{1}{\pi}\int_{-\pi}^{\pi} f(x)\cos nx\,dx = \frac{2}{b-a}\int_a^b F(y)\cos\frac{n\pi(2y-b-a)}{b-a}\,dy \tag{5.2.16}$$

and

$$b_n = \frac{1}{\pi}\int_{-\pi}^{\pi} f(x)\sin nx\,dx = \frac{2}{b-a}\int_a^b F(y)\sin\frac{n\pi(2y-b-a)}{b-a}\,dy. \tag{5.2.17}$$

5.3 Exponential Fourier series

Let us rewrite the full trigonometric Fourier series as follows:

$$f(x) = \frac{a_o}{2} + \sum_{n=1}^{\infty}\frac{a_n}{2}\left(e^{inx} + e^{-inx}\right) + \frac{b_n}{2i}\left(e^{inx} - e^{-inx}\right)$$

$$= \frac{a_o}{2} + \sum_{n=1}^{\infty}\frac{1}{2}(a_n - ib_n)e^{inx} + \frac{1}{2}(a_n + ib_n)e^{-inx}.$$

Let

$$c_o = \frac{a_o}{2}, \qquad c_n = \frac{1}{2}(a_n - ib_n), \qquad c_{-n} = \frac{1}{2}(a_n + ib_n);$$

then

$$f(x) = c_o + \sum_{n=1}^{\infty} c_n e^{inx} + \sum_{n=-1}^{-\infty} c_n e^{inx} = \sum_{n=-\infty}^{\infty} c_n e^{inx} \qquad (5.3.1)$$

with

$$c_n = \frac{1}{2}(a_n - ib_n) = \frac{1}{2\pi}\int_{-\pi}^{\pi} dx\, f(x)(\cos nx - i\sin nx)$$

$$= \frac{1}{2\pi}\int_{-\pi}^{\pi} f(x) e^{-inx}\, dx, \quad n = 0, \pm 1, \pm 2, \ldots. \qquad (5.3.2)$$

Equation (5.3.1) is the exponential Fourier series of f with the coefficients given by (5.3.2). Note the orthogonality property

$$\int_{-\pi}^{\pi} e^{i(m-n)x}\, dx = \begin{cases} 0 & \text{if } m \neq n \\ 2\pi & \text{if } m = n. \end{cases} \qquad (5.3.3)$$

If (5.3.2) is substituted into (5.3.1), the Fourier theorem is obtained

$$f(x) = \frac{1}{2\pi} \sum_{n=-\infty}^{\infty} \int_{-\pi}^{\pi} f(\xi) e^{in(x-\xi)}\, d\xi. \qquad (5.3.4)$$

If $F(y)$ is defined in $-L \leq y \leq L$, then by rescaling $x = \pi y/L$, (5.3.1) and (5.3.2) become

$$F(y) = \sum_{n=-\infty}^{\infty} c_n e^{in\pi y/L} \qquad (5.3.5)$$

with

$$c_n = \frac{1}{2L}\int_{-L}^{L} F(y') e^{-in\pi y'/L}\, dy', \qquad (5.3.6)$$

and the Fourier theorem becomes

$$F(y) = \frac{1}{2L} \sum_{n=-\infty}^{\infty} \int_{-L}^{L} F(y') e^{in\pi(y-y')/L}\, dy', \qquad (5.3.7)$$

which will be used later.

5.4 Convergence of Fourier series

If f is a periodic function with period 2π in $(-\pi, \pi)$, is piecewise continuous, and has piecewise continuous derivatives, then at any x the Fourier series

$$\frac{a_o}{2} + \sum_{n=1}^{\infty}(a_n \cos nx + b_n \sin nx)$$

with

$$a_n = \frac{1}{\pi}\int_{-\pi}^{\pi} f(x)\cos nx\, dx, \qquad b_n = \frac{1}{\pi}\int_{-\pi}^{\pi} F(x)\sin nx\, dx$$

converges to

$$\frac{1}{2}\lim_{\epsilon\to 0}[f(x+\epsilon)-f(x-\epsilon)] \equiv \frac{1}{2}[f(x+0)+f(x-0)].$$

Thus, if f is continuous at x, the Fourier series converges to f. If f is discontinuous at x, the Fourier series converges to the average. The proof of this fundamental result is given in almost every mathematical text on Fourier series (e.g., Tolstov, 1962) and is omitted here.

For practical calculations it is useful to know how fast a given Fourier series converges. The speed of convergence depends on the smoothness of the function f and is governed by the following theorem:

Theorem *If the first m derivatives of $f, f', f'', \ldots, f^{(m)}$ exist, the first $m-1$ derivatives are all continuous everywhere, but $f^{(m)}$ is piecewise continuous, then for large n the Fourier coefficients a_n and b_n diminish to zero as $O(1/n^{m+1})$.*

Corollary 1 *If f is piecewise continuous, then as n increases to infinity, a_n and b_n diminish to zero as fast as l/n.*

Corollary 2 *If $f'(x)$ exists everywhere and is piecewise continuous, then a_n, b_n diminish to zero as $1/n^2$.*

Thus the smoother the function, the faster its Fourier series converges. The proof is as follows.

Let $f(x)$ be continuous and $f'(x)$ be piecewise continuous in the interval $(-\pi, \pi)$, with $f(-\pi) = f(\pi)$ and $f'(-\pi) = f'(\pi)$. At some point $x_o \in (-\pi, \pi)$, $f'(x)$ is discontinuous, i.e., $f'(x_o-) \neq f'(x_o+)$, then

$$a_n = \frac{1}{\pi}\int_{-\pi}^{\pi} f(x)\cos nx\, dx = \frac{1}{\pi n}\int_{-\pi}^{\pi} f(x)\, d\sin nx$$

$$= \frac{1}{\pi n}\left[f(x)\frac{\sin nx}{n}\right]_{-\pi}^{\pi} - \frac{1}{\pi n}\int_{-\pi}^{\pi} f'(x)\sin nx\, dx.$$

5.4 Convergence of Fourier series

The integrated term vanishes. The remaining integral can be split into two parts and partially integrated

$$a_n = -\frac{1}{\pi n}\left[\int_{-\pi}^{x_o-} + \int_{x_o+}^{\pi}f'(x)\sin nx\right]dx$$

$$= \frac{1}{\pi n}\int_{-\pi}^{x_o-} + \int_{x_o+}^{\pi}f'(x)\frac{d\cos nx}{n}$$

$$= \frac{1}{\pi n^2}\left\{[(f'(x)\cos nx)]_{-\pi}^{x_o-} + [(f'(x)\cos nx)]_{x_o+}^{\pi}\right\}$$

$$-\frac{1}{\pi n^2}\left(\int_{-\pi}^{x_o-} + \int_{x_o+}^{\pi}\right)f''(x)\cos nx\,dx. \qquad (5.4.1)$$

Since the integrated terms do not vanish, we have

$$a_n = -\frac{1}{\pi n^2}[f'(x)]_{x_o-}^{x_o+}\cos nx_o + O\left(\frac{1}{n^2}\right) = O\left(\frac{1}{n^2}\right).$$

Clearly, if $f'(x)$ is continuous everywhere but $f''(x)$ is discontinuous at some point x_o, the integrated terms in (5.4.1) vanish and the remaining integral can be integrated once more to give

$$a_n \sim O\left(\frac{1}{n^3}\right).$$

Finally, if $f, f', f'', \ldots, f^{(m-1)}$ are continuous but $f^{(m)}$ is piecewise continuous, it follows by induction that

$$a_n \sim O\left(\frac{1}{n^{m+1}}\right).$$

Let us demonstrate the theorem by a few examples.

Example 1

$$f(x) = \begin{cases} -\pi - x & \text{if } \pi < x < 0 \\ \pi - x & \text{if } 0 < x < \pi. \end{cases}$$

Thus f is odd and discontinuous at $x = 0$, and the coefficients of the cosine series are

$$b_n = \frac{2}{\pi}\int_0^{\pi}(\pi - x)\sin nx\,dx$$

$$= -\frac{2}{\pi}\left[-\frac{\pi}{n}\cos nx - \frac{1}{n^2}(\sin nx - nx\cos nx)\right]_0^{\pi}$$

$$= \frac{2}{n}.$$

Therefore,
$$f = \sum_{n=1}^{\infty} \frac{2}{n} \sin nx,$$
where the nth term diminishes as $1/n$.

Example 2.
$$f(x) = \pi - |x|, \quad \pi < x < \pi.$$

Here f is continuous but $f'(x)$ is only piecewise continuous. The coefficients of the Fourier cosine series are
$$a_o = \frac{2}{\pi} \int_0^{\pi} (\pi - x) dx = \pi,$$
$$a_n = \frac{2}{\pi} \left[\frac{\pi}{n} \sin nx - \frac{1}{n^2} (\cos nx + nx \sin nx) \right]_0^{\pi}$$
$$= \frac{2}{\pi n^2} [1 - (-1)^n].$$

Hence,
$$f(x) = \frac{\pi}{2} + \frac{2}{\pi} \sum_{n=1}^{\infty} \frac{1}{n^2} [1 - (-1)^n] \cos nx,$$
where the nth term indeed diminishes as $1/n^2$.

The ease with which a Fourier series can be computed numerically depends on the rate of convergence. If one can extract the slowly convergent part and sum it up in closed form, the convergence of the remaining series is accelerated, and the nature of discontinuity or singularity of the solution can be revealed. Exercise (5.5) gives the reader some idea of the tactic for accelerating a Fourier series. More on these tricks is discussed in Kantorovich and Krylov (1964).

Exercises

5.1 For each of the functions defined in the interval $0 < x < L$, sketch its odd and even extensions in the interval $-L < x < 0$ and expand the functions in both sine and cosine series. Compare the convergence rate of the nth coefficient for large n, and check your observation with the theorem in §5.4.

(i) $f(x) = x$
(ii) $f(x) = L - x$
(iii) $f(x) = x, \, 0 < x < L/2; \quad f(x) = L - x, \, -L/2 < x < L$

(iv) $f(x) = x(L-x)$
(v) $f(x) = x^2(L-x)^2$
(vi) $f(x) = x^3(L-x)^3$.

5.2 Expand $f(x) = \sin \frac{n\pi x}{L}$ defined in $0 < x < L$ in a cosine series.

5.3 Expand $f(x) = \cos \frac{n\pi x}{L}$ defined in $0 < x < L$ in a sine series.

5.4 Verify the following Fourier expansions, all of which are slowly convergent (Tolstov, 1962). Use partial integration, or term-wise integration of a known expansion, if necessary.

(i)
$$-\ln\left(2\sin\frac{x}{2}\right) = \sum_{n=1}^{\infty} \frac{\cos nx}{n}, \quad 0 < x < 2\pi$$

(ii)
$$\frac{\pi - x}{2} = \sum_{n=1}^{\infty} \frac{\sin nx}{n}, \quad 0 < x < 2\pi$$

(iii)
$$\ln\left(2\cos\frac{x}{2}\right) = \sum_{n=1}^{\infty} (-1)^{n+1} \frac{\cos nx}{n}, \quad -\pi < x < \pi$$

(iv)
$$\frac{x}{2} = \sum_{n=1}^{\infty} (-1)^{n+1} \frac{\sin nx}{n}, \quad 0 < x < 2\pi$$

(v)
$$\frac{3x^2 - 6\pi x + 2\pi^2}{12} = \sum_{n=1}^{\infty} \frac{\cos nx}{n^2}, \quad 0 \leq x \leq 2\pi$$

(vi)
$$-\int_0^x \ln\left(2\sin\frac{x}{2}\right) dx = \sum_{n=1}^{\infty} \frac{\sin nx}{n^2}, \quad 0 \leq x \leq 2\pi$$

(vii)
$$\frac{\pi^2 - 3x^2}{12} = \sum_{n=1}^{\infty} (-1)^{n+1} \frac{\cos nx}{n^2}, \quad -\pi \leq x \leq \pi$$

(viii)
$$\int_0^x \ln\left(2\cos\frac{x}{2}\right) dx = \sum_{n=1}^{\infty} (-1)^{n+1} \frac{\sin nx}{n^2}, \quad -\pi \leq x \leq \pi.$$

5.5 A Fourier series converges slowly if the function it represents is not sufficiently smooth, as has been shown by the theorem in §5.4. In general a function can be decomposed into the sum of a simple discontinuous function and a smoother function. The Fourier series of the former should be responsible for the slowness of convergence. Hence, a tactic to accelerate the convergence of a series is to extract the slow part and attempt to sum it up in closed form. The remaining series must then converge very fast.

To get some ideas, consider
$$S(x) = \sum_{n=1}^{\infty} \frac{n \sin nx}{n^2+1}.$$
Decompose the nth coefficient
$$a_n = \frac{n}{n^2+1} = \frac{1}{n} - \frac{1}{n^2} + b_n$$
and rewrite the original sum as
$$S(x) = \sum_{n=1}^{\infty} \frac{\sin nx}{n} - \sum_{n=1}^{\infty} \frac{\sin nx}{n^2} + \sum_{n=1}^{\infty} b_n \sin nx.$$
Identify the discontinuous functions representing the first two series above and the coefficient b_n of the remaining series. How fast does the third series converge?

6
Introduction to Green's functions

As has been demonstrated in previous chapters, the advantage of linearity is that new solutions can be constructed by superposition of old ones. The reader is already familiar with superposition of eigenfunctions, which are the basic bricks and timber for building solutions to boundary-value problems. We now introduce Green's functions, which are more advanced modular units easily assembled to yield the final construction. We shall point out the physical significance of Green's functions as the response to unit forcing at a point, along a line, or at an instant. The δ function will be used to represent these highly localized forcing functions. The mathematical relation between Green's functions and Green's theorem will be explained. Examples are given first for ordinary differential equations, then for partial differential equations. Further extensions will be discussed in subsequent chapters.

6.1 The δ function

The one-dimensional δ function is a mathematical artifice for representing an extremely localized function with a finite total area. Specifically, $\delta(x-\xi)$ is the limit of a spike-like function of x, which is zero almost everywhere except very near $x = \xi$, where the δ function has an extremely sharp peak in such a way that its area above the x axis is unity, i.e.,

$$\int_a^b \delta(x-\xi)\,dx = \begin{cases} 1 & \text{if } \xi \in (a,b), \\ 0 & \text{if } \xi \notin (a,b), \end{cases} \quad (6.1.1)$$

where (a, b) stands for the open interval between a and b, excluding the end points. As long as the point of concentration ξ lies between a and b, the upper and lower limits can be replaced by $(-\infty, \infty)$. Therefore, an

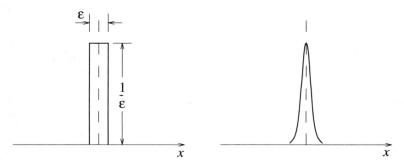

Fig. 6.1. Two models of the δ function.

alternative definition of the δ function is

$$\int_{-\infty}^{\infty} \delta(x-\xi)\,dx = 1. \tag{6.1.2}$$

In mechanics $\delta(x-\xi)$ may symbolize a concentrated force, i.e., the limit of a pressure distribution with a sharply peaked intensity around $x = x_o$ and a unit total force.

One of the many conceptual models of the δ function is the limit of a rectangle with width ϵ and height $1/\epsilon$; the other model is a Gaussian function, with the standard deviation ϵ and peak height $1/\epsilon\sqrt{\pi}$, i.e.,

$$\frac{1}{\epsilon\sqrt{\pi}} \exp\left(-\frac{x^2}{\epsilon^2}\right);$$

see Figure 6.1.

As in (6.1.1), we shall define all properties of the δ function in terms of integral operations. Thus, if $f(x)$ is a bounded and continuous function, the integral of its product with $\delta(x-\xi)$ is $f(\xi)$ if ξ lies within the domain of integration, i.e.,

$$\int_a^b \delta(x-\xi)f(x)\,dx = \begin{cases} f(\xi) & \text{if } \xi \in (a,b), \\ 0 & \text{if } x \notin (a,b). \end{cases} \tag{6.1.3}$$

The above equation is often taken as an alternative definition of the δ function.

From the preceding definition a number of properties of the δ function can be deduced. By a change of variables, we have

$$\int_{-\infty}^{\infty} \delta(-x)f(x)\,dx = \int_{-\infty}^{\infty} \delta(\xi)f(-\xi)\,d\xi = f(0) = \int_{-\infty}^{\infty} \delta(x)f(x)\,dx.$$

6.1 The δ function

This integral equality will be stated symbolically as

$$\delta(x) = \delta(-x), \qquad (6.1.4)$$

i.e., in the integral sense the δ function is even in its argument.

All equalities involving the δ function will be symbolic and interpreted in the integral sense.

Consider $\delta(cx)$ for a constant $c > 0$. Since

$$\int_{-\infty}^{\infty} \delta(cx) f(x)\, dx = \int_{-\infty}^{\infty} \delta(\xi) f\left(\frac{\xi}{c}\right) d\frac{\xi}{c} = \frac{f(0)}{c}$$

and

$$\int_{-\infty}^{\infty} \delta(x) f(x)\, dx = f(0),$$

it follows, symbolically, that

$$\delta(cx) = \frac{1}{c}\delta(x).$$

On the other hand, if $c < 0$,

$$\delta(cx) = \delta(-|c|x) = \delta(|c|x) = \frac{\delta(x)}{|c|}.$$

Hence for any constant c, we have

$$\delta(cx) = \frac{\delta(x)}{|c|}. \qquad (6.1.5)$$

The derivative of a δ function, which may represent a dipole or a concentrated torque, has the following property

$$\int_{-\infty}^{\infty} \delta' f(x)\, dx = [\delta f]_{-\infty}^{\infty} - \int_{-\infty}^{\infty} f' \delta(x)\, dx = -f'(0), \qquad (6.1.6)$$

where $f(x)$ is a function with a continuous first derivative at $x = 0$.

Finally, if

$$\phi(x) = 0 \quad \text{and} \quad \phi'(x) \neq 0 \quad \text{at} \quad x = \alpha_i, \quad i = 1, 2, 3, \ldots,$$

then the δ function of an implicit function $\phi(x)$ is

$$\delta[\phi(x)] = \sum_i \frac{\delta(x - \alpha_i)}{|\phi'(\alpha_i)|}. \qquad (6.1.7)$$

To see this result, we expand $\phi(x)$ near the zero α_i

$$\phi(x) = (x - \alpha_i)\phi'(\alpha_i) + \frac{1}{2}(x - \alpha_i)^2 \phi''(\alpha_i) + \cdots$$

so that

$$\delta(\phi(x)) = \delta\left[(x-\alpha_i)\phi'(\alpha_i)\right] = \delta\left((x-\alpha_i)|\phi'(\alpha_i)|\right) = \frac{\delta(x-\alpha_i)}{|\phi'(\alpha_i)|}.$$

Another useful generalized function is the Heaviside step function defined by

$$H(x) = \begin{cases} 0 & \text{if } x < 0, \\ 1 & \text{if } x > 0. \end{cases} \tag{6.1.8}$$

It is related to the δ function symbolically by

$$\frac{dH(x)}{dx} = \delta(x) \tag{6.1.9}$$

or

$$\int_{-\infty}^{x} \delta(\xi)\, d\xi = \begin{cases} H(x) & \text{if } x > 0, \\ 0 & \text{if } x < 0. \end{cases} \tag{6.1.10}$$

6.2 Static deflection of a string under a concentrated load

Let us use an elementary example to demonstrate the typical aspects of Green's functions: derivation, symmetry, physical significance, and application.

Consider the static deflection of a string fixed at two ends at $x = 0$ and L and under a continuously distributed load $p(x)$ per unit length. The deflection V satisfies

$$TV'' = p(x), \quad 0 < x < L \tag{6.2.1}$$

$$V(0) = V(L) = 0. \tag{6.2.2}$$

Instead of the general load, let us consider first a concentrated load applied at $x = \xi$. The corresponding deflection, denoted by $G(x, \xi)$, is governed by

$$TG'' = \delta(x - \xi), \quad 0 < x, \xi < L \tag{6.2.3}$$

and by the boundary conditions

$$G(0, \xi) = G(L, \xi) = 0, \tag{6.2.4}$$

where primes denote derivatives with respect to x. G is called the Green function of the problem defined by (6.2.1–6.2.2).

The standard way of treating the singular forcing in the governing

6.2 Static deflection of a string under a concentrated load

equation is to avoid direct contact by considering the two sides. Thus we define

$$G = G_-, \quad 0 < x < \xi \quad \text{and} \quad G = G_+, \quad \xi < x < L.$$

Away from the point load, the δ function has no value and the governing equation is homogeneous, i.e.,

$$G''_- = 0, \quad 0 < x < \xi \quad \text{and} \quad G''_+ = 0, \quad \xi < x < L. \quad (6.2.5)$$

At the two ends of the string, the boundary conditions are

$$G_-(0, \xi) = 0 \quad \text{and} \quad G_+(L, \xi) = 0. \quad (6.2.6)$$

Matching conditions are needed at $x = \xi$. First, the string should be continuous

$$G_+(\xi, \xi) = G_-(\xi, \xi). \quad (6.2.7)$$

By integrating the differential equation across the δ function, i.e., from slightly below the point $x = \xi$ to slightly above, we find

$$\int_{\xi-0}^{\xi+0} TG'' \, dx = T\left[G'_+(\xi, \xi) - G'_-(\xi, \xi)\right] = \int_{\xi-0}^{\xi+0} \delta(x - \xi) \, dx = 1$$

or

$$G'_+(\xi, \xi) - G'_-(\xi, \xi) = 1/T. \quad (6.2.8)$$

As can be verified by balancing the vertical forces, this result simply means that the effect of a concentrated force causes a slope discontinuity in the string.

The original boundary-value problem for G with a singular forcing is now replaced by (6.2.5), subject to the end conditions (6.2.6) and the matching conditions (6.2.7) and (6.2.8). The solution is easily obtained

$$G(x, \xi) = \begin{cases} \dfrac{(\xi-L)x}{TL}, & 0 < x < \xi, \\ \dfrac{\xi(x-L)}{TL}, & \xi < x < L. \end{cases} \quad (6.2.9)$$

Note that this Green function is symmetric with respect to the interchange of x and ξ:

$$G(\alpha, \beta) = G(\beta, \alpha)$$

since, by putting $x = \alpha$ and $\xi = \beta$ in the first of (6.2.9),

$$G(\alpha, \beta) = \frac{(\beta - L)\alpha}{TL}$$

and, by putting $\xi = \alpha$ and $x = \beta$ in the second of (6.2.9),
$$G(\beta, \alpha) = \frac{\alpha(\beta - L)}{TL} = G(\alpha, \beta).$$

Physically, the deflection at α due to a unit load at β is the same as the response at β due to a unit load at α.

The solution to the original problem (6.2.1–6.2.2) for continuous loading will now be obtained by superposition. Specifically, we multiply $G(x, \xi)$ by $p(\xi)d\xi$ and integrate the result with respect to ξ from 0 to L:

$$\begin{aligned} V(x) &= \int_0^L G(x, \xi) p(\xi) \, d\xi \\ &= \int_0^x p(\xi) \frac{\xi(x - L)}{TL} \, d\xi + \int_x^L p(\xi) \frac{x(\xi - L)}{TL} \, d\xi. \end{aligned} \quad (6.2.10)$$

It is straightforward to verify that this solution satisfies (6.2.1–6.2.3).

In structural engineering, the Green function has long been called the *influence function*, and the symmetry property is called the *principle of reciprocity*.

6.3 String under a simple harmonic point load

For any distributed load, the governing equation for the string displacement V is

$$T \frac{\partial^2 V}{\partial x^2} = \rho \frac{\partial^2 V}{\partial t^2} + p(x, t). \quad (6.3.1)$$

Let us confine ourselves to a distributed load that is simple harmonic in time

$$p = P(x) \cos \omega t.$$

It is convenient to replace the cosine term by the exponential function

$$p = \text{Re}\,[P(x) e^{-i\omega t}],$$

where $\text{Re}\, f$ stands for *the real part of f*.

Since all coefficients in the equation are independent of time, the response is also simple harmonic long after the passage of the initial transients, i.e.,

$$V = \text{Re}\,[Y(x) e^{-i\omega t}]. \quad (6.3.2)$$

For brevity it is customary to omit the operator Re, but its presence

6.3 String under a simple harmonic point load

must always be implied. From the governing equation we get

$$Y'' + k^2 Y = P(x), \quad k^2 = \rho\omega^2/T. \tag{6.3.3}$$

For fixed ends, the boundary conditions are

$$Y(0) = Y(L) = 0. \tag{6.3.4}$$

Now let us define the Green function G for this boundary-value problem by the governing equation

$$G'' + k^2 G = \delta(x - \xi) \tag{6.3.5}$$

and the boundary conditions

$$G = 0, \quad x = 0, L. \tag{6.3.6}$$

Integrating with respect to x across the δ function, we get

$$[G']_{\xi-0}^{\xi+0} + \int_{\xi-0}^{\xi+0} k^2 G\, dx = \int_{\xi-0}^{\xi+0} \delta(x - \xi)\, dx = 1$$

or

$$G'_+(\xi,\xi) - G'_-(\xi,\xi) = 1. \tag{6.3.7}$$

Now we can restate the boundary-value problem as follows:

$$G''_- + k^2 G_- = 0, \quad 0 < x < \xi$$

$$G''_+ + k^2 G_+ = 0, \quad \xi < x < L,$$

with the boundary conditions

$$G_-(0,\xi) = 0, \quad G_+(L,\xi) = 0$$

$$G_-(\xi,\xi) = G_+(\xi,\xi)$$

$$G'_+(\xi,\xi) - G'_-(\xi,\xi) = 1.$$

The homogeneous solutions on two sides of the load are

$$G_- = a \sin kx \quad \text{and} \quad G_+ = b \sin k(L - x),$$

which have been chosen to satisfy the boundary conditions at $x = 0$ and L. To satisfy the two coupling conditions at $x = \xi$, we must have

$$a \sin k\xi - b \sin k(L - \xi) = 0$$

$$ak \cos k\xi + bk \cos(L - \xi) = -1.$$

The coefficients are obtained as

$$a = -\frac{\sin k(L-\xi)}{k \sin kL}, \qquad b = -\frac{\sin k\xi}{k \sin kL}.$$

The Green function is, therefore,

$$G_+ = -\frac{1}{k}\frac{\sin k(L-x)\sin k\xi}{\sin kL}, \qquad x > \xi \qquad (6.3.8)$$

and

$$G_- = -\frac{1}{k}\frac{\sin k(L-\xi)\sin kx}{\sin kL}, \qquad x < \xi. \qquad (6.3.9)$$

A more compact way of writing the result is

$$G(x_<, x_>) = -\frac{1}{k}\frac{\sin k(L-x_>)\sin kx_<}{\sin kL}, \qquad 0 < x_< < x_> < L, \qquad (6.3.10)$$

where $(x_<, x_>)$ denotes the (smaller, larger) member of the pair (x, ξ), i.e.,

$$x_< = \xi, \quad x_> = x, \quad \text{if} \quad \xi < x$$

$$x_< = x, \quad x_> = \xi, \quad \text{if} \quad x > \xi.$$

Directly beneath the point load the deflection is

$$G_+ = G_- = -\frac{1}{k}\frac{\sin k(L-\xi)\sin k\xi}{\sin kL}.$$

Note that when $kL = n\pi, n = 1, 2, 3, \ldots$, $G \to \infty$. These special values of kL correspond to the natural modes, which are the eigensolutions of the homogeneous boundary-value problem.

6.4 Sturm–Liouville boundary-value problem

Consider the problem defined by the Sturm–Liouville equation

$$\mathcal{L}y \stackrel{\text{def}}{=} \frac{d}{dx}\left(p(x)\frac{dy}{dx}\right) + qy = f(x), \qquad a < x < b \qquad (6.4.1)$$

and the boundary conditions

$$a_1 y + a_2 \frac{dy}{dx} = g, \qquad x = a \qquad (6.4.2)$$

$$b_1 y + b_2 \frac{dy}{dx} = h, \qquad x = b, \qquad (6.4.3)$$

6.4 Sturm–Liouville boundary-value problem

where $\mathcal{L}$ denotes the Sturm–Liouville operator

$$\mathcal{L} = \frac{d}{dx}p\frac{d}{dx} + q.$$

Together, (6.4.1–6.4.3) define a Sturm–Liouville boundary-value problem.

Let us define the Green function $G(x,\xi)$ for the Sturm–Liouville problem by

$$\mathcal{L}G = \frac{d}{dx}\left(p(x)\frac{dG}{dx}\right) + qG = \delta(x-\xi), \quad a < x < b \quad (6.4.4)$$

$$a_1 G + a_2 \frac{dG}{dx} = 0, \quad x = a \quad (6.4.5)$$

$$b_1 G + b_2 \frac{dG}{dx} = 0, \quad x = b. \quad (6.4.6)$$

Note that in this definition, the governing equations for G are similar to those for y, except that the interior forcing is a δ function, and boundary forcing is zero. Recall from (4.5.10) the Green theorem for a pair of twice-differentiable functions Y and Z. Now let $Z = G$, and Y be the solution to (6.4.1–6.4.3). The left-hand side of (4.5.10) gives

$$\int_a^b (G\mathcal{L}Y - Y\mathcal{L}G)\,dx = \int_a^b [Gf - Y\delta(x-\xi)]\,dx$$

$$= \int_a^b G(x,\xi)f(x)\,dx - Y(\xi).$$

The right-hand side of (4.5.10) gives

$$\left[pG\frac{dY}{dx} - pY\frac{dG}{dx}\right]_a^b = p(b)\left[G\left(-\frac{b_1}{b_2}Y + \frac{h}{b_2}\right) - Y\left(-\frac{b_1}{b_2}\right)G\right]_{x=b}$$

$$-p(a)\left[G\left(-\frac{a_1}{a_2}Y + \frac{g}{a_2}\right) - Y\left(-\frac{a_1}{a_2}\right)G\right]_{x=a}$$

$$= p(b)\frac{h}{b_2}G(b,\xi) - p(a)\frac{g}{a_2}G(a,\xi),$$

provided that $a_2, b_2 \neq 0$. It follows that

$$Y(\xi) = \int_a^b G(x,\xi)f(x)\,dx + p(b)\frac{h}{b_2}G(b,\xi) - p(a)\frac{g}{a_2}G(a,\xi). \quad (6.4.7)$$

Thus, in principle, if the Green function of the Sturm–Liouville problem can be found, the solution to the most general inhomogeneous boundary-value problem can be obtained by quadrature through Green's formula.

An important general property of Green's function defined by (6.4.4–6.4.6) is its symmetry with respect to the interchanges of the points of cause and effect

$$G(x,\xi) = G(\xi,x). \tag{6.4.8}$$

To prove this, we put $Y = G(x,\xi_1)$ and $Z = G(x,\xi_2)$ in the left-hand side of Green's formula; then

$$\int_a^b [G(x,\xi_2)\mathcal{L}G(x,\xi_1) - G(x,\xi_1)\mathcal{L}G(x,\xi_2)]\,dx$$
$$= \left[pG(x,\xi_2)\frac{d}{dx}G(x,\xi_1) - pG(x,\xi_1)\frac{d}{dx}G(x,\xi_2)\right]_{x=a}^{x=b}.$$

The right-hand side is zero by virtue of the boundary conditions. The left-hand side

$$\int_a^b [G(x,\xi_2)\delta(x,\xi_1)dx - G(x,\xi_1)\delta(x-\xi_2)]\,dx = G(\xi_1,\xi_2) - G(\xi_2,\xi_1)$$

must also vanish, hence G is symmetric. This result generalizes an earlier observation made in §6.2.

Using symmetry, we may interchange x and ξ and rewrite (6.4.7) as

$$Y(x) = \int_a^b G(x,\xi)f(\xi)d\xi + p(b)\frac{h}{b_2}G(x,b) - p(a)\frac{g}{a_2}G(x,a). \tag{6.4.9}$$

While the integral represents superposition of internal sources, the last two terms represent contributions from forcing at the boundary.

Green's function is inseparable from Green's theorem, hence the name.

6.5 Bending of an elastic beam on an elastic foundation

6.5.1 Formulation of the beam problem

As an example involving a fourth-order differential equation, let us consider the vibration of a beam under lateral loading. Referring to Figure 6.2, we first describe the momentum conservation of a beam of rectangular cross section with width a and depth h. Assume the beam to have a horizontal axis when it is not loaded and to have uniform material properties. Let $V(x,t)$ denote the upward deflection of the beam axis. If the thickness is small compared to the length and the deflection small

6.5 Bending of an elastic beam on an elastic foundation

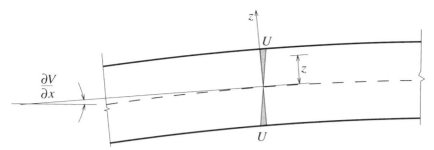

Fig. 6.2. Deflection of a beam.

compared to the thickness, a plane cross section remains approximately plane and perpendicular to the deformed axis. Hence, the longitudinal displacement U at section x and height z above the axis is proportional to z and to the tilt angle $\partial V/\partial x$

$$U \cong -z\frac{\partial V}{\partial x}.$$

Consequently, the longitudinal strain is

$$\varepsilon_x = \frac{\partial U}{\partial x} = -z\frac{\partial^2 V}{\partial x^2}$$

and the longitudinal stress is

$$\sigma_x = E\frac{\partial U}{\partial x} = -Ez\frac{\partial^2 V}{\partial x^2}, \quad (6.5.1)$$

where E is Young's modulus. The total moment about the mid-section $z = 0$ due to the stress distribution across the section is

$$M = -a\int_{-h/2}^{h/2} \sigma_x z\, dz = aE\frac{\partial^2 V}{\partial x^2}\int_{-h/2}^{h/2} z^2 dz = EI\frac{\partial^2 V}{\partial x^2}, \quad (6.5.2)$$

where

$$I = a\int_{-h/2}^{h/2} z^2 dz = \frac{ah^3}{12}$$

is the moment of inertia of the cross section with respect to its midsection $z = 0$. Consider a length element of the beam from x to $x + dx$, as sketched in Figure 6.3. The balance of angular momentum about the center of the element requires that

$$M + \frac{\partial M}{\partial x}dx - M + \left(S + \frac{\partial S}{\partial x}dx\right)\frac{dx}{2} + S\frac{dx}{2} = \rho I dx \frac{\partial^3 V}{\partial x \partial t^2},$$

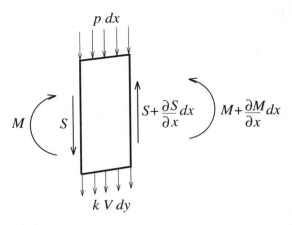

Fig. 6.3. Forces and moments on a beam segment from x to $x + dx$.

where ρ is the mass per unit volume and the right-hand-side represents the angular or rotatory inertia per unit length of the beam. Thus,

$$S = -\frac{\partial M}{\partial x} + \rho I \frac{\partial^3 V}{\partial x \partial t^2} = -EI \frac{\partial^3 V}{\partial x^3} + \rho I \frac{\partial^3 V}{\partial x \partial t^2}. \tag{6.5.3}$$

On the other hand, balance of vertical forces requires that

$$S + \frac{\partial S}{\partial x} dx - S = p\, dx + kV\, dx + \rho A \frac{\partial^2 V}{\partial t^2} dx,$$

where $A = ah$ is the cross-sectional area, $p(x,t)$ is the external load per unit length, and k the elastic constant of the lateral support. Making use of (6.5.3), we get

$$\rho \left(A - I \frac{\partial^2}{\partial x^2} \right) \frac{\partial^2 V}{\partial t^2} + EI \frac{\partial^4 V}{\partial x^4} + kV = -p(x,t), \tag{6.5.4}$$

which is a fourth-order partial differential equation derived first by Lord Rayleigh. Consistent with the original assumption of a shallow beam, $h/L \ll 1$, the term representing rotatory inertia is smaller than A by the ratio $O(h^2/L^2)$ and is usually omitted. Equation (6.5.4) may then be simplified to the form due to Bernoulli and Euler,

$$\rho A \frac{\partial^2 V}{\partial t^2} + EI \frac{\partial^4 V}{\partial x^4} + kV = -p(x,t). \tag{6.5.5}$$

Only this simplified equation will be discussed here. The role of rotatory inertia and other refinements are discussed in Graf (1975). Now the boundary conditions. Because of the fourth-order derivative, a total of

6.5 Bending of an elastic beam on an elastic foundation

four conditions is needed. For a beam of finite length, each end can be free, clamped, or supported on a hinge. At a free end, there is neither torque nor shear

$$\frac{\partial^2 V}{\partial x^2} = 0, \qquad \frac{\partial^3 V}{\partial x^3} = 0. \qquad (6.5.6)$$

At a clamped end, the deflection and slope must vanish

$$V = 0, \qquad \frac{\partial V}{\partial x} = 0. \qquad (6.5.7)$$

At a hinged end both the deflection and the torque are zero

$$V = 0, \qquad \frac{\partial^2 V}{\partial x^2} = 0. \qquad (6.5.8)$$

For an infinitely long beam, the boundary conditions at infinity depend on the loading, as discussed in the next example.

6.5.2 Beam under a sinusoidal concentrated load

Let there be a concentrated load at $x = 0$, oscillating at the frequency ω. By assuming negligible rotatory inertia, the governing equation for the Green function $G(x;0)$ is

$$EI \frac{d^4 G}{dx^4} + (k - \rho A \omega^2) G = -P_0 \delta(x), \qquad -\infty < x < \infty. \qquad (6.5.9)$$

The solution should be symmetric about the load, hence G and its even derivatives are even in x, while the odd derivatives are odd in x. The deflection and its derivatives must not be unbounded at infinities. At $x = 0$, the deflection G, the slope G', and the moment G'' must be continuous where primes denote derivatives with respect to x. If the load is at any other point $x = \xi$, a new coordinate may be chosen so that the load is at the new origin, hence the present formulation suffices.

Let us integrate the governing equation across the concentrated load

$$EI \int_{0-}^{0+} G'''' dx + (k - \rho A \omega^2) \int_{0-}^{0+} G \, dx = -P_0 \int_{0-}^{0+} \delta(x) dx = -P_0,$$

where primes denote derivatives with respect to x. Thus

$$EI \left[G'''\right]_{0-}^{0+} = -P_0,$$

which means physically that the difference of shear forces on both sides of the concentrated load is equal to the concentrated load. Hence, we can reformulate the boundary-value problem as follows:

$$G_-'''' + 4\beta^4 G_- = 0, \quad -\infty < x < 0 \tag{6.5.10}$$

$$G_+'''' + 4\beta^4 G_+ = 0, \quad 0 < x < \infty, \tag{6.5.11}$$

where

$$4\beta^4 = (k - \rho A \omega^2)/EI.$$

At $x = 0$ we have the matching conditions

$$G_- = G_+, \quad G_-' = G_+', \quad G_-'' = G_+'' \tag{6.5.12}$$

and

$$G_+''' - G_-''' = -\frac{P_o}{EI}. \tag{6.5.13}$$

The conditions at infinities must now be made more specific according to the sign of β^4. We now distinguish two cases:

Case (i): $k > \rho A \omega^2$

The homogeneous differential equation

$$G'''' + 4\beta^4 G = 0$$

has four independent solutions of the type e^{Dx} with

$$D^4 + 4\beta^4 = 0 \quad \text{or} \quad D = \sqrt{2}\beta(-1)^{1/4}.$$

There are four distinct roots for D, each corresponding to a quartic root of -1:

$$(-1)^{1/4} = e^{i(2n+1)\frac{\pi}{4}} = e^{i\frac{\pi}{4}}, e^{i\frac{3\pi}{4}}, e^{i\frac{5\pi}{4}}, e^{i\frac{7\pi}{4}}$$

$$= \frac{1}{\sqrt{2}}(1+i, -1+i, -1-i, 1-i).$$

All four roots lie on a unit circle in the complex plane; see Figure 6.4. Thus the most general solution for either G_- or G_+ is a linear combination of all four solutions

$$\exp\{\beta(\pm 1 \pm i)x\} \quad \text{or} \quad e^{\pm \beta x}(\cos \beta x, \sin \beta x).$$

To ensure boundedness at infinities and symmetry in x, we take

$$G_\pm = e^{\mp \beta x}(A \cos \beta x \pm B \sin \beta x)$$
$$= \tfrac{1}{2} e^{\mp \beta x}\left[(A \mp iB)e^{i\beta x} + (A \pm iB)e^{-i\beta x}\right].$$

6.5 Bending of an elastic beam on an elastic foundation

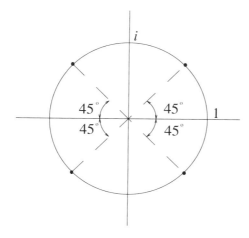

Fig. 6.4. Quartic roots of -1 in the complex plane.

In this symmetric form $G_+(0) = G_-(0)$ and $G''_+(0) = G''_-(0)$ are guaranteed. For $G'_+(0) = G'_-(0)$, it is necessary that $A = B$ so that

$$G_\pm(x) = \frac{A}{2} e^{\mp \beta x} \left[(1 \mp i) e^{i\beta x} + (1 \pm i) e^{-i\beta x} \right].$$

Finally, the matching condition (6.5.13) demands that

$$A = -\frac{P_o}{8\beta^3 EI}.$$

The solution is, therefore,

$$G_\pm(x) = -\frac{P_o e^{\mp \beta x}}{8\beta^3 EI} (\cos \beta x \pm \sin \beta x) \qquad x \gtrless 0$$

or

$$\operatorname{Re} G(x) e^{-i\omega t} = -\frac{P_o e^{-\beta |x|}}{8\beta^3 EI} (\cos \beta |x| + \sin \beta |x|) \cos \omega t. \qquad (6.5.14)$$

Thus for low-frequency forcing, the oscillations are localized near the load.

Case (ii): $k < \rho A \omega^2$

We denote $\alpha^4 = -4\beta^4 > 0$ so that

$$G'''''_\pm - \alpha^4 G_\pm = 0, \qquad x \lessgtr 0.$$

The homogeneous solution in the form e^{Dx} exists only if $D^4 - \alpha^4 = 0$ or $D = \pm\alpha, \pm i\alpha$. The independent solutions are $e^{\pm\alpha x}$ and $e^{\pm i\alpha x}$. To avoid unboundedness at infinities we choose from the first pair $e^{-\alpha x}$ for $x > 0$ and $e^{\alpha x}$ for $x < 0$. Both $e^{i\alpha x}$ and $e^{-i\alpha x}$ of the second pair are bounded, and a stronger condition is needed to decide between them. Note that

$$e^{i\alpha x - i\omega t}, \quad e^{-i\alpha x - i\omega t}$$

represent, respectively, rightward- and leftward-going waves. For a disturbance originated from a local region, waves should only propagate away from the region, hence we must take $e^{i\alpha x}$ for $x > 0$ and $e^{-i\alpha x}$ for $x < 0$. This requirement that localized disturbance can only radiate waves outward is called the *radiation condition*, about which more discussion will be given later. In summary we take

$$G_\pm = C_1 e^{\mp\alpha x} + C_2 e^{\pm i\alpha x} \quad \text{for} \quad x \gtrless 0.$$

By symmetry G and G'' are continuous at $x = 0$. For the continuity of G' at $x = 0$ it is necessary that $C_1 = iC_2$ so that

$$G_\pm = C_2 \left(i e^{\mp\alpha x} + e^{\pm i\alpha x} \right).$$

Finally, (6.5.13) can be satisfied if

$$2i^3 C_2 = \frac{P_o}{\alpha^3 EI}.$$

Hence,

$$G_\pm(x) = \frac{P_o}{2\alpha^3 EI} \left(e^{\mp\alpha x} - i e^{\pm i\alpha x} \right)$$

and

$$\operatorname{Re} G(x) e^{-i\omega t} = \frac{P_o}{2\alpha^3 EI} \left[e^{\mp\alpha x} \cos\omega t + \sin(\pm\alpha x - \omega t) \right]. \tag{6.5.15}$$

Thus at high enough frequencies the response consists of a localized oscillation and a radiating wave.

If the concentrated load is applied at any point $x = \xi$, the expression for $G(x, \xi)$ can be obtained by replacing x by $x - \xi$ in $G_-(x)$ and $G_+(x)$.

We now use the Green function to construct the solution for a distributed load of $p(x)$ per unit length. The deflection $\operatorname{Re}\left[V(x) \exp(-i\omega t) \right]$ is governed by

$$\mathcal{L} V = \frac{p(x)}{EI}, \tag{6.5.16}$$

where

$$\mathcal{L} \stackrel{\text{def}}{=} \frac{d^4}{dx^4} + 4\beta^4,$$

and V satisfies the same boundary conditions as G at infinities. Because of these homogeneous boundary conditions, it is expected that

$$V(x) = \int_{-\infty}^{\infty} G(x,\xi)p(\xi)d\xi. \qquad (6.5.17)$$

Let us deduce the result more formally through the use of Green's theorem, which we must first derive for the fourth-order differential operator

$$\int_{-\infty}^{\infty} (V\mathcal{L}G - G\mathcal{L}V)\,dx$$

$$= \int_{-\infty}^{\infty} \left[VG'''' - GV'''' + 4\beta^4(VG - GV)\right]dx$$

$$= \int_{-\infty}^{\infty} \left[(VG''')' - V'G''' - (GV''')' + G'V'''\right]dx$$

$$= [VG''' - GV''']_{-\infty}^{\infty} - \int_{-\infty}^{\infty}(V'G''' - G'V''')\,dx$$

$$= -\int_{-\infty}^{\infty}\left[(V'G'' - G'V'')' + (V''G'' - G''V'')\right]dx$$

$$= -[V'G'' - G'V'']_{-\infty}^{\infty}.$$

So far this result holds for any pair of functions V and G, which are differentiable four times. Because of the governing equations of V and G and the boundary conditions, in particular the radiation condition at $\pm\infty$, it follows that

$$V(\xi) = \int_{-\infty}^{\infty} G(x,\xi)p(x)\,dx.$$

The reader should prove the symmetry of G. Equation (6.5.17) then follows.

6.6 Fundamental solutions

We now turn to an example for partial differential equations with two independent variables, say x and y. The Green function $G(x,y;0,0)$ in an infinite domain defined by

$$\mathcal{L}G = \delta(x)\delta(y), \qquad -\infty < x, y < \infty \qquad (6.6.1)$$

is often called the *fundamental solution*. Note that the two-dimensional δ function $\delta(x)\delta(y)$ has the property

$$\iint_{-\infty}^{\infty} \delta(x)\delta(y)\,dx\,dy = 1. \tag{6.6.2}$$

To be specific we choose $\mathcal{L}$ to be Laplacian

$$\mathcal{L} = \frac{\partial^2}{\partial x^2} + \frac{\partial^2}{\partial y^2}.$$

The problem is clearly isotropic about the origin and suggests the use of cylindrical polar coordinates with $x = r\cos\theta$ and $y = r\sin\theta$. Let us rewrite the two dimensional δ function as

$$\delta(x)\delta(y) = \frac{\delta(r)}{2\pi r}, \tag{6.6.3}$$

where $\delta(r)$ is narrow and sharply peaked near $r = 0$. When r refers to the radial coordinate, we redefine the δ function by

$$\int_0^\infty \delta(r)\,dr = 1$$

so that

$$\iint_{-\infty}^\infty \delta(x)\delta(y)\,dx dy = \int_0^{2\pi} d\theta \int_0^\infty \frac{\delta(r)}{2\pi r} r\,dr = 1.$$

In polar coordinates the Green function must satisfy

$$\frac{1}{r}\frac{\partial}{\partial r}\left(r\frac{\partial G}{\partial r}\right) = \frac{\delta(r)}{2\pi r}, \qquad 0 \le r < \infty. \tag{6.6.4}$$

Let us multiply the above equation by $2\pi r$ and integrate the result from $r = 0$ to $r = 0+$, yielding

$$\lim_{r\to 0+} 2\pi r \frac{\partial G}{\partial r} = 1, \tag{6.6.5}$$

which becomes a boundary condition for G. For all $r > 0$ the Green function is governed by the homogeneous equation

$$\frac{1}{r}\frac{\partial}{\partial r}\left(r\frac{\partial G}{\partial r}\right) = 0, \qquad 0 < r < \infty. \tag{6.6.6}$$

Physically, G defined by (6.6.5) and (6.6.6) can represent the velocity potential in a homogeneous porous medium due to a concentrated source. The mathematical solution is simply $G = C\ln r$. From the boundary

6.6 Fundamental solutions

condition (6.6.5) the coefficient is determined to be $C = 1/2\pi$, so that the fundamental solution is

$$G(x, y; 0, 0) = \frac{1}{2\pi} \ln r. \qquad (6.6.7)$$

If the source is located at any other point (x', y'), the fundamental solution is obtained by a simple change of coordinates

$$G(x, y; x', y') = G(\mathbf{r}; \mathbf{r}') = \frac{1}{2\pi} \ln(|\mathbf{r} - \mathbf{r}'|), \qquad (6.6.8)$$

where $\mathbf{r} = (x, y)$, $\mathbf{r}' = (x', y')$, and

$$|\mathbf{r} - \mathbf{r}'| = \sqrt{(x - x')^2 + (y - y')^2 + (z - z')^2}.$$

For half-plane problems with homogeneous Dirichlet or Neuman condition on the straight boundary, the Green function can be constructed from the fundamental solution by the so-called *method of images*. For example, let us seek the Green function defined in the upper half plane by

$$\frac{\partial^2 G}{\partial x^2} + \frac{\partial^2 G}{\partial y^2} = \delta(x - x')\delta(y - y'), \qquad 0 < x, x', y, y' < \infty, \qquad (6.6.9)$$

and by the Neuman condition

$$\frac{\partial G}{\partial y} = 0 \quad \text{on} \quad y = 0. \qquad (6.6.10)$$

The solution is obtained by adding two fundamental solutions: one at the source point (x', y') and one at the image point $(x', -y')$

$$G(x, y; x', y') = \frac{1}{2\pi} \ln r + \frac{1}{2\pi} \ln r' = \frac{1}{2\pi} \ln(rr') \qquad (6.6.11)$$

with

$$r = \sqrt{(x - x')^2 + (y - y')^2}, \qquad r' = \sqrt{(x - x')^2 + (y + y')^2}.$$

By extending Green's theorem to two dimensions, this Green function can be used to solve the general boundary-value problem for the half plane. This is left as an exercise.

The fundamental solution of Laplace's equation in three dimensions can be obtained in much the same way. The three-dimensional δ function $\delta(x)\delta(y)\delta(z)$ has the property that

$$\iiint_{-\infty}^{\infty} \delta(x)\delta(y)\delta(z)\, dx\, dy\, dz = 1. \qquad (6.6.12)$$

Isotropy suggests the use of the spherical polar coordinates in terms of which $\delta(r)$ can be defined by

$$\delta(x)\delta(y)\delta(z) = \frac{\delta(r)}{4\pi r^2} \qquad (6.6.13)$$

so that (6.6.12) becomes

$$\int_0^{2\pi} d\theta \int_0^{\pi} \sin\phi \, d\phi \int_0^{\infty} \frac{\delta(r)}{4\pi r^2} r^2 \, dr = \int_0^{\infty} \frac{\delta(r)}{4\pi r^2} 4\pi r^2 \, dr = 1.$$

We leave it to the reader to verify that the fundamental solution defined by

$$\frac{1}{r^2}\frac{\partial}{\partial r}\left(r^2 \frac{\partial G}{\partial r}\right) = \frac{\delta(r)}{4\pi r^2}, \qquad 0 \leq r < \infty$$

is

$$G = -\frac{1}{4\pi r}. \qquad (6.6.14)$$

If the source is any other point (x', y'), the Green function is simply

$$G(x,y,z;x',y',z') = -\frac{1}{4\pi\sqrt{(x-x')^2 + (y-y')^2 + (z-z')^2}}. \qquad (6.6.15)$$

6.7 Green's function in a finite domain

As demonstrated so far, the Green function, once found, is useful in constructing new solutions to general boundary-value problems. However, finding the Green function explicitly may itself be a challenging task and depends on the equation, the boundary geometry, and the boundary condition. For example, even for the Laplace equation, the task is not simple except for a circular boundary and for the Dirichlet or Neuman boundary condition. We demonstrate the case for a rectangle.

Let $G(x,y;x'y')$ be defined by

$$\frac{\partial^2 G}{\partial x^2} + \frac{\partial^2 G}{\partial y^2} = \delta(x-x')\delta(y-y'), \quad 0 < x, x' < a, \quad 0 < y, y' < b \quad (6.7.1)$$

with the boundary conditions

$$G = 0, \quad x = 0, a; \qquad G = 0, \quad y = 0, b. \qquad (6.7.2)$$

The Green function can be found in terms of known eigenfunctions. Let

$$G(x,y;x'y') = \sum_{m=1}^{\infty} \sum_{n=1}^{\infty} a_{mn} \sin\frac{m\pi x}{a} \sin\frac{n\pi y}{b}.$$

Each of the series terms satisfies the Dirichlet boundary condition but not Laplace's equation. Now to satisfy the inhomogeneous (Poisson) equation, we substitute the expansion and get

$$\sum_{m=1}^{\infty}\sum_{n=1}^{\infty}\left[\left(\frac{m\pi}{a}\right)^2+\left(\frac{n\pi}{b}\right)^2\right]a_{mn}\sin\frac{m\pi x}{a}\sin\frac{n\pi y}{b}=\delta(x-x')\delta(y-y').$$

The double Fourier series must be inverted. Using orthogonality of the sine functions, we get,

$$\left[\left(\frac{m\pi}{a}\right)^2+\left(\frac{n\pi}{b}\right)^2\right]a_{mn}\frac{ab}{4}=\sin\frac{m\pi x'}{a}\sin\frac{n\pi y'}{b}.$$

Hence,

$$G=\sum_{m=1}^{\infty}\sum_{n=1}^{\infty}\frac{4/(ab)}{\left(\frac{m\pi}{a}\right)^2+\left(\frac{n\pi}{b}\right)^2}\sin\frac{m\pi x'}{a}\sin\frac{n\pi y'}{b}\sin\frac{m\pi x}{a}\sin\frac{n\pi y}{b}. \quad (6.7.3)$$

This double series converges rather slowly due to the expected singularity at the source point. This is a disadvantage of eigenfunction expansions and can be remedied by further manipulations to separate the singularity from the series so that the remainder converges fast (Exercise 6.12). In any case it is clear that the solution procedure for G is almost as demanding as that for the more general boundary-value problem itself.

6.8 Adjoint operator and Green's function

The special feature of the Sturm–Liouville operator $\mathcal{L}$ is that the integral of the difference of two scalar products $\langle Y, \mathcal{L}Z\rangle - \langle Z, \mathcal{L}Y\rangle$ reduces to a boundary integral, i.e., Y and Z satisfy Green's formula. Any operator with this property is called *self-adjoint*. For the self-adjoint Sturm–Liouville operator, the proper Green function can help solve an inhomogeneous boundary-value problem by superposition. What can we do for a non-self-adjoint operator?

Let us start with a simple boundary-value problem defined by a non-self-adjoint operator

$$\mathcal{L}u=f, \quad 0<x<L, \quad \text{where} \quad \mathcal{L}=\frac{d^2}{dx^2}+3\frac{d}{dx}+4, \quad (6.8.1)$$

with the boundary conditions

$$u(0)=a \quad \text{and} \quad u(L)=b. \quad (6.8.2)$$

Consider the product

$$v\mathcal{L}u = v\left(\frac{d^2u}{dx^2} + 3\frac{du}{dx} + 4u\right)$$

$$= \frac{d}{dx}\left(v\frac{du}{dx}\right) - \left(\frac{dv}{dx}\frac{du}{dx}\right) + 3\frac{duv}{dx} - 3u\frac{dv}{dx} + 4uv$$

$$= \frac{d}{dx}\left(v\frac{du}{dx} - u\frac{dv}{dx} + 3uv\right) + u\left(\frac{d^2v}{dx^2} - 3\frac{dv}{dx} + 4v\right). \quad (6.8.3)$$

We now define the adjoint operator

$$\mathcal{L}^* = \frac{d^2}{dx^2} - 3\frac{d}{dx} + 4.$$

Then the integral of (6.8.3) gives

$$\int_0^L (v\mathcal{L}u - u\mathcal{L}^*v)\,dx = \left[v\frac{du}{dx} - u\frac{dv}{dx} + 3uv\right]_0^L. \quad (6.8.4)$$

We have now extended Green's theorem to a non-self-adjoint operator.

Now let us define a Green function for the adjoint operator in order to help solve for u. Let $v = G$ in (6.8.4)

$$\int_0^L (G\mathcal{L}u - u\mathcal{L}^*G)\,dx = \left[G\frac{du}{dx} - u\frac{dG}{dx} + 3uG\right]_0^L.$$

Clearly, if G is defined by

$$\mathcal{L}^*G = \delta(x - \xi), \quad 0 < x, \xi < L \quad (6.8.5)$$

$$G = 0, \quad x = 0, L, \quad (6.8.6)$$

then u is related to G by

$$u = \int_0^L G(x, \xi) f(x)\,dx, \quad (6.8.7)$$

which is the usual form signifying the superposition of a distributed forcing.

Thus, if $\mathcal{L}$ is not self-adjoint, the method of Green's function still works. The procedure is to find first the adjoint operator by partial integration, then find G for the adjoint boundary-value problem with boundary conditions of the same type, but homogeneous, and with singular forcing.

These ideas can, in principle, be extended to more general partial

6.8 Adjoint operator and Green's function

differential equations (as in, e.g., Koshlyakov et al., 1964). Let us take for example the second-order equation with variable coefficients

$$\mathcal{L}u \equiv A_{ij}\frac{\partial^2 u}{\partial x_i \partial x_j} + B_i \frac{\partial u}{\partial x_i} + Cu = 0, \qquad (6.8.8)$$

where A_{ij}, B_i and C are functions of x_i, with $i = 1, 2, 3$. An alternative form is

$$\frac{\partial}{\partial x_i}\left(A_{ij}\frac{\partial u}{\partial x_j}\right) + D_i \frac{\partial u}{\partial x_i} + Cu = 0, \qquad (6.8.9)$$

where

$$D_i = B_i - \frac{\partial A_{ij}}{\partial x_j}. \qquad (6.8.10)$$

Consider the product

$$v\mathcal{L}u = v\frac{\partial}{\partial x_i}\left(A_{ij}\frac{\partial u}{\partial x_j}\right) + D_i v \frac{\partial u}{\partial x_i} + Cuv$$

$$= \frac{\partial}{\partial x_i}\left(A_{ij} v \frac{\partial u}{\partial x_j}\right) + A_{ij}\frac{\partial v}{\partial x_i}\frac{\partial u}{\partial x_j} + D_i v \frac{\partial u}{\partial x_i} + Cuv$$

$$= \frac{\partial}{\partial x_i}\left[A_{ij}\left(v\frac{\partial u}{\partial x_j} - u\frac{\partial v}{\partial x_j}\right) + D_i uv\right]$$

$$+ u\left[\frac{\partial}{\partial x_i}\left(A_{ij}\frac{\partial v}{\partial x_i}\right) - D_i\frac{\partial v}{\partial x_i} + \left(C - \frac{\partial D_i}{\partial x_i}\right)v\right].$$

Let us define the adjoint operator by

$$\mathcal{L}^* v = \frac{\partial}{\partial x_i}\left(A_{ij}\frac{\partial v}{\partial x_i}\right) - D_i\frac{\partial v}{\partial x_i} + \left(C - \frac{\partial D_i}{\partial x_i}\right)v \qquad (6.8.11)$$

so that

$$v\mathcal{L}u - u\mathcal{L}^* x = \frac{\partial}{\partial x_i}\left[A_{ij}\left(v\frac{\partial u}{\partial x_j} - u\frac{\partial v}{\partial x_j}\right) + D_i uv\right].$$

Taking the volume integral and applying Gauss' theorem, we get a surface integral from the right-hand side

$$\iiint_\mathcal{V} (v\mathcal{L}u - u\mathcal{L}^* v)\, d\mathcal{V} = \iint_\mathcal{S}\left[A_{ij}\left(v\frac{\partial u}{\partial x_j} - u\frac{\partial v}{\partial x_j}\right) + D_i uv\right] n_i\, d\mathcal{S}, \qquad (6.8.12)$$

where $\mathcal{S}$ is the boundary of $\mathcal{V}$. This result is the generalization of Green's formula. Obviously, if $D_i = 0$, $\mathcal{L} = \mathcal{L}^*$ is self-adjoint.

Green's function G can be defined by requiring

$$\mathcal{L}^* G(x_i, x_i') = \delta(x - x')\delta(y - y')(\delta(z - z'), \qquad (6.8.13)$$

subject to homogeneous boundary conditions of the same type satisfied by u.

In this chapter we have only discussed problems for which Green's functions can be found analytically. Many other interesting examples can be found in Morse and Feshbach (1953) and Zauderer (1983). If either the boundary condition or boundary geometry is less simple, however, such analytical solutions may no longer be feasible. One then must relax one or more of the boundary conditions so that the incomplete Green function can be found explicitly. When such a result is used in Green's formula along with the unknown $u(x, y)$ of a boundary-value problem, one usually gets an integral equation for u that must be solved numerically. This technique is called the *boundary integral method*, which is now an important tool for numerical computations in engineering applications. A classical instrument has been given a new life! Interested readers are referred to Gipson (1987) for further information.

Exercises

6.1 Two rods of length L_1 and L_2 and constant thermal conductivities k_1 and k_2 are joined end to end at $x = 0$. The ends at $-L_1$ and $x = L_2$ are kept at zero temperature. What is the steady-state temperature in the rods if there is a point source at x'? Distinguish two cases where the point source is located on the left or on the right of the point of contact.

6.2 A heavy elastic rod of length L, density ρ, and Young's modulus E is hung from the ceiling where $x = 0$. If a small but heavy steel collar is attached to the spring at $x = L'$, how far downward is the vertical extension of the lower end of the rod?

6.3 A tightrope is stretched along the x axis from $x = 0$ to $x = L$. The left end is fixed. The right end is attached to an elastic spring that can only be deformed in the y direction. Verify that the boundary condition at $x = L$ is

$$T\frac{\partial V}{\partial x} + kV = 0.$$

Derive the Green function due to an oscillating point load at $x = x'$.

6.4 Derive the Green function for a beam of infinite length surrounded by a uniform elastic medium and forced by a point

Exercises

load oscillating at the frequency $\omega = \sqrt{k/\rho}$. Ignore rotatory inertia.

6.5 Derive the Green function for an infinitely long beam forced by a point load oscillating at the frequency ω. Include rotatory inertia but omit the lateral support. Discuss the role of rotatory inertia as the driving frequency varies.

6.6 Derive the fundamental solution for the one-dimensional Helmholtz equation

$$\frac{d^2 G}{dx^2} + k^2 G = \delta(x).$$

G represents waves generated by an oscillating point source and must be as outgoing as $x \to \pm\infty$. The time factor is $e^{-i\omega t}$.

6.7 Derive the fundamental solution for the three-dimensional Helmholtz equation in the infinite space

$$\nabla^2 G + k^2 G = -\frac{\delta(r)}{4\pi r}.$$

This fundamental solution represents sound generated by a point source oscillating at the frequency $\omega = kC$. With the time factor $e^{-i\omega t}$, G must behave as an outgoing wave at infinity.

6.8 Use the method of images to derive the Green function for the upper half plane, where

$$\nabla^2 G = 0, \quad y > 0, \quad -\infty < x < \infty.$$

Assume a homogeneous Dirichlet condition along the x axis.

6.9 Use the method of images to derive the Green function for the quarter plane $x > 0, y > 0$. Assume Laplace's equation in the region and a homogeneous Neuman condition along the plane boundaries.

6.10 Let $F(x,y)$ and $H(x,y)$ be two functions defined in the domain $\mathcal{A}$ bounded by the closed curve $\mathcal{S}$. Derive the following Green's theorem:

$$\iint_{\mathcal{A}} \left(F \nabla^2 H - H \nabla^2 F \right) dx\, dy = \oint_{\mathcal{S}} (F \nabla H - H \nabla F) \cdot \mathbf{n}\, ds.$$

6.11 In the half plane $y > 0$, u is governed by

$$\nabla^2 u = h(x,y),$$

subject to the boundary condition

$$u(x,0) = f(x), \quad y = 0, \quad |x| < \infty,$$

where h and f are nonzero only in a finite domain. Use the Green function for the half plane and Green's formula to construct the solution for u.

6.12 In the quarter plane $x > 0, y > 0$, u is governed by Poisson's equation as in the previous problem. The boundary conditions on the straight boundaries are inhomogeneous

$$u(x,0) = f(x), \quad 0 < x < \infty; \qquad u(0,y) = g(y), \quad 0 < y < \infty.$$

Find the solution by using Green's theorem and Green's function in Problem 6.9.

6.13 Find the solution G defined in the rectangle by $0 < x < a$, $0 < y < b$ if

$$\nabla^2 G + k^2 G = 0, \qquad 0 < x < a, \ 0 < y < b$$

$$\frac{\partial G}{\partial y} = 0, \qquad 0 < x < a, \ y = 0 \text{ and } b$$

$$\frac{\partial G(0,y)}{\partial x} = 0, \quad 0 < y < b$$

and

$$\frac{\partial G(a,y)}{\partial x} = \delta(y - y'), \quad 0 < y < b.$$

G represents the oscillation in a rectangular sound chamber forced by a line source on a wall.

Show first that

$$G = \sum_{n=0}^{\infty} X_n(x) \cos \frac{n\pi y}{b} \cos \frac{n\pi y'}{b},$$

where

$$X_n = \frac{\epsilon_n \cos(K_n(x-a))}{bK_n \sin K_n a}$$

and

$$K_n^2 = k^2 - \left(\frac{n\pi}{b}\right)^2,$$

with ϵ_n being the Jacobi symbol ($\epsilon_0 = 1; \epsilon_n = 2$, for $n = 1, 2, 3, \ldots$).

Next, show that for large n

$$X_n \sim \hat{X}_n(x) \equiv -\frac{2}{n\pi} e^{-n\pi x/b}.$$

Express the cosines as exponential functions and sum the series

$$\sum_{1}^{\infty} \hat{X}_n(x) \cos \frac{n\pi y}{b} \cos \frac{n\pi y'}{b}$$

by using the result

$$\sum_{1}^{\infty} \frac{e^{-nx}}{n} = -\ln(1 - e^{-x}).$$

Examine the logarithmic behavior near the source point $x = 0, y = y'$ and check the strength of the source. How fast does the remaining series in G converge?

6.14 Consider the second-order partial differential operator $\mathcal{L}$ and its adjoint $\mathcal{L}^*$ defined, respectively, by (6.8.8) and (6.8.11). Let $G(x_i, \xi_i)$ be the Green function of $\mathcal{L}$ defined in the region $\mathcal{V}$ by

$$\mathcal{L} G = \delta(x - \xi)\delta(y - \eta)\delta(z - \zeta), \quad x_i, \xi_i \in \mathcal{V}$$

and by the homogeneous condition on the bounding surface $\mathcal{S}$

$$G(x_i, \xi_i) = 0, \quad x_i \in \mathcal{S}.$$

Similarly let G^* be the Green function of $\mathcal{L}^*$ defined by

$$\mathcal{L}^* G^* = \delta(x - \xi')\delta(y - \eta')\delta(z - \zeta'), \quad x_i, \xi'_i \in \mathcal{V}$$

and by

$$G^*(x_i, \xi'_i) = 0, \quad x_i \in \mathcal{S}.$$

Show that

$$G(\xi_i, \xi'_i) = G^*(\xi'_i, \xi_i).$$

7
Unbounded domains and Fourier transforms

For finite domains of simple geometry, whenever the method of separation of variables is applicable partial differential equations are reduced to ordinary differential equations. The general solution can be constructed by superposition of eigenfunctions. When the domain is infinitely large, is it possible to achieve a similar reduction? The technique of Fourier transform can often be useful toward attaining that goal.

There is a large variety of integral transforms associated with the names Fourier, Laplace, Hankel, Mellin, etc. Among them the Fourier and Laplace transforms are the most basic and useful. In this chapter we only introduce the Fourier transforms of three varieties (exponential, sine and cosine). The Laplace transform is postponed until the tools of complex analysis are presented in Chapter 9.

7.1 Exponential Fourier transform
7.1.1 From Fourier series to Fourier transform

With a trivial change of notation, the exponential Fourier integral theorem (5.3.7) may be rewritten

$$f(x) = \frac{1}{2L} \sum_{n=-\infty}^{\infty} \int_{-L}^{L} f(\xi) e^{in\pi(x-\xi)/L} \, d\xi \qquad (7.1.1)$$

for $-L \leq x \leq L$. Consider the limit as $L \to \infty$. By the change of variable

$$\alpha_n = \frac{n\pi}{L},$$

which implies

$$\Delta \alpha = \alpha_{n+1} - \alpha_n = \frac{\pi}{L} \quad \text{or} \quad \frac{\Delta \alpha}{\pi} = \frac{1}{L},$$

7.1 Exponential Fourier transform

(7.1.1) may be written

$$f(x) = \lim_{L\to\infty} \sum_{n=-\infty}^{\infty} \Delta\alpha \widehat{f}(\alpha_n, x), \qquad (7.1.2)$$

where

$$\widehat{f}(\alpha_n, x) = \frac{1}{2\pi} \int_{-L}^{L} f(\xi) e^{i\alpha_n(x-\xi)} \, d\xi. \qquad (7.1.3)$$

In the limit of $L \to \infty$ or $\Delta\alpha \to 0$, the sum becomes an integral in the sense of Riemann, so that (7.1.2) becomes

$$f(x) = \int_{-\infty}^{\infty} \widehat{f}(\alpha, x) \, d\alpha \qquad (7.1.4)$$

while (7.1.3) becomes

$$\widehat{f}(\alpha, x) = \frac{1}{2\pi} \int_{-\infty}^{\infty} f(\xi) e^{i\alpha(x-\xi)} \, d\xi, \qquad (7.1.5)$$

provided that $f(x)$ is absolutely integrable

$$\int_{-\infty}^{\infty} |f(x)| \, dx < \infty.$$

The combination of (7.1.4) and (7.1.5) gives the Fourier integral theorem

$$f(x) = \frac{1}{2\pi} \int_{-\infty}^{\infty} e^{i\alpha x} \, d\alpha \int_{-\infty}^{\infty} f(\xi) e^{-i\alpha\xi} \, d\xi. \qquad (7.1.6)$$

If we define

$$F(\alpha) = \int_{-\infty}^{\infty} f(x) e^{-i\alpha x} \, dx, \qquad (7.1.7)$$

then

$$f(x) = \frac{1}{2\pi} \int_{-\infty}^{\infty} F(\alpha) e^{i\alpha x} \, dx. \qquad (7.1.8)$$

$F(\alpha)$ is called the exponential Fourier transform of $f(x)$, while $f(x)$ is called the inverse exponential Fourier transform of $F(\alpha)$. Equations (7.1.7) and (7.1.8) form the Fourier transform pair. We adopt the convention that a function and its transform be written in lower- and upper-case letters, respectively.

In the literature, there are various equivalent definitions. For example, the constant factor $1/2\pi$ can be split between the transform and its inverse

$$F(\alpha) = \frac{1}{\sqrt{2\pi}} \int_{-\infty}^{\infty} f(x) e^{-i\alpha x} \, dx \qquad (7.1.9)$$

$$f(x) = \frac{1}{\sqrt{2\pi}} \int_{-\infty}^{\infty} F(\alpha) e^{i\alpha x} \, dx, \qquad (7.1.10)$$

or the sign of the dummy variable α can be reversed so that

$$F(\alpha) = \int_{-\infty}^{\infty} f(x) e^{i\alpha x} \, dx \qquad (7.1.11)$$

and

$$f(x) = \frac{1}{2\pi} \int_{-\infty}^{\infty} F(\alpha) e^{-i\alpha x} \, d\alpha. \qquad (7.1.12)$$

Let us derive some useful formulas.

7.1.2 Transforms of derivatives

Under the assumption that u and its derivatives vanish sufficiently fast at infinities, it can be shown by repeated partial integration and induction that

$$\int_{-\infty}^{\infty} \frac{\partial^n u}{\partial x^n} e^{-i\alpha x} \, dx = (i\alpha)^n \, U. \qquad (7.1.13)$$

For example, for the first- and second-order derivatives

$$\int_{-\infty}^{\infty} \frac{\partial u}{\partial x} e^{-i\alpha x} \, dx = \left[u \, e^{-i\alpha x} \right]_{-\infty}^{\infty} + i\alpha \int_{-\infty}^{\infty} u \, e^{-i\alpha x} \, dx$$
$$= i\alpha U \qquad (7.1.14)$$

and

$$\int_{-\infty}^{\infty} \frac{\partial^2 u}{\partial x^2} e^{-i\alpha x} \, dx = \left[\frac{\partial u}{\partial x} e^{-i\alpha x} \right]_{-\infty}^{\infty} + i\alpha \int_{-\infty}^{\infty} \frac{\partial u}{\partial x} e^{-i\alpha x} \, dx$$
$$= (i\alpha)^2 \, U. \qquad (7.1.15)$$

Thus, each differentiation of u with respect to x brings out a factor $i\alpha$ to its Fourier transform.

7.1.3 Convolution theorem

If

$$F(\alpha) = \int_{-\infty}^{\infty} f(x) \, e^{-i\alpha x} \, dx \quad \text{and} \quad G(\alpha) = \int_{-\infty}^{\infty} g(x) \, e^{-i\alpha x} \, dx,$$

then

$$\frac{1}{2\pi} \int_{-\infty}^{\infty} F(\alpha)\,G(\alpha)\,e^{i\alpha x}\,d\alpha$$
$$= \int_{-\infty}^{\infty} f(x-\xi)g(\xi)\,d\xi = \int_{-\infty}^{\infty} f(\xi)g(x-\xi)\,d\xi. \quad (7.1.16)$$

The integrals on the right are called the convolution integrals. Thus, if the inverse transforms of $F(\alpha)$ and $G(\alpha)$ are known, the inverse transform of their product is given by the convolution integral of their inverses. This result can be used to facilitate the inversion of a transform, which can be factorized into two parts with known inverses.

To prove this theorem we take the Fourier transform of the convolution integral

$$\int_{-\infty}^{\infty} dx\, e^{-i\alpha x} \int_{-\infty}^{\infty} d\xi\, f(x-\xi) g(\xi) =$$
$$\int d\xi\, e^{-i\alpha \xi} g(\xi) \int_{-\infty}^{\infty} dx\, e^{-i\alpha(x-\xi)} f(x-\xi)$$
$$= G(\alpha)\, F(\alpha).$$

The inversion of this result is the convolution theorem.

7.2 One-dimensional diffusion

7.2.1 General solution in integral form

Consider the initial-value problem of diffusion in an infinite medium

$$\frac{\partial u}{\partial t} = k \frac{\partial^2 u}{\partial x^2}, \quad |x| < \infty, \quad t > 0 \quad (7.2.1)$$

$$u(x,0) = f(x), \quad |x| < \infty \quad (7.2.2)$$

$$u \to 0 \quad \text{as} \quad |x| \to \infty, \quad t > 0. \quad (7.2.3)$$

Let us define $U(\alpha, t)$ to be the Fourier transform of $u(x,t)$ and take the Fourier transform of the diffusion equation. On the left-hand side of the transformed equation the order of spatial integration and time differentiation is interchanged to give

$$\int_{-\infty}^{\infty} \frac{\partial u}{\partial t} e^{-i\alpha x}\, dx = \frac{\partial}{\partial t} \int_{-\infty}^{\infty} u e^{-i\alpha x}\, dx = \frac{dU}{dt}.$$

On the right-hand side we use (7.1.15) to get

$$\int_{-\infty}^{\infty} \frac{\partial^2 u}{\partial x^2} e^{-i\alpha x}\, dx = -\alpha^2 U.$$

It follows that

$$\frac{dU}{dt} + k\alpha^2 U = 0, \tag{7.2.4}$$

which is a first-order ordinary differential equation. An initial condition is needed and is found from the Fourier transform of the original initial condition (7.2.2)

$$U(\alpha, 0) = \int_{-\infty}^{\infty} f(x)\, e^{-i\alpha x}\, dx = F(\alpha), \tag{7.2.5}$$

where $F(\alpha)$ is the Fourier transform of the initial temperature (or concentration). Now the partial differential equation is reduced to an ordinary differential equation. This reduction of order is the typical advantage of the Fourier transform, similar to the method of separation of variables. Clearly, this reduction works if all coefficients in the governing equations are independent of x.

The solution to (7.2.4) and (7.2.5) is immediate

$$U(\alpha, t) = F(\alpha) e^{-k\alpha^2 t}. \tag{7.2.6}$$

Using the definition of the inverse transform, we get the solution in integral form

$$u = \frac{1}{2\pi} \int_{-\infty}^{\infty} d\alpha\, F(\alpha)\, e^{-k\alpha^2 t}\, e^{i\alpha x}$$

or, more explicitly,

$$\begin{aligned}
u &= \frac{1}{2\pi} \int_{-\infty}^{\infty} e^{i\alpha x - k\alpha^2 t}\, d\alpha \int_{-\infty}^{\infty} f(\xi)\, e^{-i\alpha \xi}\, d\xi \\
&= \int_{-\infty}^{\infty} f(\xi)\, d\xi\, \frac{1}{2\pi} \int_{-\infty}^{\infty} e^{i\alpha(x-\xi) - k\alpha^2 t}\, d\alpha \\
&= \int_{-\infty}^{\infty} f(\xi)\, d\xi\, \frac{1}{\pi} \int_{0}^{\infty} \cos\alpha(x - \xi)\, e^{-k\alpha^2 t}\, d\alpha. \tag{7.2.7}
\end{aligned}$$

An integral solution is often called a closed-form solution, in contrast to a series or numerical solution. We shall now explore the physical significance of the above general result by examining specific initial conditions.

7.2 One-dimensional diffusion

7.2.2 A localized source

Consider first the initial concentration to be localized at the origin

$$u(x,0) = f(x) = \delta(x). \qquad (7.2.8)$$

The corresponding total concentration at $t = 0$ in the entire domain $x \in (-\infty, \infty)$ is equal to unity. The Fourier transform of the initial data is

$$F(\alpha) = \int_{-\infty}^{\infty} \delta(x) e^{-i\alpha x}\, dx = 1, \qquad (7.2.9)$$

hence,

$$u = \frac{1}{2\pi} \int_{-\infty}^{\infty} d\alpha\, e^{i\alpha x} e^{-k\alpha^2 t} = \frac{1}{\pi} \int_{0}^{\infty} d\alpha\, \cos \alpha x\, e^{-k\alpha^2 t}. \qquad (7.2.10)$$

Introducing the change of variables $\alpha = z/\sqrt{kt}$, we get

$$u = \frac{1}{\pi\sqrt{kt}} \int_0^\infty e^{-z^2} \cos \frac{xz}{\sqrt{kt}}\, dz = \frac{1}{\pi\sqrt{kt}} I(\mu),$$

where

$$I(\mu) \equiv \int_0^\infty e^{-z^2} \cos \mu z\, dz \quad \text{and} \quad \mu = \frac{x}{\sqrt{kt}}.$$

The integration in I can be carried out by first differentiating it with respect to the parameter μ

$$\frac{dI}{d\mu} = -\int_0^\infty z e^{-z^2} \sin \mu z\, dz = \frac{1}{2} \int_0^\infty \sin \mu z\, d(e^{-z^2})$$

$$= \left[\frac{1}{2} e^{-z^2} \sin \mu z\right]_0^\infty - \int_0^\infty \frac{1}{2} e^{-z^2} \mu \cos \mu z\, dz$$

$$= -\frac{1}{2}\mu I.$$

This result is a first-order differential equation for $I(\mu)$ with the initial condition

$$I(0) = \int_0^\infty e^{-z^2}\, dz = \frac{\sqrt{\pi}}{2}. \qquad (7.2.11)$$

The solution for $I(\mu)$ is easily obtained

$$I(\mu) = I(0) e^{-\frac{\mu^2}{4}} = \frac{\sqrt{\pi}}{2} e^{-\frac{\mu^2}{4}}.$$

Finally,

$$u = \frac{1}{\pi} \int_0^\infty d\alpha\, \cos \alpha x\, e^{-k\alpha^2 t} = \frac{1}{2\sqrt{\pi kt}} e^{-\frac{x^2}{4kt}}. \qquad (7.2.12)$$

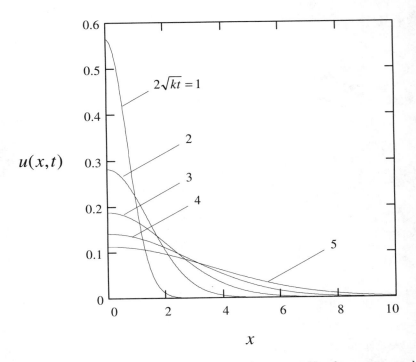

Fig. 7.1. Temperature distribution due to a point source. Number next to each curve represents the value of $2(kt)^{1/2}$.

The evolution of $u(x,t)$ is plotted in Figure 7.1. At any time, the spatial distribution of u is Gaussian. The peak height of u diminishes inversely with $\sqrt{kt}$, while the width of the peak increases with $\sqrt{kt}$. These features are the earmarks of diffusion phenomena.

A few additional remarks are appropriate here. First, the initial condition (7.2.2) can be transformed to an impulsive internal source. Consider the following problem

$$\frac{\partial u}{\partial t} - k\frac{\partial^2 u}{\partial x^2} = f(x)\,\delta(t), \quad |x| < \infty, \qquad (7.2.13)$$

subject to the initial condition

$$u(x,-\epsilon) = 0, \quad |x| < \infty, \qquad (7.2.14)$$

where $\epsilon > 0$ is a small number. Equations (7.2.13) and (7.2.14) define the diffusion due to impulsive release of heat at the instant $t = 0$. Integrating

(7.2.13) with respect to t from $-\epsilon$ to ϵ, we get

$$\int_{-\epsilon}^{\epsilon} \left(\frac{\partial u}{\partial t} - k\frac{\partial^2 u}{\partial x^2}\right) dt = u(x, \epsilon) - u(x, -\epsilon)$$

$$= \int_{-\epsilon}^{\epsilon} \delta(t) f(x)\, dt = f(x).$$

Hence, (7.2.13) and (7.2.14) can be equivalently stated as

$$\frac{\partial u}{\partial t} = k\frac{\partial^2 u}{\partial t^2}, \quad |x| < \infty, \quad t > \tau$$

$$u(x, \epsilon) = f(x), \quad |x| < \infty.$$

Thus, the problem for an impulsive source released at $t = 0$ is equivalent to the initial-value problem with $t = \epsilon$ as the initial instant.

If the source in (7.2.8) is at $x = \xi \neq 0$, the solution can be written by replacing x in (7.2.12) by $x - \xi$,

$$u(x - \xi, t) = \frac{1}{\pi}\int_0^\infty d\alpha \cos\alpha(x - \xi)\, e^{-k\alpha^2 t} = \frac{1}{2\sqrt{\pi k t}} e^{-\frac{(x-\xi)^2}{4kt}}, \quad (7.2.15)$$

i.e., by a shift of origin.

For general initial data $f(x)$, the temperature at any time is

$$u(x, t) = \int_{-\infty}^{\infty} d\xi\, f(\xi)\, \frac{e^{-\frac{(x-\xi)^2}{4kt}}}{\sqrt{4\pi k t}}. \quad (7.2.16)$$

This integral means that the temperature at any x and t is the sum of contributions by the initial source distribution of intensity $f(x)$ per unit length.

7.2.3 Discontinuous initial temperature

As another special case of the general result, we take the initial temperature to be a step function in x, i.e., $f(x) = f_o H(x)$, where $H(x)$ is the Heaviside function. We then have from (7.2.16)

$$u = \int_{-\infty}^{\infty} d\xi\, f_o H(\xi)\, \frac{e^{-\frac{(x-\xi)^2}{4kt}}}{\sqrt{4\pi k t}} = \frac{f_o}{\sqrt{4\pi k t}}\int_0^\infty d\xi\, e^{-\frac{(x-\xi)^2}{4kt}}.$$

Let $\zeta = x - \xi$, then

$$u = -\frac{f_o}{\sqrt{4\pi k t}}\int_x^{-\infty} d\zeta\, e^{-\frac{\zeta^2}{4kt}}.$$

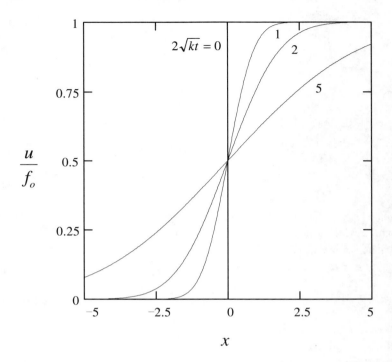

Fig. 7.2. Diffusion due to a discontinuous initial temperature at different instants. Number next to each curve represents the value of $2(kt)^{1/2}$.

With the further change of variable

$$\eta = \frac{\zeta}{\sqrt{4kt}},$$

the solution becomes

$$u = \frac{f_o}{\sqrt{\pi}} \int_{-\infty}^{\frac{x}{\sqrt{4kt}}} d\eta\, e^{-\eta^2} = \frac{f_o}{\sqrt{\pi}} \left[\int_{-\infty}^{0} + \int_{0}^{\frac{x}{\sqrt{4kt}}} e^{-\eta^2}\, d\eta \right]$$

$$= \frac{f_o}{\sqrt{\pi}} \frac{\sqrt{\pi}}{2} \left(1 + \operatorname{erf} \frac{x}{\sqrt{4kt}}\right) = \frac{f_o}{2} \left(1 + \operatorname{erf} \frac{x}{\sqrt{4kt}}\right), \qquad (7.2.17)$$

where

$$\operatorname{erf} z = \frac{2}{\sqrt{\pi}} \int_{0}^{z} e^{-\eta^2}\, d\eta$$

is called the error function. The implied temperature variation is shown in Figure 7.2. As t increases, the discontinuity is smoothed out, while the width of the transition zone expands as $\sqrt{kt}$.

7.3 Forced waves in one dimension

The initial-value problem of the inhomogeneous one-dimensional wave equation in an infinite domain was solved in §3.4 with considerable effort. We now show that the same result can be found more straightforwardly by Fourier transform. Recall the governing field equation

$$\frac{\partial^2 u}{\partial t^2} = c^2 \frac{\partial^2 u}{\partial x^2} + h(x,t), \quad t > 0, \quad |x| < \infty, \tag{7.3.1}$$

the initial conditions

$$u(x,0) = f(x), \quad \frac{\partial u}{\partial t}(x,0) = g(x), \tag{7.3.2}$$

and the boundary conditions

$$u \to 0, \quad |x| \to \infty. \tag{7.3.3}$$

Let

$$H(\alpha, t) = \int_{-\infty}^{\infty} h(x,t) e^{-i\alpha x} \, dx$$

denote the Fourier transform of the forcing term. The transformed wave equation is now an ordinary differential equation

$$\frac{d^2 U}{dt^2} + c^2 \alpha^2 U = H, \quad t > 0$$

with the initial data

$$U(\alpha, 0) = F(\alpha), \quad \frac{dU(\alpha, 0)}{dt} = G(\alpha).$$

The general solution to the inhomogeneous second-order ordinary differential equation is

$$U = C_1 U_1 + C_2 U_2 + \int_0^t \frac{H(\tau)}{W} \left[U_1(\tau) U_2(t) - U_2(\tau) U_2(t) \right] d\tau,$$

where U_1 and U_2 are the homogeneous solutions of the differential equation governing U, and W is the Wronskian

$$W = U_1 \frac{dU_2}{dt} - U_2 \frac{dU_1}{dt}.$$

The parametric dependence on α has been suppressed for brevity. In the present case

$$U_1 = e^{-i\alpha ct}, \quad U_2 = e^{i\alpha ct}$$

and

$$W = U_1 U_2' - U_2 U_1' = 2i\alpha c = \text{constant},$$

hence,
$$U = C_1 e^{-i\alpha ct} + C_2 e^{i\alpha ct} + \int_0^t \frac{H(\alpha,\tau)}{2i\alpha c}\left[e^{i\alpha c(t-\tau)} - e^{-i\alpha c(t-\tau)}\right] d\tau.$$

Applying the initial conditions, we get
$$C_1 + C_2 = F$$
$$i\alpha c\,(C_1 - C_2) = G.$$

Hence,
$$\begin{pmatrix} C_1 \\ C_2 \end{pmatrix} = \frac{1}{2}\left(F \mp \frac{G}{i\alpha c}\right)$$

and
$$U(\alpha,t) = \frac{F}{2}\left(e^{-i\alpha ct} + e^{i\alpha ct}\right) - \frac{G}{2i\alpha c}\left(e^{-i\alpha ct} - e^{i\alpha ct}\right)$$
$$+ \frac{1}{2\pi}\int_0^t d\tau \frac{H(\alpha,\tau)}{2i\alpha c}\left[e^{i\alpha c(t-\tau)} - e^{-i\alpha c(t-\tau)}\right].$$
(7.3.4)

The inverse transform of $Fe^{\pm i\alpha ct}$ is
$$\frac{1}{2\pi}\int_{-\infty}^{\infty} F(\alpha)e^{i\alpha(x\pm ct)}\,d\alpha = f(x \pm ct),$$

and the inverse of $G/(2i\alpha c)$ is
$$-\frac{1}{2\pi}\int_{-\infty}^{\infty} \frac{G}{2i\alpha c}\left(e^{-i\alpha ct} - e^{i\alpha ct}\right)e^{i\alpha x}\,d\alpha = \frac{1}{4\pi c}\int_{-\infty}^{\infty} d\alpha\, G \int_{x-ct}^{x+ct} d\xi\, e^{i\alpha\xi}$$
$$= \frac{1}{4\pi c}\int_{x-ct}^{x+ct} d\xi \int_{-\infty}^{\infty} Ge^{i\alpha\xi}\,d\alpha = \frac{1}{2c}\int_{x-ct}^{x+ct} d\xi\, g(\xi).$$

To invert the integral in (7.3.4), observe that
$$\int_a^b d\xi\, h(\xi,\tau) = \frac{1}{2\pi}\int_a^b d\xi \int_{-\infty}^{\infty} d\alpha\, H\, e^{i\alpha\xi}$$
$$= \frac{1}{2\pi}\int_{-\infty}^{\infty} d\alpha\, H(\alpha,\tau)\frac{e^{i\alpha b} - e^{i\alpha a}}{i\alpha}$$

after changing the order of integration. If we let $b = x + c(t-\tau)$ and $a = x - c(t-\tau)$, the following
$$\frac{1}{2c}\int_0^t d\tau \int_{x-c(t-\tau)}^{x+c(t-\tau)} d\xi\, h(\xi,\tau)$$

7.4 Seepage flow into a line drain

is easily seen to be the inverse transform of the double integral. In summary, the final result is the full solution of d'Alembert

$$u(x,t) = \frac{1}{2}[f(x+ct) + f(x-ct)] + \frac{1}{2c}\int_{x-ct}^{x+ct} g(\xi)\,d\xi$$

$$+ \frac{1}{2c}\int_0^t d\tau \int_{x-c(t-\tau)}^{x+c(t-\tau)} h(\xi,\tau)\,d\xi, \qquad (7.3.5)$$

which agrees with (3.4.10).

7.4 Seepage flow into a line drain

Water is pumped from a line drain on the covered surface of a wet ground. What is the flow pattern in the ground?

Let us model the wet ground as a porous medium in which the seepage flow velocity obeys Darcy's law (1.4.5)

$$\mathbf{u} = \nabla \phi, \quad |x| < \infty, y < 0, \qquad (7.4.1)$$

where ϕ is proportional to the sum of the pore pressure and the potential energy. For an incompressible pore fluid, conservation of mass leads to

$$\nabla^2 \phi = \frac{\partial^2 \phi}{\partial x^2} + \frac{\partial^2 \phi}{\partial y^2} = 0, \quad |x| < \infty, y < 0. \qquad (7.4.2)$$

On the ground surface $y = 0$ the prescribed flux into the line drain implies the boundary condition

$$v = \frac{\partial \phi}{\partial y} = Q\delta(x), \quad -\infty < x < \infty. \qquad (7.4.3)$$

Far from the drain

$$\mathbf{u} = \nabla \phi \to 0, \quad x^2 + y^2 \to \infty. \qquad (7.4.4)$$

Since the ground surface occupies the entire x axis, we apply the exponential Fourier transform with respect to x. The transformed Laplace equation is

$$\frac{d^2 \Phi}{dy^2} + \alpha^2 \Phi = 0 \qquad (7.4.5)$$

and the transformed surface condition is

$$\frac{d\Phi}{dy} = Q, \quad y = 0. \qquad (7.4.6)$$

At great depths, the transform of (7.4.4) implies

$$\Phi, \frac{d\Phi}{dy} \to 0, \quad y \to -\infty. \tag{7.4.7}$$

The solution is

$$\Phi = \frac{Q}{|\alpha|} e^{|\alpha|y}.$$

Note that $|\alpha|$ is used to ensure boundedness at $y \to \infty$ for both positive and negative α. The inverse transform is

$$\phi = \frac{Q}{2\pi} \int_{-\infty}^{\infty} e^{i\alpha x} \frac{e^{|\alpha|y}}{|\alpha|} d\alpha = \frac{Q}{\pi} \int_0^{\infty} \frac{e^{\alpha y}}{\alpha} \cos \alpha x \, d\alpha.$$

To evaluate this integral, we consider

$$\begin{aligned}
\frac{\partial \phi}{\partial y} &= \frac{Q}{\pi} \int_0^{\infty} e^{\alpha y} \cos \alpha x \, d\alpha \\
&= \frac{Q}{\pi} \int_0^{\infty} \frac{d\alpha}{2} \left[e^{i\alpha x + \alpha y} + e^{-i\alpha x + \alpha y} \right] \\
&= \frac{Q}{\pi} \int_0^{\infty} \frac{d\alpha}{2} \left[e^{i\alpha(x - iy)} + e^{i\alpha(-x - iy)} \right] \\
&= -\frac{Q}{2\pi} \left[\frac{1}{i(x - iy)} - \frac{1}{i(x + iy)} \right] \\
&= -\frac{Q}{\pi} \frac{y}{x^2 + y^2} = -\frac{Q}{\pi} \frac{y}{r^2},
\end{aligned}$$

where $r = \sqrt{x^2 + y^2}$. By integrating with respect to y from $-\infty$ to y, we get

$$\phi = -\frac{Q}{\pi} \ln r + \text{constant}. \tag{7.4.8}$$

The equipotential lines are circles centered at the drain. In cylindrical polar coordinates the velocity field is purely radial with the radial component

$$u_r = \frac{\partial \phi}{\partial r} = -\frac{Q}{\pi r}. \tag{7.4.9}$$

This potential solution is the lower half of a sink flow in the entire plane of x, y.

7.5 Surface load on an elastic ground

One of the most basic problems in soil mechanics is to predict the stresses in the ground under surface load. Based on the linear elasticity theory, Boussinesq (1885) gave the classic solution for a concentrated load on the surface of a homogeneous half space. His two-dimensional solution for a line surface load is most simply derived in terms of polar coordinates. We shall, however, use it for demonstrating the use of Fourier transform. First, some background of static elasticity in a plane (Fung, 1965).

7.5.1 Field equations for plane elasticity

Consider a rectangular element of a solid with two sides of dimension dx, dy, as sketched in Figure 7.3. On the side at x, the normal stress is σ_x and the shear stress τ_{xy}. On the side at y, the normal stress is σ_y and the shear stress τ_{yx}. In static equilibrium the inertial force vanishes. Momentum conservation requires that the net force in both x and y

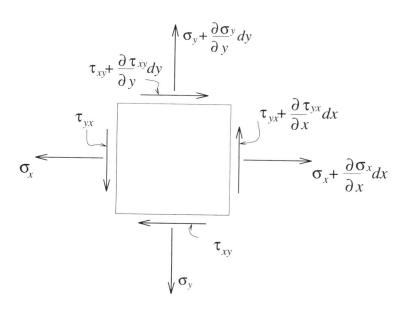

Fig. 7.3. Stresses on an elastic element.

directions must vanish; accordingly,

$$\left(\sigma_x + \frac{\partial \sigma_x}{\partial x}dx - \sigma_x\right)dy + \left(\tau_{xy} + \frac{\partial \tau_{xy}}{\partial y}dy - \tau_{xy}\right)dx = 0$$

and

$$\left(\sigma_y + \frac{\partial \sigma_y}{\partial y}dy - \sigma_y\right)dx + \left(\tau_{yx} + \frac{\partial \tau_{yx}}{\partial y}dx - \tau_{yx}\right)dy = 0.$$

Conservation of angular momentum requires that the net torque on the element also vanishes

$$\left(\tau_{yx} + \frac{\partial \tau_{yx}}{\partial x}dx + \tau_{yx}\right)dy\frac{dx}{2} - \left(\tau_{xy} + \frac{\partial \tau_{xy}}{\partial y}dy + \tau_{xy}\right)dx\frac{dy}{2} = 0.$$

Keeping only leading-order terms for vanishingly small dx and dy, we get

$$\frac{\partial \sigma_x}{\partial x} + \frac{\partial \tau_{xy}}{\partial y} = 0, \qquad (7.5.1)$$

$$\frac{\partial \tau_{yx}}{\partial x} + \frac{\partial \sigma_y}{\partial y} = 0 \qquad (7.5.2)$$

and

$$\tau_{xy} = \tau_{yx}. \qquad (7.5.3)$$

Let us define the elastic strain components by

$$\varepsilon_x = \frac{\partial u}{\partial x}, \qquad \varepsilon_y = \frac{\partial v}{\partial y}, \qquad \gamma_{xy} = \frac{\partial u}{\partial y} + \frac{\partial v}{\partial x}. \qquad (7.5.4)$$

It follows by straightforward differentiation that

$$\frac{\partial^2 \varepsilon_x}{\partial y^2} + \frac{\partial^2 \varepsilon_y}{\partial x^2} = \frac{\partial^2 \gamma_{xy}}{\partial x \partial y}, \qquad (7.5.5)$$

which is called the *compatibility condition*.

For small enough deformation, the linear law of Hooke between stresses and strains applies. For strain components in the x, y plane, the law, for an isotropic material, gives

$$E\varepsilon_x = \sigma_x - \nu(\sigma_y + \sigma_z), \qquad E\varepsilon_y = \sigma_y - \nu(\sigma_x + \sigma_z),$$

$$E\gamma_{xy} = 2(1+\nu)\tau_{xy}, \qquad (7.5.6)$$

with ν being the Poisson ratio. In the general case of three-dimensional deformation, there are other similar equations involving the z direction. For example, Hooke's law for the z component of strain reads

$$E\varepsilon_z = \sigma_z - \nu(\sigma_x + \sigma_y). \qquad (7.5.7)$$

7.5 Surface load on an elastic ground

If along the z direction the dimension is large and the deformation is constrained, we then have a case of plane strain with $\varepsilon_z = 0$, which implies

$$\sigma_z = \nu(\sigma_x + \sigma_y). \tag{7.5.8}$$

In terms of the stresses, the compatibility condition (7.5.5) reads

$$\frac{\partial^2}{\partial y^2}[\sigma_x - \nu(\sigma_x + \sigma_z)] + \frac{\partial^2}{\partial x^2}[\sigma_y - \nu(\sigma_x + \sigma_z)] = 2(1+\nu)\frac{\partial^2 \tau_{xy}}{\partial x \partial y},$$

which can be simplified by making use of (7.5.1), (7.5.2) and (7.5.8) to give

$$\nabla^2(\sigma_x + \sigma_y) = 0. \tag{7.5.9}$$

Equations (7.5.1) and (7.5.2) can be formally satisfied if we define Airy's stress function F by

$$\sigma_x = \frac{\partial^2 F}{\partial y^2}, \quad \sigma_y = \frac{\partial^2 F}{\partial x^2}, \quad \tau_{xy} = -\frac{\partial^2 F}{\partial x \partial y}. \tag{7.5.10}$$

It follows from (7.5.9) that

$$\nabla^2 \nabla^2 F = 0, \tag{7.5.11}$$

which is called the biharmonic equation.

We now have the basic equations of two-dimensional elasticity.

7.5.2 Half plane under surface load

Consider first a boundary-value problem for the lower half plane $y < 0$ with the following normal and shear stresses applied on the ground surface

$$\sigma_y = \frac{\partial^2 F}{\partial x^2} = -P_o(x), \quad \tau_{xy} = -\frac{\partial^2 F}{\partial x \partial y} = 0, \quad y = 0, \quad |x| < \infty. \tag{7.5.12}$$

Far away from the external load there should be no stresses

$$F \to 0, \quad \sqrt{x^2 + y^2} \to \infty. \tag{7.5.13}$$

Let us take the Fourier transform of the biharmonic equation

$$\left[\frac{d^2}{dy^2} - \alpha^2\right]^2 \overline{F} = 0, \tag{7.5.14}$$

where
$$\overline{F} = \int_{-\infty}^{\infty} dx\, F\, e^{-i\alpha x}$$

denotes the Fourier transform of F. The transforms of the stress components and the stress function are related by

$$\overline{\sigma}_y = \int_{-\infty}^{\infty} \sigma_y\, e^{-i\alpha x}\, dx = \int_{-\infty}^{\infty} \frac{\partial^2 F}{\partial x^2}\, e^{-i\alpha x}\, dx = -\alpha^2 \overline{F} \qquad (7.5.15)$$

$$\overline{\sigma}_x = \int_{-\infty}^{\infty} \sigma_x\, e^{-i\alpha x}\, dx = \int_{-\infty}^{\infty} \frac{\partial^2 F}{\partial y^2}\, e^{-i\alpha x}\, dx = \frac{d^2 \overline{F}}{dy^2} \qquad (7.5.16)$$

$$\overline{\tau}_{xy} = \int_{-\infty}^{\infty} \tau_{xy}\, e^{-i\alpha x}\, dx = -\int_{-\infty}^{\infty} \frac{\partial^2 F}{\partial x \partial y}\, e^{-i\alpha x}\, dx = -i\alpha \frac{d\overline{F}}{dy}. \qquad (7.5.17)$$

It is convenient to express the general solution as follows:

$$\overline{F} = (A + By)\, e^{-|\alpha|y} + (C + Dy)\, e^{|\alpha|y},$$

where A, B, C, D are unknown coefficients. To ensure boundedness of the solution at $y \sim -\infty$ for all α, we insist that $A = B = 0$. From the Fourier transforms of the boundary conditions (7.5.12) on the ground surface, we have

$$-\alpha^2 C = -\int_{-\infty}^{\infty} P_o(x)\, e^{-i\alpha x}\, dx = -\overline{P}_o(\alpha)$$

and

$$D + |\alpha| C = 0.$$

Hence, the transformed stress function is

$$\overline{F} = \frac{1}{\alpha^2} \overline{P}_o(\alpha)\, (1 - |\alpha|y)\, e^{|\alpha|y},$$

which has the following inverse transform:

$$F = \frac{1}{2\pi} \int_{-\infty}^{\infty} \frac{d\alpha}{\alpha^2}\, \overline{P}_o(\alpha)(1 - |\alpha|y)\, e^{|\alpha|y}\, e^{i\alpha x}. \qquad (7.5.18)$$

The stress components are found either by taking the inverse transforms of (7.5.15–7.5.17), or by applying (7.5.10) to (7.5.18),

$$\sigma_x = -\frac{1}{2\pi} \int_{-\infty}^{\infty} d\alpha\, \overline{P}_o(\alpha)\, e^{|\alpha|y}\, e^{i\alpha x}\, (1 + |\alpha|y) \qquad (7.5.19)$$

$$\sigma_y = -\frac{1}{2\pi} \int_{-\infty}^{\infty} d\alpha\, \overline{P}_o(\alpha)\, e^{|\alpha|y}\, e^{i\alpha x}\, (1 - |\alpha|y) \qquad (7.5.20)$$

7.5 Surface load on an elastic ground

$$\tau_{xy} = \frac{i}{2\pi} \int_{-\infty}^{\infty} d\alpha\, \alpha y\, \overline{P}_o(\alpha)\, e^{|\alpha|y}\, e^{i\alpha x}. \qquad (7.5.21)$$

We must now evaluate the inverse transforms.

7.5.3 Response to a line load

Let us examine the detailed response to a line surface load of unit strength

$$P_o(x) = \delta(x),$$

whose transform is

$$\overline{P}_o(\alpha) = \int_{-\infty}^{\infty} \delta(x)\, e^{i\alpha x}\, dx = 1.$$

The stress distributions are

$$\sigma_x = -\frac{1}{2\pi} \int_{-\infty}^{\infty} d\alpha\, e^{|\alpha|y + i\alpha x}\, (1 + |\alpha|y)$$

$$\sigma_y = -\frac{1}{2\pi} \int_{-\infty}^{\infty} d\alpha\, e^{|\alpha|y + i\alpha x}\, (1 - |\alpha|y)$$

$$\tau_{xy} = \frac{i}{2\pi} \int_{-\infty}^{\infty} d\alpha\, \alpha y\, e^{|\alpha|y + i\alpha x}.$$

Using the evenness of $|\alpha|$, we obtain

$$\sigma_x = -\frac{1}{\pi} \int_0^{\infty} d\alpha\, e^{\alpha y} \cos \alpha x\, (1 + \alpha y)$$

$$\sigma_y = -\frac{1}{\pi} \int_0^{\infty} d\alpha\, e^{\alpha y} \cos \alpha x\, (1 - \alpha y)$$

$$\tau_{xy} = -\frac{y}{\pi} \int_0^{\infty} d\alpha\, \alpha e^{\alpha y} \sin \alpha x.$$

All these integrals may be easily evaluated, for example,

$$\int_0^{\infty} d\alpha\, e^{\alpha y} \cos \alpha x = \frac{1}{2} \int_0^{\infty} d\alpha\, e^{\alpha y} \left(e^{-i\alpha x} + e^{i\alpha x}\right)$$

$$= \frac{1}{2} \int_0^{\infty} e^{\alpha(y+ix)} + e^{\alpha(y-ix)}\, d\alpha$$

$$= \frac{1}{2} \left[\frac{e^{\alpha(y+ix)}}{y+ix} + \frac{e^{\alpha(y-ix)}}{y-ix} \right]_{\alpha=0}^{\alpha=\infty}$$

$$= -\frac{1}{2} \left(\frac{1}{y+ix} + \frac{1}{y-ix} \right) = -\frac{y}{x^2 + y^2}$$

and
$$\int_0^\infty d\alpha\, \alpha\, e^{\alpha y} \cos \alpha x = \frac{d}{dy} \int_0^\infty d\alpha\, e^{\alpha y} \cos \alpha x.$$

Therefore, the stresses are

$$\sigma_x = \frac{1}{\pi} \frac{y}{x^2 + y^2} + \frac{y}{\pi} \frac{d}{dy}\left(\frac{y}{x^2+y^2}\right) = \frac{2x^2 y}{\pi(x^2+y^2)^2} \quad (7.5.22)$$

$$\sigma_y = \frac{2y^3}{\pi(x^2+y^2)^2} \quad (7.5.23)$$

$$\tau_{xy} = -\frac{2xy^2}{\pi(x^2+y^2)^2}. \quad (7.5.24)$$

In polar coordinates, $x = r\cos\theta, y = r\sin\theta$, the stress components can be expressed as

$$\sigma_x = \frac{2}{\pi r} \cos^2\theta \sin\theta \quad (7.5.25)$$

$$\sigma_y = \frac{2}{\pi r} \sin^3\theta \quad (7.5.26)$$

$$\tau_{xy} = -\frac{2}{\pi r} \cos\theta \sin^2\theta. \quad (7.5.27)$$

In solid mechanics a useful tool for finding the maximum stresses at a point is Mohr's circle, which can be constructed by first marking σ_x and σ_y on the horizontal axis and then drawing two vertical line segments of height τ_{xy} in opposite directions from these two points. The circle centered at the mid-point $(\sigma_x - \sigma_y)/2$ and passing the tips of the vertical line segments is Mohr's circle; see Figure 7.4.

It is easy to show by simple geometry that the maximum shear stress is, in general,

$$\tau_{max} = \left[\tau_{xy}^2 + \frac{1}{4}(\sigma_x - \sigma_y)^2\right]^{\frac{1}{2}}.$$

In the present case the maximum shear stress is

$$\tau_{max} = \frac{|\sin\theta|}{\pi r} = \frac{|y|}{\pi r^2}. \quad (7.5.28)$$

The contours of τ_{max} are circles tangent to the ground surface at the loading point

$$x^2 + \left(y + \frac{1}{2\pi\tau_{max}}\right)^2 = \frac{1}{(2\pi\tau_{max})^2},$$

as shown in Figure 7.5.

7.5 Surface load on an elastic ground   151

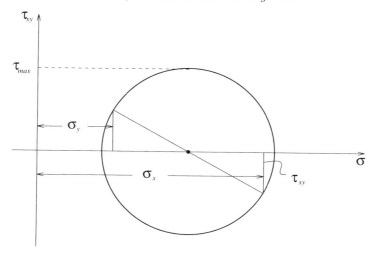

Fig. 7.4. Mohr's circle.

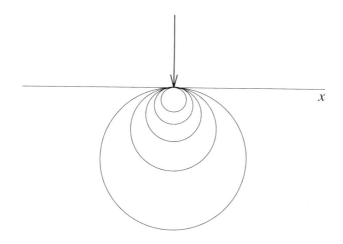

Fig. 7.5. Contours of maximum shear stress.

An issue that should be of practical importance in foundation engineering but rarely mentioned in soil mechanics texts is the soil deformation implied by the present solution. The solid displacements are again best described in polar coordinates after integrating Hooke's law

in polar coordinates (Fung, 1965). We shall only consider the vertical displacement
$$E\frac{\partial v}{\partial y} = (1 - \nu^2)\sigma_y - \nu(1+\nu)\sigma_x.$$

In particular, along the vertical line directly beneath the load, the normal stresses are
$$\sigma_x = 0, \qquad \sigma_y = \frac{2}{\pi y}.$$

Let $y = y_o$ be a reference point, where $v = v_o$ is measured; then by integration we have
$$E[v(y) - v(y_o)] = \frac{2}{\pi}(1 - \nu^2)\ln\left(\frac{y}{y_o}\right). \tag{7.5.29}$$

The displacement at great depth y/y_o would be unbounded even though the stresses are finite. This is of course physically absurd!

Consider next the horizontal displacement
$$E\frac{\partial u}{\partial x} = E\varepsilon_x = (1 - \nu^2)\sigma_x - \nu(1+\nu)\sigma_y.$$

In particular, for any $x > 0$ and $y = 0$
$$E[u(x) - u(-x)] = E\int_{-x}^{x}\frac{\partial u}{\partial x}dx = \int_{-x}^{x}\left[(1-\nu^2)\sigma_x - \nu(1+\nu)\sigma_y\right]dx$$
$$= \nu(1+\nu)\int_{-x}^{x}\delta(\xi)d\xi = \nu(1+\nu).$$

By virtue of symmetry $u(-x) = -u(x)$, hence
$$Eu(x) = \frac{1}{2}\nu(1+\nu), \qquad |x| > 0. \tag{7.5.30}$$

Thus, the ground surface everywhere, including infinity, would be moved!

As is usual in physical theories, a dilemma is resolved by reassessing the original assumptions. The culprit here is the geometrical oversimplification of the half plane in which both the length of the line loading in the z direction and the soil depth are assumed to be infinite. In reality neither length is truly infinite. If the length of the load is much greater than the soil depth, a theory for a finite soil depth, though more complex, is needed. On the other hand, if the soil depth is much greater than the length of the load, a three-dimensional theory for a line load of finite length is needed in order to describe the regions near the two ends of, and far away from, the line load. The present theory is only valid close to the center portion of the line load.

7.6 Fourier sine and cosine transforms

If the physical problem is defined for a semi-infinite domain, the exponential transform is inappropriate as it stands. Instead, half-sided sine or cosine transforms can be more useful.

Let us recall the Fourier integral theorem

$$f(x) = \frac{1}{2\pi} \int_{-\infty}^{\infty} d\alpha \int_{-\infty}^{\infty} d\xi\, f(\xi) e^{i\alpha(x-\xi)}$$

$$= \frac{1}{2\pi} \int_{-\infty}^{\infty} d\alpha \int_{-\infty}^{\infty} d\xi\, f(\xi)[\cos\alpha(x-\xi) + i\sin\alpha(x-\xi)]$$

$$= \frac{1}{\pi} \int_{0}^{\infty} d\alpha \int_{-\infty}^{\infty} d\xi\, f(\xi) \cos\alpha(x-\xi)$$

since $\sin\alpha(x-\xi)$ is odd and $\cos\alpha(x-\xi)$ is even in α. By the sum formula for the cosine function, we get

$$f(x) = \frac{1}{\pi} \int_{0}^{\infty} d\alpha \int_{-\infty}^{\infty} d\xi\, f(\xi)\, [\cos\alpha x \cos\alpha\xi + \sin\alpha x \sin\alpha\xi]$$

$$= \frac{1}{\pi} \int_{0}^{\infty} d\alpha \cos\alpha x \int_{-\infty}^{\infty} f(\xi) \cos\alpha\xi\, d\xi$$

$$+ \frac{1}{\pi} \int_{0}^{\infty} d\alpha \sin\alpha x \int_{-\infty}^{\infty} f(\xi) \sin\alpha\xi\, d\xi.$$

The sine transform: If $f(x)$ is odd in x for $-\infty < x < \infty$, or $f(x)$ is first defined in $0 < x < \infty$ but is extended to all $x \in (-\infty, \infty)$ such that f is odd in x, then

$$\int_{-\infty}^{\infty} f(\xi) \cos\alpha\xi\, d\xi = 0$$

and

$$f(x) = \frac{2}{\pi} \int_{0}^{\infty} d\alpha \sin\alpha x \int_{0}^{\infty} d\xi\, f(\xi) \sin\alpha\xi.$$

Let us define

$$F_s(\alpha) = \int_{0}^{\infty} d\alpha\, f(x) \sin\alpha x \qquad (7.6.1)$$

to be the Fourier sine transform of f. Then

$$f(x) = \frac{2}{\pi} \int_{0}^{\infty} d\alpha\, F_s(\alpha) \sin\alpha x \qquad (7.6.2)$$

is the inverse sine transform of F_s.

The cosine transform: On the other hand, if $f(x)$ is even in x for $-\infty < x < \infty$, or if it is only defined in $0 < x < \infty$ but its extension in $-\infty < x < \infty$ is even in x, then

$$f(x) = \frac{2}{\pi} \int_0^\infty d\alpha \cos \alpha x \int_0^\infty d\xi\, f(\xi) \cos \alpha \xi. \tag{7.6.3}$$

We define

$$F_c(\alpha) = \int_0^\infty dx\, f(x) \cos \alpha x \tag{7.6.4}$$

to be the Fourier cosine transform of $f(x)$ and

$$f(x) = \frac{2}{\pi} \int_0^\infty d\alpha\, F_c(\alpha) \cos \alpha x \tag{7.6.5}$$

the inverse Fourier cosine transform of $F_c(\alpha)$.

Both sine and cosine transforms are useful for problems involving half-infinite domains, but when to use which depends on the boundary condition at $x = 0$. In the next section we illustrate this issue for a diffusion problem.

7.7 Diffusion in a semi-infinite domain

Consider heat diffusion in a semi-infinitely long rod. Let the temperature $u(x,t)$ be governed by

$$\frac{\partial u}{\partial t} = k \frac{\partial^2 u}{\partial x^2}, \quad 0 < x < \infty \tag{7.7.1}$$

with

$$u(\infty, t) = 0 \tag{7.7.2}$$

$$u(x, 0) = h(x). \tag{7.7.3}$$

Three different boundary conditions at the left end will now be discussed.

Case (i): Flux rate is given at the left end

$$\frac{\partial u}{\partial x} = f(t), \quad x = 0. \tag{7.7.4}$$

Let us first try the Fourier cosine transform. From the left-hand side of (7.7.1) we get

$$\int_0^\infty \frac{\partial u}{\partial t} \cos \alpha x\, dx = \frac{\partial}{\partial t} \int_0^\infty u \cos \alpha x\, dx = \frac{dU_c}{dt},$$

7.7 Diffusion in a semi-infinite domain

where U_c denotes the cosine transform of u. For the right-hand side we perform partial integration repeatedly

$$k \int_0^\infty dx \, \frac{\partial^2 u}{\partial x^2} \cos \alpha x = k \left\{ \left[\frac{\partial u}{\partial x} \cos \alpha x \right]_0^\infty + \alpha \int_0^\infty \frac{\partial u}{\partial x} \sin \alpha x \, dx \right\}$$

$$= k \left\{ \left[-\frac{\partial u}{\partial x} \right]_0 + [\alpha u \sin \alpha x]_0^\infty - \alpha^2 \int_0^\infty u \cos \alpha x \, dx \right\}$$

$$= k \left\{ -\frac{\partial u(0,t)}{\partial x} - \alpha^2 U_c \right\}.$$

Although $u(0,t)$ is unknown, it is nevertheless not needed in the second line above. Hence,

$$\frac{dU_c}{dt} + k\alpha^2 U_c = -k \frac{\partial u(0,t)}{\partial x} = -kf(t).$$

The initial value of U_c can be found from the initial condition

$$U_c(\alpha, 0) = H_c(\alpha) \stackrel{\text{def}}{=} \int_0^\infty \cos \alpha x \, h(x) \, dx.$$

The problem can now readily be solved by finding an integrating factor

$$e^{-k\alpha^2 t} \left(U_c e^{k\alpha^2 t} \right)' = -k \, f(t)$$

$$U_c = H_c e^{-k\alpha^2 t} - k \int_0^t d\tau \, f(\tau) \, e^{-k\alpha^2 (t-\tau)}.$$

The inverse transform is

$$u = \frac{2}{\pi} \int_0^\infty d\alpha \, \cos \alpha x \, H_c \, e^{-k\alpha^2 t}$$

$$- \frac{2k}{\pi} \int_0^\infty d\alpha \, \cos \alpha x \int_0^t d\tau \, f(\tau) \, e^{-k\alpha^2 (t-\tau)}$$

$$= \int_0^\infty d\xi \, h(\xi) \frac{2}{\pi} \int_0^\infty d\alpha \, \cos \alpha \xi \, \cos \alpha x \, e^{-k\alpha^2 t}$$

$$- k \int_0^t d\tau \, f(\tau) \frac{2}{\pi} \int_0^\infty d\alpha \, \cos \alpha x \, e^{-k\alpha^2 (t-\tau)}. \qquad (7.7.5)$$

The first integral above represents the effect of initial data, while the second represents the effect of boundary data. Let us rewrite the first integral as

$$\int_0^\infty d\xi\, h(\xi) \frac{1}{\pi} \int_0^\infty d\alpha\, e^{-k\alpha^2 t} [\cos\alpha(x-\xi) + \cos\alpha(x+\xi)]$$

$$= \int_0^\infty d\xi\, h(\xi) \frac{1}{\sqrt{4\pi kt}} \left\{ e^{\frac{-(x-\xi)^2}{4kt}} + e^{\frac{-(x+\xi)^2}{4kt}} \right\}$$

$$= \int_0^\infty d\xi\, \frac{h(\xi)}{\sqrt{4\pi kt}} e^{\frac{-(x-\xi)^2}{4kt}} + \int_{-\infty}^0 d\xi\, \frac{h(-\xi)}{\sqrt{4\pi kt}} e^{\frac{-(x-\xi)^2}{4kt}}$$

$$= \int_{-\infty}^\infty d\xi\, \frac{h(|\xi|)}{\sqrt{4\pi kt}} e^{\frac{-(x-\xi)^2}{4kt}},$$

which represents the effect of heat sources distributed symmetrically with respect to the origin along the entire x axis. The second integral after the second equality in (7.7.5) can be written as

$$-\sqrt{\frac{k}{\pi}} \int_0^t d\tau\, \frac{f(\tau)}{\sqrt{(t-\tau)}} e^{-\frac{x^2}{4k(t-\tau)}}.$$

Thus the final form of the solution is

$$u = \int_{-\infty}^\infty d\xi\, \frac{h(|\xi|)}{\sqrt{4\pi kt}} e^{\frac{-(x-\xi)^2}{4kt}}$$

$$-\sqrt{\frac{k}{\pi}} \int_0^t d\tau\, \frac{f(\tau)}{\sqrt{t-\tau}} e^{\frac{-x^2}{4k(t-\tau)}}. \qquad (7.7.6)$$

What if we had chosen the sine transform instead? Partial integration would lead to

$$\int_0^\infty dx\, \sin\alpha x\, \frac{\partial^2 u}{\partial x^2} = \left[\frac{\partial u}{\partial x} \sin\alpha x\right]_0^\infty - \alpha \int_0^\infty \cos\alpha x\, \frac{\partial u}{\partial x} dx$$

$$= \left[\frac{\partial u}{\partial x} \sin\alpha x\right]_0^\infty - \alpha u\cos\alpha x|_0^\infty - \alpha^2 \int_0^\infty \sin\alpha x\, u\, dx.$$

In the first integrated term the flux condition prescribed at the boundary $x = 0$ is rendered useless, while in the second $u(0,t)$ is still unknown. Clearly, the sine transform is inappropriate.

7.7 Diffusion in a semi-infinite domain

Case (ii): Temperature is prescribed at the left end
Let the rod be brought to contact with a heat bath at the left end

$$u(0,t) = g(t), \quad t > 0, \tag{7.7.7}$$

then the sine transform is just what is needed. The transformed problem is governed by

$$\frac{dU_s}{dt} + k\alpha^2 U_s = k\alpha g(t), \quad t > 0$$

with the initial condition

$$U_s(\alpha, 0) = H_s(\alpha) \equiv \int_0^\infty h(x) \sin \alpha x \, dx.$$

The solution for U_s can be obtained as before with the inverse transform

$$u = \frac{2}{\pi} \int_0^\infty d\alpha \sin \alpha x \, H_s \, e^{-k\alpha^2 t}$$

$$+ \frac{2k}{\pi} \int_0^\infty d\alpha \, \alpha \sin \alpha x \int_0^t d\tau \, g(\tau) \, e^{-k\alpha^2(t-\tau)}. \tag{7.7.8}$$

The first and second integrals represent the effects of initial and boundary data, respectively.

Let us work out the details only for $h = H_s = 0$. The second integral above can be written

$$u = \frac{2k}{\pi} \int_0^t d\tau \, g(\tau) \left(-\frac{\partial}{\partial x}\right) \int_0^\infty d\alpha \cos \alpha x \, e^{-k\alpha^2(t-\tau)}$$

$$= 2k \int_0^t d\tau \, g(\tau) \left(-\frac{\partial}{\partial x}\right) \frac{e^{-\frac{x^2}{4k(t-\tau)}}}{\sqrt{4\pi k(t-\tau)}}$$

$$= \frac{x}{\sqrt{4\pi k}} \int_0^t \frac{d\tau \, g(\tau)}{(t-\tau)^{3/2}} e^{-\frac{x^2}{4k(t-\tau)}}. \tag{7.7.9}$$

In particular, if $g(t) = \text{constant} = U_0$, then

$$u = \frac{U_0 x}{\sqrt{4\pi k}} \int_0^t \frac{d\tau}{(t-\tau)^{3/2}} e^{-x^2/4k(t-\tau)}.$$

A change of variable

$$\eta = \frac{x}{\sqrt{4k(t-\tau)}}, \quad \frac{\partial \eta}{\partial \tau} = \frac{x}{4\sqrt{k}(t-\tau)^{3/2}}$$

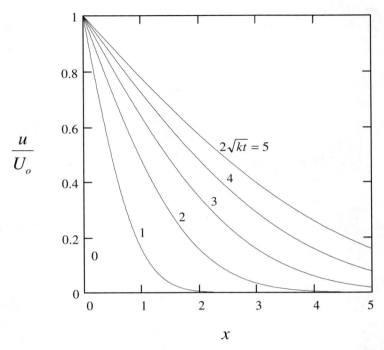

Fig. 7.6. Diffusion in a rod in contact with a heat bath. Number next to each curve represents the value of $2(kt)^{1/2}$.

finally yields

$$u = \frac{U_0}{\sqrt{\pi}} \int_0^\infty \frac{x\,d\tau}{2\sqrt{k}(t-\tau)^{3/2}} e^{-x^2/4k(t-\tau)}$$

$$= \frac{2U_0}{\sqrt{\pi}} \int_{x/2\sqrt{kt}}^\infty e^{-\eta^2}\,d\eta = U_0 \left(1 - \frac{2}{\sqrt{\pi}} \int_0^{x/\sqrt{4kt}} e^{-\eta^2}\,d\eta\right)$$

$$= U_0 \left[1 - \mathrm{erf}\left(\frac{x}{\sqrt{4kt}}\right)\right] \equiv U_0 \,\mathrm{erfc}\left(\frac{x}{\sqrt{4kt}}\right), \qquad (7.7.10)$$

where $\mathrm{erfc}(z)$ is called the complimentary error function of z. The result is plotted in Figure 7.6.

Case (iii): Partially insulated left end

If the boundary condition is

$$\frac{\partial u}{\partial x} - \beta u = f(t), \quad x = 0, \qquad (7.7.11)$$

then there is no clear advantage of using either sine or cosine transform.

7.8 Potential problem in a semi-infinite strip

Let us rewrite the above condition as

$$\frac{\partial u}{\partial x}(0,t) = f(t) + \beta u(0,t)$$

and apply cosine transform formally. From (7.7.5) the formal solution is

$$u(x,t) = -k \int_0^t d\tau \, [f(\tau) + \beta u(0,\tau)] \frac{2}{\pi} \int_0^\infty d\alpha \, \cos \alpha x \, e^{-k\alpha^2(t-\tau)}. \tag{7.7.12}$$

At the end $x = 0$ the boundary condition requires

$$u(0,t) = -k \int_0^t d\tau \, [f(\tau) + \beta u(0,t)] \frac{2}{\pi} \int_0^\infty d\alpha \, e^{-k\alpha^2(t-\tau)}, \tag{7.7.13}$$

which is an integral equation for $u(0,t)$. Let $z = \alpha\sqrt{k(t-\tau)}$, then the inner integral becomes

$$\frac{2}{\pi} \frac{1}{\sqrt{k(t-\tau)}} \int_0^\infty dz \, e^{-z^2} = \sqrt{\frac{2}{\pi}} \frac{1}{\sqrt{k(t-\tau)}}$$

and (7.7.13) may be written

$$u(0,t) = \beta \sqrt{\frac{2k}{\pi}} \int_0^t \frac{u(0,\tau)d\tau}{\sqrt{(t-\tau)}} + \sqrt{\frac{2k}{\pi}} \int_0^t \frac{f(\tau)d\tau}{\sqrt{t-\tau}}. \tag{7.7.14}$$

This is a Voltera integral equation of the second kind, which can in principle be solved by the method of Laplace transform, a subject of Chapter 10. After $u(0,t)$ is solved from the integral equation the full solution for $u(x,t)$ is given by (7.7.12). The subject of integral equations is beyond the scope of this book; the interested reader may consult one of many outstanding texts such as Tricomi (1957).

7.8 Potential problem in a semi-infinite strip

As the last example, we apply the Fourier sine transform to a potential problem in a semi-infinite strip governed by

$$\nabla^2 u = \frac{\partial^2 u}{\partial x^2} + \frac{\partial^2 u}{\partial y^2} = 0, \quad x > 0, \quad 0 < y < a \tag{7.8.1}$$

$$u(0,y) = 0, \quad 0 < y < a \tag{7.8.2}$$

$$u(x,0) = f(x), \quad 0 < x < \infty \tag{7.8.3}$$

$$u(x,a) = 0, \quad x > 0. \tag{7.8.4}$$

Taking the Fourier sine transform with respect to x, we get

$$\frac{d^2 U_s}{dy^2} - \alpha^2 U_s = 0, \quad 0 < y < a,$$

where $U_s(\alpha, y)$ denotes the Fourier sine transform of $u(x, y)$ and the boundary condition at $x = 0$ has been used. The sine transforms of the boundary conditions (7.8.3) and (7.8.4) are

$$U_s(\alpha, 0) = F_s(\alpha) = \int_0^\infty f(x) \sin \alpha x \, dx$$

$$U_s(\alpha, a) = 0,$$

which now serve as the boundary conditions for $U_s(\alpha, y)$. The solution is

$$U_s = F_s \frac{\sinh \alpha(a - y)}{\sinh \alpha a}$$

and the inverse transform is

$$u = \frac{2}{\pi} \int_0^\infty d\xi \, f(\xi) \int_0^\infty d\alpha \, \sin \alpha \xi \, \sin \alpha x \, \frac{\sinh \alpha(a - y)}{\sinh \alpha a}. \qquad (7.8.5)$$

In previous examples the inverse transforms can be evaluated explicitly by simple means. In general, the task is not so straightforward. Indeed, to calculate the integral in (7.8.5) one needs the technique of complex variables, to be described in Chapter 9. The reader is invited to obtain a series solution to the present problem by the familiar method of separation of variables.

The limiting case of $a \to \infty$ is easier to invert since

$$\frac{\sinh \alpha(a-y)}{\sinh \alpha a} \to \frac{e^{\alpha(a-y)} - e^{-\alpha(a-y)}}{e^{\alpha a} - e^{\alpha y}} \to e^{-\alpha y}$$

so that

$$u(x, y) = \frac{2}{\pi} \int_0^\infty d\xi \, f(\xi) \int_0^\infty d\alpha \, \sin \alpha \xi \, \sin \alpha x \, e^{-\alpha y}.$$

Since

$$2 \int_0^\infty \sin \lambda \xi \, \sin \lambda x \, e^{-\lambda y} \, d\lambda$$

$$= \int_0^\infty e^{-\lambda y} \{\cos \lambda(x - \xi) - \cos \lambda(x + \xi)\} \, d\lambda$$

$$= \frac{1}{2} \int_0^\infty e^{-\lambda y} \left\{ \left(e^{i\lambda(x-\xi)} + e^{-i\lambda(x-\xi)}\right) - \left(e^{i\lambda(x+\xi)} + e^{-i\lambda(x+\xi)}\right) \right\} d\lambda$$

$$= \frac{y}{y^2 + (x - \xi)^2} - \frac{y}{y^2 + (x + \xi)^2},$$

7.8 Potential problem in a semi-infinite strip

it follows that

$$u(x,y) = \frac{y}{\pi} \int_0^\infty d\xi\, f(\xi) \left[\frac{1}{(x-\xi)^2 + y^2} - \frac{1}{(x+\xi)^2 + y^2} \right]. \quad (7.8.6)$$

This solution is for a harmonic function $u(x,y)$ in the quarter plane $x > 0, y > 0$ subject to the Dirichlet conditions that $u(x,0) = f(x)$, $x > 0$ and $u(0,y) = 0$, $y > 0$. This result can also be written as

$$u(x,y) = \frac{y}{\pi} \left[\int_0^\infty \frac{f(\xi)d\xi}{(x-\xi)^2 + y^2} - \int_{-\infty}^0 \frac{f(-\xi)d\xi}{(x-\xi)^2 + y^2} \right]$$

$$= \frac{y}{\pi} \int_{-\infty}^\infty \frac{\text{sgn}(\xi) f(|\xi|) d\xi}{(x-\xi)^2 + y^2},$$

where $\text{sgn}(\xi)$ is the sign of ξ, being 1 if $\xi > 0$ and -1 if $\xi < 0$. In this form $u(x,y)$ represents the solution for Laplace's equation in the upper half plane $-\infty < x < \infty, y > 0$, with the boundary value being antisymmetric about the origin.

For a spatially unbounded domain, a virtue of the Fourier transform with respect to one variable is to reduce the number of independent variables by one. Thus an ordinary differential equation is reduced to an algebraic equation, a one-dimensional heat or wave equation to an ordinary differential equation in time, and a two-dimensional Laplace equation to an ordinary differential equation in space, etc. For a higher-dimensional problem one may try to apply the Fourier transform to two or more of the spatial coordinates repeatedly.

It may occur that the range of one space coordinate, say x, is infinite while the others are finite. The Fourier transform may be applied to x, and the transformed problem may be treated by separation of variables or other means. The solution will then involve a mixture of an infinite integral and a series. Examples on repeated transforms and on mixed use of series and transforms will be found in the following exercises.

In most practical problems the inverse Fourier transform can be difficult to evaluate exactly. Aside from strictly numerical means, there are asymptotic techniques applicable when a certain parameter is large. A systematic exposition of these techniques (method of stationary phase, steepest descent, etc.) is beyond the scope of this book and may be found in Carrier, Krook and Pearson (1966), Erdelyi (1953b), Jeffreys and Jeffreys (1950), Morse and Feshbach (1953), Nayfeh (1981), and other treatises.

Exercises

7.1 A boat carrying toxic waste cruises at the constant speed V along a long and narrow canal of negligible flow. After passing the station $x = 0$ at $t = 0$, the boat begins to leak. Let the concentration of waste in the canal be governed by

$$\frac{\partial C}{\partial t} = k\frac{\partial^2 C}{\partial x^2} + q(t)\delta(x - Vt), \quad t > 0,$$

where $q(t) = 0$ for $t < 0$. Find $C(x,t)$ for all x and $t > 0$.

7.2 In a straight and uniform river $-\infty < x < \infty$, water is flowing at the constant velocity U. A line source of contaminant with strength $f(x)g(t)$ is released in the domain $a < x < b$. Show first that the concentration C satisfies

$$\frac{\partial C}{\partial t} + U\frac{\partial C}{\partial x} = k\frac{\partial^2 C}{\partial x^2} + f(x)g(t).$$

If the source is highly localized in space and time so that $f(x) = \delta(x)$ and $g(t) = \delta(t)$, find $C(x,t)$. Show the equivalence of this exercise with the preceding one.

7.3 Along the southern bank of a long river of width a, a sewage plant emits pollutant at the steady rate of Q_o from $x = 0$ while the rest of the bank is nonabsorbing

$$-K\frac{\partial C}{\partial y} = Q_o\delta(x), \quad -\infty < x < \infty, \quad y = 0.$$

The flow rate in the river is negligible. Find $C(x,y)$. What must be the rate of cleanup along the entire northern bank in order to keep it free of pollution, i.e., $C(x,a) = 0$? Sketch and discuss the results.

7.4 In two dimensions, diffusion induced by a source is governed by

$$\frac{\partial C}{\partial t} = k\left(\frac{\partial^2 C}{\partial x^2} + \frac{\partial^2 C}{\partial y^2}\right) + q(x,y,t), \quad -\infty < x, y < \infty$$

for $t > 0$. Let the initial condition be

$$C(x,y,0) = 0.$$

Use the two-dimensional Fourier transform

$$U(\alpha, \beta, t) = \iint_{-\infty}^{\infty} e^{-i(\alpha x + \beta y)} u(x,y,t) dx dy$$

to show that

$$C(x,y,t) = \int_0^t d\tau \iint_{-\infty}^{\infty} \exp\left[-\frac{(x-\xi)^2 + (y-\eta)^2}{4k(t-\tau)}\right] \frac{q(\xi,\eta,\tau)}{\sqrt{4\pi k(t-\tau)}} d\xi d\eta.$$

7.5 Ignoring rotatory inertia, the forced vibration of a uniform beam obeys

$$\frac{\partial^4 u}{\partial x^4} + \frac{1}{c^2}\frac{\partial^2 u}{\partial t^2} = \frac{P(x,t)}{EI},$$

where u denotes the lateral deflection and $c^2 = EI/\rho S$. Let the beam be infinitely long, the forcing be impulsive and localized

$$P = \delta(x)\delta(t),$$

and the initial state is calm. Find the transient response for all x and t. Use the following identity

$$\frac{1}{\sqrt{2\pi}} \int_{-\infty}^{\infty} \exp(-\xi^2 a - i\xi x) d\xi = \frac{1}{\sqrt{2a}} e^{-x^2/4a},$$

where a can be complex, and evaluate all the integrals explicitly. Interpret the results physically. Speculate on the possible role of rotatory inertia in this problem.

7.6 A long estuary is invaded by tides. The spreading and transport of pollutant from a point source is affected by the time-dependent flow and by dispersion (diffusion augmented by shear). In the one-dimensional case, the depth-averaged transport equation for the concentration is

$$\frac{\partial C}{\partial t} + U\cos\omega t \frac{\partial C}{\partial x} = D\frac{\partial^2 C}{\partial x^2} + q(x,t), \quad t > 0,$$

where $U\cos\omega t$ represents the tidal velocity and D the dispersion coefficient. Solve for C in an infinitely long estuary $-\infty < x < \infty$ subject to the initial condition $C(x,t) = 0$.

7.7 A fault line along $x = x'$ is parallel to a long and straight seacoast $x = 0$. Apply the linearized long-wave theory to find the water waves in the region $0 < x < \infty$ due to a sudden jolt of the sea bottom along the fault. For a movable bottom use the following conservation laws

$$\frac{\partial \zeta}{\partial t} + \frac{\partial b}{\partial t} + h\frac{\partial u}{\partial x} = 0$$

and
$$\frac{\partial u}{\partial t} + g\frac{\partial \zeta}{\partial x} = 0,$$

where $b(x,t)$ denotes the upward displacement of the seabed. Assume
$$b = A\delta(x-x')\delta(t)$$

and impose the boundary conditions that $u(0,t) = 0$, and that u and ζ vanish at $x = \infty$ for $t > 0$. Initially, both u and ζ are zero. Find $\zeta(x,t)$.

7.8 Contaminant is released from a chemical plant at the dead end of infinitely long river $0 < x < \infty, 0 < y < a$. Let there be no absorption along the two banks $y = 0, a$, and the flux along $x = 0$ be given by

$$-K\frac{\partial C}{\partial x} = \begin{cases} f(y,t) = y(a-y), & 0 < t < T \\ 0, & t > T. \end{cases}$$

Find the concentration $C(x,y,t)$ in the river for all t.

7.9 A landslide occurs along the south bank of a long river for a limited duration T. What is the water surface elevation at any later time and any (x,y)? Use shallow-water approximation and represent the boundary condition on the south bank as

$$g\frac{\partial \zeta}{\partial y} = \begin{cases} \omega V \delta(x) \cos \omega t, & 0 < \omega t < \pi \\ 0, & \omega t > \pi. \end{cases}$$

7.10 A semi-infinite string is kept taut at tension T. The left end at $x = 0$ is fixed at all times. At time $t = 0$ a point force is applied impulsively at $x = x' > 0$. There is otherwise no disturbance anywhere. Find the string displacement for $t > 0$. Use the formula

$$\int_0^\infty \frac{d\alpha}{\alpha} \sin \alpha a \cos \alpha b = \frac{\pi}{2} H(a-b),$$

where H is the Heaviside function.

7.11 Referring to §7.5, find the elastic stress field in the lower half plane $y < 0$ if there is a distributed shear stress acting on the free surface. Get the solution explicitly if the applied stress is localized at the point $x = 0$.

8
Bessel functions and circular boundaries

Beyond elementary functions such as exponential, logarithmic, sinusoidal and hyperbolic functions, there are a host of so-called special functions that arise frequently in physical problems. Examples are Bessel functions, Legendre polynomials, Mathieu functions, hypergeometric functions, etc. Often these special functions emerge from the solution of partial differential equations when the boundary possesses a certain special geometry. For example, Bessel functions are associated with circular boundaries, while Legendre polynomials are associated with spherical boundaries, etc. In this chapter we choose to acquaint the readers only with the basic properties of the Bessel functions, and with applications in wave propagation and fluid flow. Certain essential facts such as series definitions, recursion formulas, orthogonality and asymptotic approximations will be discussed. Though far from exhaustive, these facts can already go a long way toward many applications, and can prepare the reader for further study of advanced aspects and other special functions. For quick access to further properties the reader should take advantage of some of the popular handbooks of special functions such as Erdelyi (1953a) and Abramowitz and Stegun (1964). For thorough theoretical expositions the reader must consult more advanced treatises such as Watson (1958).

8.1 Circular region and Bessel's equation

In this section we give a practical motivation for the need of Bessel functions by examining wave motion in a circular domain. For either the vibration of a membrane stretched over a circular rim, or oscillations of water in a shallow pond of constant depth h, the vertical displacement is governed by (1.6.12), which reads, in cylindrical polar coordinates,

$$\frac{1}{c^2}\frac{\partial^2 \zeta}{\partial t^2} = \frac{\partial^2 \zeta}{\partial r^2} + \frac{1}{r}\frac{\partial \zeta}{\partial r} + \frac{1}{r^2}\frac{\partial^2 \zeta}{\partial \theta^2}, \quad 0 < r < a, \quad 0 < \theta \leq 2\pi. \quad (8.1.1)$$

For the membrane the displacement vanishes along the rim

$$\zeta(a, \theta) = 0, \quad r = a. \quad (8.1.2)$$

For water in a pond the no-flux condition along the bank at radius a implies

$$\frac{\partial \zeta}{\partial r} = 0, \quad r = a. \quad (8.1.3)$$

The initial conditions on the displacement and velocity of the surface are

$$\zeta(r, \theta, 0) = f(r, \theta) \quad (8.1.4)$$

and

$$\frac{\partial \zeta}{\partial t}(r, \theta, 0) = g(r, \theta). \quad (8.1.5)$$

Let us apply the method of separation of variables and assume $\zeta = R(r)\Theta(\theta)T(t)$ in (8.1.1)

$$\frac{1}{c^2} R\Theta T'' = \left(R'' + \frac{R'}{r} \right) \Theta T + \frac{R}{r^2} \Theta'' T,$$

where primes denote differentiation with respect to the argument. Dividing throughout by $R\Theta T$, we get

$$\frac{1}{c^2}\frac{T''}{T} = \frac{R'' + R'/r}{R} + \frac{\Theta''}{r^2 \Theta} = -k^2,$$

which leads to

$$T'' + k^2 c^2 T = 0 \quad (8.1.6)$$

and

$$\frac{R'' + R'/r}{R} + \frac{\Theta''}{r^2 \Theta} = -k^2,$$

where k^2 is a separation constant. The last equation can be further written as

$$\frac{r^2 (R'' + R'/r)}{R} + k^2 r^2 = -\frac{\Theta''}{\Theta} = \alpha,$$

where α is another separation constant and gives in turn

$$\Theta'' + \alpha\Theta = 0 \qquad (8.1.7)$$

and

$$r^2 R'' + rR' + \left(k^2 r^2 - \alpha\right) R = 0. \qquad (8.1.8)$$

The general solution for T is the linear combination of $\exp(\pm ikct)$, and Θ is a linear combination of $\cos\sqrt{\alpha}\theta$ and $\sin\sqrt{\alpha}\theta$. Since the physical domain covers the entire circle, Θ and Θ' must be periodic in θ. Consequently, $\sqrt{\alpha}$ must be an integer n, so that Θ is a linear combination of $\cos n\theta$ and $\sin n\theta$. Equation (8.1.8) then becomes

$$r^2 R'' + rR' + \left(k^2 r^2 - n^2\right) R = 0, \qquad (8.1.9)$$

which is called the Bessel equation, whose solutions are needed before the oscillation problem can be analyzed. First, we make the following change of variables

$$kr = x, \qquad R = y,$$

where x, y no longer represent the rectangular coordinates, so as to rewrite the Bessel equation in its canonical form

$$x^2 y'' + xy' + \left(x^2 - n^2\right) y = 0. \qquad (8.1.10)$$

This second-order equation must have two independent homogeneous solutions, which we now seek.

8.2 Bessel function of the first kind

For generality let us replace in (8.1.10) the integer order n by any positive real number ν

$$x^2 y'' + xy' + \left(x^2 - \nu^2\right) y = 0. \qquad (8.2.1)$$

Let us seek a homogeneous solution by using the Frobenius method of power series and assume

$$y = \sum_{n=0}^{\infty} a_n x^{s+n}, \qquad (8.2.2)$$

where n is a non-negative integer and s is to be determined, then

$$y' = \sum_{n=0}^{\infty} a_n(s+n)x^{s+n-1}$$

$$xy' = \sum_{n=0}^{\infty} a_n(s+n)x^{s+n}$$

$$x^2y'' = \sum_{n=0}^{\infty} a_n(s+n)(s+n-1)x^{s+n}.$$

Hence,

$$x^2y'' + xy' + (x^2 - \nu^2)y = 0 = \left[a_0(s-1)s + a_0 s - \nu^2 a_0\right]x^s$$
$$+ \left[a_1(s+1)s + a_1(s+1) - \nu^2 a_1\right]x^{s+1}$$
$$+ \sum_{n=2}^{\infty} \left[a_n(s+n)(s+n-1) + a_n(s+n) + a_{n-2} - \nu^2 a_n\right]x^{s+n}.$$

Let us collect all terms of the same power of x and rewrite the right-hand side

$$a_0\left(s^2 - \nu^2\right)x^s + a_1\left[(s+1)^2 - \nu^2\right]x^{s+1}$$

$$+ \sum_{n=2}^{\infty}\left\{\left[(s+n)^2 - \nu^2\right]a_n + a_{n-2}\right\}x^{s+n} = 0.$$

Equating the coefficient of x^s to zero and insisting that $a_0 \neq 0$, we get the *indicial equation* for s: $s^2 = \nu^2$ so that $s = \pm\nu$. For the coefficient of x^{s+1} to vanish it is necessary that

$$a_1 = 0,$$

since $(s+1)^2 - \nu^2 = (\nu+1)^2 - \nu^2 \neq 0$ except for $\nu = -1/2$. For any integer $n > 2$, the coefficient of x^{s+n} vanishes if

$$a_n = -\frac{a_{n-2}}{(s+n)^2 - \nu^2} = -\frac{a_{n-2}}{2n\nu + n^2}.$$

This result is a recursion relation between a_n and a_{n-2} and can be used repeatedly to give, for $n = $ odd,

$$a_1 = a_3 = a_5 = a_{2n-1} = \cdots = 0,$$

8.2 Bessel function of the first kind

and for $n = $ even,

$$a_2 = -\frac{a_0}{2(2\nu+2)} = \frac{(-1)a_0}{2^2(\nu+1)}$$

$$a_4 = -\frac{a_2}{4(2\nu+4)} = \frac{(-1)^2 a_0}{2^4 \cdot 2(\nu+2)(\nu+1)}$$

$$a_6 = -\frac{a_4}{6(2\nu+6)} = \frac{(-1)^3 a_0}{2^6 \cdot 3 \cdot 2(\nu+3)(\nu+2)(\nu+1)}$$

$$\vdots$$

$$a_{2p} = \frac{(-1)^p a_0}{2^{2p} p!(\nu+p)(\nu+p-1)\ldots(\nu+2)(\nu+1)}$$

$$\vdots$$

The symbol $p! = p(p-1)(p-2)\cdots 3\cdot 2\cdot 1$ is called the factorial function of p and is a special case of the Gamma function $\Gamma(z)$ when the argument $z = n+1$ is an integer, $\Gamma(n+1) = n!$. Properties of these functions are summarized in Appendix C. In terms of Gamma functions, we may write

$$a_{2p} = \frac{(-1)^p 2^\nu \Gamma(\nu+1) a_0}{2^{2p+\nu} p! \Gamma(\nu+p+1)}. \tag{8.2.3}$$

Since a homogeneous solution needs only to be determined up to a constant factor, we may choose

$$a_0 = \frac{1}{2^\nu \Gamma(\nu+1)}$$

so that

$$a_{2p} = \frac{(-1)^p}{2^{2p+\nu} p! \Gamma(\nu+p+1)}.$$

With this normalization the homogeneous solution is denoted by $J_\nu(x)$

$$y = J_\nu(x) \stackrel{\text{def}}{=} \sum_{p=0}^{\infty} \frac{(-1)^p x^{2p+\nu}}{2^{2p+\nu} p! \Gamma(\nu+p+1)}, \tag{8.2.4}$$

which is called the Bessel function of the first kind and order ν.

In particular, if $\nu = 0$ and 1, the Bessel functions are

$$J_0(x) = 1 - \frac{x^2}{2^2} + \frac{x^4}{2^4 (2!)^2} - \frac{x^6}{2^6 (3!)^2} + \cdots \tag{8.2.5}$$

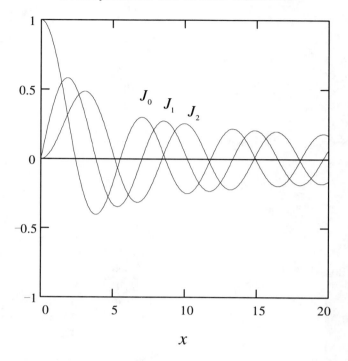

Fig. 8.1. Bessel functions $J_n(x)$ with $n = 0, 1, 2$.

and
$$J_1(x) = \frac{x}{2} - \frac{x^3}{2^3 2!} + \frac{x^5}{2^5 2! 3!} - \cdots. \tag{8.2.6}$$

The variations of $J_0(x)$, $J_1(x)$ and $J_2(x)$ are plotted in Figure 8.1.

Corresponding to the second solution $s = -\nu$ of the indicial equation, the second homogeneous solution to the Bessel equation is

$$J_{-\nu} = \sum_{p=0} \frac{(-1)^k x^{2p-\nu}}{2^{2p-\nu} p! \Gamma(-\nu + p + 1)} \tag{8.2.7}$$

if ν is not an integer. The general solution to (8.2.1) is then

$$y = a J_\nu(x) + b J_{-\nu}(x), \quad \nu \neq \text{integer}. \tag{8.2.8}$$

If $\nu = n = $ integer, however, then

$$\Gamma(-n + p + 1) = \infty \quad \text{for} \quad p = 0, 1, 2, \ldots, n - 1;$$

8.3 Bessel function of the second kind for integer order

see Appendix C. The first n coefficients of the series in J_n vanish and

$$J_{-n} = (-1)^n \sum_{p=n}^{\infty} \frac{(-1)^{p-n} x^{2(p-n)+n}}{2^{2(p-n)+n} p! \Gamma(p-n+1)}$$

$$= (-1)^n \sum_{K=0}^{\infty} \frac{(-1)^K \left(\frac{x}{2}\right)^{2K+n}}{\Gamma(K+1)(K+n)!}$$

$$= (-1)^n \sum_{K=0}^{\infty} \frac{(-1)^K \left(\frac{x}{2}\right)^{2K+n}}{K! \Gamma(K+n+1)} = (-1)^n J_n(x), \qquad (8.2.9)$$

where a change of variables $p - n = K$ has been made. Hence, J_{-n} is not independent of J_n, and we must look for another independent solution.

8.3 Bessel function of the second kind for integer order

For $\nu = n$ the search for the second independent solution of the Bessel equation is lengthy. We only illustrate the special case of $\nu = 0$

$$xy'' + y' + xy = 0 \qquad (8.3.1)$$

for which the indicial equation has a double root $s = 0$.

As a clue we note that the origin $x = 0$ is a singular point of the differential equation. The second derivative term drops out at the origin, and the reduced first-order equation cannot be used to generate two independent solutions. Near $x = 0$ the singular solution is expected to be large, hence $xy \ll y', xy''$. The dominant part must be

$$xy'' + y' \sim 0,$$

which implies that $y' \propto 1/x$ or $y \propto \ln x$. Accordingly, we shall try a solution of the form

$$y = J_0(x) \ln x + \sum_{m=1}^{\infty} A_m x^m, \qquad (8.3.2)$$

where $J_0(x)$ and the second series are both regular at $x = 0$. Now let us find the coefficients A_m by first inserting (8.3.2) into (8.3.1). Since

$$y' = J_0' \ln x + \frac{J_0}{x} + \sum_{m=1}^{\infty} m A_m x^{m-1}$$

$$y'' = J_0'' \ln x + \frac{2J_0'}{x} - \frac{J_0}{x^2} + \sum_{m=1}^{\infty} m(m-1) A_m x^{m-2},$$

we get from (8.3.1)

$$2J_0' + \sum_{m=1}^{\infty} m(m-1)A_m x^{m-1} + \sum_{m=1}^{\infty} m A_m x^{m-1} + \sum_{m=1}^{\infty} A_m x^{m+1} = 0, \tag{8.3.3}$$

where use has been made of the fact that

$$xJ_0'' + J_0' + xJ_0 = 0.$$

Since from (8.2.4)

$$J_0 = \sum_{m=0}^{\infty} \frac{(-1)^m x^{2m}}{2^{2m} m! m!},$$

the derivative is

$$J_0' = \sum_{m=1}^{\infty} \frac{(-1)^m x^{2m-1}}{2^{2m-1} m!(m-1)!}.$$

Hence, (8.3.3) may be rewritten

$$\sum_{m=1}^{\infty} \frac{(-1)^m x^{2m-1}}{2^{2m-2} m!(m-1)!} + \sum_{m=1}^{\infty} m^2 A_m x^{m-1} + \sum_{m=1}^{\infty} A_m x^{m+1} = 0. \tag{8.3.4}$$

We now collect terms with the same power of x. The coefficient of x^0 is simply

$$A_1 = 0.$$

In general the first series of (8.3.4) consists of only odd powers of x, while the second and third series contain both odd and even powers of x. Consider first the even powers. Let $m - 1 = 2s$ or $m = 2s + 1$ in the second series from which the part consisting only of the even powers can be extracted

$$\sum_{s=0}^{\infty} (2s+1)^2 A_{2s+1} x^{2s}.$$

Similarly, in the third series the part consisting only of the even powers is

$$\sum_{s=0}^{\infty} A_{2s-1} x^{2s}.$$

The coefficient of x^{2s} from all series in (8.3.4) must vanish, hence,

$$A_{2s-1} + (2s+1)^2 A_{2s+1} = 0 \quad \text{with} \quad A_{-1} = 0.$$

8.3 Bessel function of the second kind for integer order

Since $A_1 = 0$ we conclude that

$$A_3 = A_5 = A_7 = A_{\text{odd}} = 0.$$

Now consider the odd powers in (8.3.4). The first series is

$$\sum_{m=1}^{\infty} \frac{(-1)^s x^{2s-1}}{2^{2s-2} s!(s-1)!}.$$

The part with odd powers in the second series of (8.3.4) can be made explicit by letting $m = 2s$; the result is

$$\sum_{s=1}^{\infty} (2s)^2 A_{2s} x^{2s-1}.$$

To collect the odd powers from the third series, we let $m = 2s - 2$ to get

$$\sum_{s=1}^{\infty} A_{2s-2} x^{2s-1}.$$

Equating to zero the coefficients of x^{2s-1} in (8.3.4), we have

$$\frac{(-1)^{s+1}}{2^{2s}(s+1)s!} + (2s+2)^2 A_{2s+2} + A_{2s} = 0,$$

which is a recursion relation. In particular, for $s = 0$ the recursion relation is

$$-1 + 4A_2 = 0, \quad \text{i.e.,} \quad A_2 = \frac{1}{4},$$

where $A_0 = 0$ has been used. For $s = 1$ the corresponding relation is

$$\frac{1}{8} + 16 A_4 + A_2 = 0,$$

implying that

$$A_4 = -\frac{1}{16}\left(A_2 + \frac{1}{8}\right) = -\frac{1}{16}\left(\frac{1}{4} + \frac{1}{8}\right).$$

In general,

$$A_{2m} = \frac{(-1)^{m-1}}{2^{2m}(m!)^2} h_m,$$

where

$$h_m = 1 + \frac{1}{2} + \cdots \frac{1}{m}. \tag{8.3.5}$$

Without loss of generality the second solution to the Bessel equation of order zero is defined as

$$Y_0(x) = \frac{2}{\pi}\left[J_0(x)\left(\ln\frac{x}{2}+\gamma\right) + \sum_{m=1}^{\infty}\frac{(-1)^{m-1}h_m}{(m!)^2}\left(\frac{x}{2}\right)^{2m}\right], \quad (8.3.6)$$

where γ denotes the Euler constant

$$\gamma = \lim_{s\to\infty}\left(1+\frac{1}{2}+\cdots\frac{1}{s}-\ln s\right) = 0.5772157\ldots.$$

$Y_0(x)$ is called the Weber function of order zero.

The general solution to the homogeneous Bessel equation with $\nu = n = 0$ is, therefore,

$$y = aJ_0(x) + bY_0(x).$$

For any other integer order $\nu = n$, consideration of the singularity at $x = 0$ shows that in addition to logarithmic behavior, y should be as singular as x^{-n}. A lengthier analysis (see, e.g., Hildebrand, 1964) shows that in general the Weber function of order n is

$$Y_n(x) = \frac{2}{\pi}\left[J_n(x)\left(\ln\frac{x}{2}+\gamma\right) + \frac{1}{2}\sum_{m=0}^{\infty}\frac{(-1)^{m-1}(h_m+h_{m+n})}{m!(m+n)!}\left(\frac{x}{2}\right)^{2m+n}\right.$$
$$\left. - \frac{1}{2}\sum_{m=0}^{n-1}\frac{(n-m-1)!}{m!}\left(\frac{x}{2}\right)^{2m-n}\right], \quad (8.3.7)$$

where h_m is defined by (8.3.5). The variations of $Y_n(x)$ for $n = 0, 1, 2$ are shown in Figure 8.2. We mention that the Weber function can be alternatively defined by

$$Y_\nu(x) = \frac{J_\nu(x)\cos\nu\pi - J_{-\nu}(x)}{\sin\nu\pi}. \quad (8.3.8)$$

When ν is not an integer, Y_ν thus defined is certainly an independent solution of (8.2.1). When $\nu = n$ is an integer, careful use of l'Hopital's rule leads also to (8.3.7).

The general solution to the Bessel equation of integer order is

$$y = aJ_n(x) + bY_n(x). \quad (8.3.9)$$

8.4 Some properties of Bessel functions

In problems described by cylindrical polar coordinates, the Bessel function $J_n(x)$ and the Weber function $Y_n(x)$ play the roles of the trigonometric functions $\cos x$ and $\sin x$ in rectangular coordinates. Like their

8.4 Some properties of Bessel functions

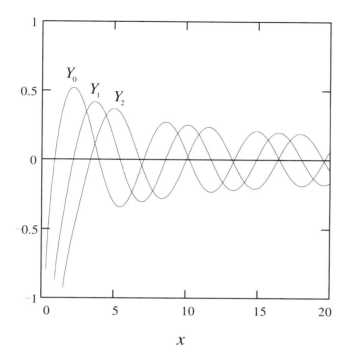

Fig. 8.2. Weber functions $Y_n(x)$.

trigonometric cousins, the Bessel functions have among them a number of identities, most of which can be derived from their series definitions. We only demonstrate such derivation for a recursion relation.

8.4.1 Recurrence relations

One of the most useful recurrence relations is

$$\nu J_\nu(x) + x J'_\nu(x) = x J_{\nu-1}(x). \qquad (8.4.1)$$

To verify this identity, we begin with the definition

$$J_\nu(x) = \sum_{p=0}^{\infty} \frac{(-1)^p \left(\frac{x}{2}\right)^{2p+\nu}}{p!(\nu+p)!}$$

and expand the left-hand side of (8.4.1) into a power series of x. Consider the rth term

$$\{\nu J_\nu + x J'_\nu\}_r = \frac{(-)^r \left(\frac{x}{2}\right)^{\nu+2r}}{r!\Gamma(\nu+r+1)}[\nu + (\nu+2r)]$$

$$= x\frac{(-1)^r \left(\frac{x}{2}\right)^{\nu-1+2r}}{r!\Gamma(\nu-1+r+1)}$$

$$= x\{J_{\nu-1}\}_r$$

after using $\Gamma(\nu+r+1) = (\nu+r)\Gamma(\nu+r)$. Thus (8.4.1) is proven.

It may also be shown that

$$xJ'_\nu(x) = -xJ_{\nu+1}(x) + \nu J_\nu(x) \tag{8.4.2}$$

and

$$\left[x^{\pm\nu} J_\nu(x)\right]' = \pm x^{\pm\nu} J_{\nu\mp1}(x). \tag{8.4.3}$$

The difference of (8.4.1) and (8.4.2) gives another recursion relation

$$J_{\nu+1}(x) + J_{\nu-1}(x) = \frac{2\nu}{x} J_\nu(x). \tag{8.4.4}$$

It may be similarly shown that all four recursion formulas (8.4.1–8.4.4) are satisfied by $Y_\nu(x)$.

These formulas are useful not only for analytical purposes, but also for computing the numerical values of Bessel functions of higher orders from the known values for low orders (see Press et al., 1985).

8.4.2 Behavior for small argument

From the series expansions the behavior of a Bessel function at small x is found by keeping the first term in the expansions. The following results are easily verified:

$$J_\nu \sim \frac{1}{2^\nu \nu!} x^\nu, \quad Y_0 \sim \frac{2}{\pi}\ln x, \quad Y_\nu \sim -\frac{2^\nu (\nu-1)!}{\pi} x^{-\nu}. \tag{8.4.5}$$

8.4.3 Behavior for large argument

Neither for numerical evaluation nor for analytical purposes is it efficient to use the series expansions to compute Bessel functions when either the argument or the order is large; asymptotic approximations are more powerful. The details, however, are complicated and beyond the scope of

8.4 Some properties of Bessel functions

this book. We shall merely give an incomplete argument to rationalize the form of such approximations for large argument.

Let us introduce a transformation to rid the first derivative in (8.2.1)

$$y = \frac{u}{\sqrt{x}}.$$

Substituting

$$y' = \frac{u'}{\sqrt{x}} - \frac{1}{2}\frac{u}{x^{3/2}}$$

$$y'' = \frac{u''}{x^{1/2}} - \frac{1}{2}\frac{u'}{x^{3/2}} - \frac{1}{2}\frac{u'}{x^{3/2}} + \frac{3}{4}\frac{u}{x^{5/2}}$$

into the Bessel equation, we get

$$u'' + \left(1 - \frac{\nu^2 - 1/4}{x^2}\right)u = 0. \tag{8.4.6}$$

For $x \gg 1$ and finite ν, the equation can be approximated by

$$u'' + u \cong 0,$$

whose general solution is

$$u \sim A\cos(x + \phi) + B\sin(x + \phi).$$

Therefore, for large x the independent solutions to the Bessel equation should be of the form

$$\frac{\cos(x+\phi)}{\sqrt{x}} \quad \text{and} \quad \frac{\sin(x+\phi)}{\sqrt{x}}.$$

A more sophisticated asymptotic analysis is needed to decide the phase ϕ, which we omit here and only cite the following results:

$$J_\nu(x) \sim \sqrt{\frac{2}{\pi x}} \cos\left[x - (2\nu + 1)\frac{\pi}{4}\right] \tag{8.4.7}$$

$$Y_\nu(x) \sim \sqrt{\frac{2}{\pi x}} \sin\left(x - (2\nu + 1)\frac{\pi}{4}\right). \tag{8.4.8}$$

These formulas further confirm that $J_\nu(x)$ and $Y_\nu(x)$ are the mathematical counterparts of $\cos x$ and $\sin x$, respectively.

A by-product of (8.4.6) is that for $\nu = \pm\frac{1}{2}$,

$$u'' + u = 0,$$

hence $u = \sin x$ or $\cos x$. It can be verified that

$$J_{\frac{1}{2}}(x) = \sqrt{\frac{2}{\pi x}} \sin x. \tag{8.4.9}$$

From the recursion formula (8.4.3)

$$\frac{d}{dx}\left(\sqrt{x}J_{\frac{1}{2}}(x)\right) = \sqrt{x}J_{-\frac{1}{2}}(x) = \sqrt{\frac{2}{\pi}}\cos x,$$

it follows that

$$J_{-\frac{1}{2}}(x) = \sqrt{\frac{2}{\pi x}}\cos x. \qquad (8.4.10)$$

By recursion formulas it can be shown that $J_{\pm n/2}$ can also be expressed in terms of sinusoidal functions (McLachlan, 1955; Erdelyi, 1953a; and Abramowitz and Stegun, 1964). The asymptotic properties of Bessel functions for large orders are also discussed in these references.

8.4.4 Wronskians

If y_1 and y_2 are two linearly independent Bessel functions, then

$$(xy_1')' + \frac{x^2 - \nu^2}{x}y_1 = 0$$

$$(xy_2')' + \frac{x^2 - \nu^2}{x}y_2 = 0.$$

Multiplying the first equation by y_2 and the second by y_1 and substracting the results, we get after partial integration

$$\frac{d}{dx}(xy_1'y_2 - xy_2y_1') = 0.$$

Hence, by integration with respect to x, we get

$$y_1'y_2 - y_1y_2' = -\frac{A}{x},$$

where A is some constant. The left side is called the Wronskian of y_1 and y_2. The constant A can be determined by using the limiting forms of Bessel functions for either small or large x. We leave it to the reader to verify that

$$J_n(x)Y_n'(x) - J_n'(x)Y_n(x) = \frac{2}{\pi x} \qquad (8.4.11)$$

for any x. There are also other Wronskian identities between other pairs of independent solutions of the Bessel equation such as

$$J_\nu(x)J_{-\nu}'(x) - J_\nu'(x)J_{-\nu}(x) = -\frac{2\sin\nu\pi}{\pi x}. \qquad (8.4.12)$$

8.4.5 Partial wave expansion

A useful result in wave theory is the expansion of the plane wave as a Fourier series of the polar angle θ. In polar coordinates the spatial factor of a plane wave of unit amplitude is

$$e^{ikx} = e^{ikr\cos\theta}.$$

Consider the following product of exponential functions

$$e^{zt/2}e^{-z/2t} = \left[\sum_{n=0}^{\infty} \frac{1}{n!}\left(\frac{zt}{2}\right)^n\right]\left[\sum_{n=0}^{\infty} \frac{1}{n!}\left(\frac{-z}{2t}\right)^n\right]$$

$$= \sum_{-\infty}^{\infty} t^n \left[\frac{(z/2)^n}{n!} - \frac{(z/2)^{n+2}}{1!(n+1)!} + \frac{(z/2)^{n+4}}{2!(n+2)!} + \cdots\right.$$

$$\left. + (-1)^r \frac{(z/2)^{n+2r}}{r!(n+r)!} + \cdots\right].$$

The coefficient of t^n is nothing but $J_n(z)$, hence

$$\exp\left[\frac{z}{2}\left(t - \frac{1}{t}\right)\right] = \sum_{-\infty}^{\infty} t^n J_n(z).$$

Now we set

$$t = ie^{i\theta}, \quad z = kr.$$

The plane wave then becomes

$$e^{ikx} = \sum_{n=-\infty}^{\infty} e^{in(\theta+\pi/2)} J_n(z).$$

Using the fact that $J_{-n} = (-1)^n J_n$, we finally get

$$e^{ikx} = e^{ikr\cos\theta} = \sum_{n=0}^{\infty} \epsilon_n i^n J_n(kr)\cos n\theta, \qquad (8.4.13)$$

where ϵ_n is the Jacobi symbol. The above result may be viewed as the Fourier expansion of the plane wave with Bessel functions being the expansion coefficients. In wave propagation theories, each term in the series represents a distinct angular variation and is called a *partial wave*.

Using the orthogonality of $\cos n\theta$, we may evaluate the Fourier coefficient

$$J_n(kr) = \frac{1}{i^n \pi} \int_0^\pi e^{ikr\cos\theta} \cos n\theta \, d\theta, \qquad (8.4.14)$$

which is one of a host of integral representations of Bessel functions.

8.5 Oscillations in a circular region

We now return to the problem started in §8.1.

8.5.1 Radial eigenfunctions and natural modes

In view of our experience with oscillation problems in rectangular domains, we shall first examine unforced, simple harmonic oscillations in a circular domain. Let the unknown displacement be of the form

$$\zeta = \mathrm{Re}\left(\eta e^{-i\omega t}\right), \qquad (8.5.1)$$

where $\omega = kc$ and $\eta = \eta(r,\theta)$ is the spatial factor. It follows from the wave equation (8.1.1) that

$$\nabla^2 \eta + k^2 \eta = 0. \qquad (8.5.2)$$

From the results of separation of variables in §8.1, we have

$$\Theta_n = C_n \sin n\theta + D_n \cos n\theta$$

and

$$R_n = A_n J_n(kr) + B_n Y_n(kr).$$

Since the displacement must be finite, the Weber function $Y_n(kr)$, which is unbounded at the origin, must be discarded. The solution corresponding to the nth angular mode is

$$\zeta_n = \mathrm{Re}\left(\eta_n e^{-i\omega t}\right) = \mathrm{Re}\left[e^{-i\omega t} J_n(kr)(A_n \cos n\theta + B_n \sin n\theta)\right].$$

For larger n, the angular variation is more rapid.

For a circular membrane clamped at the rim the boundary condition requires

$$\eta_n(a, \theta) = 0,$$

hence,

$$J_n(ka) = 0.$$

As can be seen from the graph for J_n, there are infinitely many zeros of this transcendental equation. It is customary to denote the mth zero of J_n by j_{nm}, i.e.,

$$J_n(j_{nm}) = 0, \quad m = 1, 2, 3, \ldots. \qquad (8.5.3)$$

8.5 Oscillations in a circular region

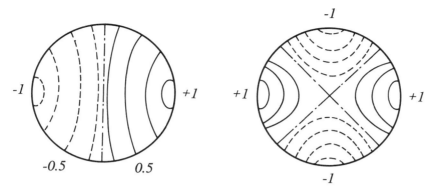

Fig. 8.3. Contour lines of two natural modes in a circular basin: $J_1(k_{11}r)\cos\theta$ and $J_2(k_{21}r)\cos 2\theta$.

The values of j_{nm} are well tabulated in handbooks such as Abramowitz and Stegun (1964). The eigen-wavenumbers and the eigenfrequencies are then

$$k_{nm} = \frac{j_{nm}}{a}, \qquad \omega_{nm} = \frac{j_{nm}c}{a}. \tag{8.5.4}$$

Thus for each n there are infinitely many modes corresponding to different values of m. For each pair of (n, m), the radial variation is proportional to $J_n(k_{nm}r)$.

For a shallow pond the boundary condition is

$$\frac{\partial \eta_n}{\partial r} = 0, \quad r = a,$$

implying

$$J'_n(ka) = 0.$$

The mth zero of $J'_n(z)$, customarily denoted by $j'_{n,m}$, is defined by

$$J'_n(j'_{n,m}) = 0, \quad m = 1, 2, \ldots. \tag{8.5.5}$$

The values of j'_{nm} are also found in standard mathematical handbooks. The corresponding eigen-wavenumbers and eigenfrequencies are

$$k_{nm} = \frac{j'_{nm}}{a}, \qquad \omega_{nm} = \frac{j'_{nm}c}{a}. \tag{8.5.6}$$

For each pair of (n, m), the radial variation is proportional to $J_n(k'_{nm}r)$. Figure 8.3 shows the contours of the surface for the first two modes of water waves in a pond.

To solve an initial-value problem we must start with the most general solution constructed by superposition

$$u = \sum_{n=0}^{\infty} \sum_{m=1}^{\infty} \{[A_{nm} \cos \omega_{nm} t + B_{nm} \sin \omega_{nm} t] \cos n\theta +$$
$$[C_{nm} \cos \omega_{nm} t + D_{nm} \sin \omega_{nm} t] \sin n\theta\} J_n(k_{nm} r). \quad (8.5.7)$$

The expansion coefficients A_{nm}, B_{nm}, C_{nm} and D_{nm} must now be found by studying first the orthogonality of the radial eigenfunctions.

8.5.2 Orthogonality of natural modes

To simplify notations we drop the double subscripts in this subsection and let $u = J_n(\alpha x)$ and $v = J_n(\beta x)$ be two eigenfunctions of the nth-order Bessel equation

$$xu'' + u' + \frac{\alpha^2 x^2 - n^2}{x} u = 0, \qquad xv'' + v' + \frac{\beta^2 x^2 - n^2}{x} v = 0, \quad (8.5.8)$$

and satisfy the same homogeneous boundary conditions at $x = a$

$$u' + hu = 0, \qquad v' + hv = 0. \quad (8.5.9)$$

The constants α and β are different eigenvalues. At $x = 0$, u and v are both finite. It will now be shown that u and v are orthogonal if $\alpha \neq \beta$.

As in the analysis for a Stürm-Liouville problem we multiply the first equation for u by v and the second for v by u, and then add the results to get

$$\int_0^a \{x(u''v - uv'') + (u'v - uv')\} dx + (\alpha^2 - \beta^2) \int_0^a xuv\,dx = 0$$

or

$$\int_0^a \frac{d}{dx} \{x(u'v - uv')\} dx = (\beta^2 - \alpha^2) \int_0^a xuv\,dx.$$

It follows that

$$(\beta^2 - \alpha^2) \int_0^a xuv\,dx = [x(u'v - uv')]_0^a. \quad (8.5.10)$$

8.5 Oscillations in a circular region

Using the behavior of Bessel functions for small x, we get from (8.2.5)

$$J_0 = 1 - \frac{x^2}{4} - \cdots, \quad J_0' = -\frac{x}{2} + \cdots,$$

hence $xJ_0'J_0 \propto x^2$. For $n = 1, 2, 3, \ldots$, (8.4.5) gives

$$J_n \sim \frac{x^n}{2^n n!}, \quad J_n' \sim \frac{x^{n-1}}{2^n(n-1)!}$$

so that $xJ_n'J_n \propto x^{2n}$. Thus the integrated term in (8.5.10) vanishes at the lower limit $x = 0$. At the upper limit $x = a$, the integrated term vanishes also because of the boundary condition (8.5.9). Therefore,

$$\int_0^a xuv\,dx = 0, \quad \alpha \neq \beta,$$

i.e., u and v are orthogonal with respect to the weighting function x. More explicitly, we have

$$\int_0^a xJ_n(\alpha x)J_n(\beta x)dx = 0 \qquad (8.5.11)$$

if

$$\frac{dJ_n(cx)}{dx} + hJ_n(cx) = 0 \quad c = \alpha, \beta \quad \text{at} \quad x = a. \qquad (8.5.12)$$

Thus Bessel solutions of the same order but with different eigenvalues are orthogonal.

Let us evaluate the scalar product of the eigenfunction with itself

$$\int_0^a xJ_n^2(\alpha x)dx.$$

Consider

$$2u'\left(x^2 u'' + xu' + (\alpha^2 x^2 - n^2)u\right) = 0,$$

which may be rewritten as

$$2x^2 u'u'' + 2xu'^2 + 2\left(\alpha^2 x^2 - n^2\right)uu' = 0$$

or

$$\frac{d}{dx}\{x^2 u'^2 - n^2 u^2 + \alpha^2 x^2 u^2\} = 2\alpha^2 xu^2.$$

Integrating both sides from 0 to a, we get

$$2\alpha^2 \int_0^a xu^2 dx = \left[x^2 u'^2 + \left(\alpha^2 x^2 - n^2\right) u^2\right]_0^a$$

or, more explicitly,

$$\int_0^a x J_n^2(\alpha x) dx = \left\{\frac{x^2}{2}\left[\frac{1}{\alpha^2}\left(\frac{dJ_n}{dx}\right)^2 + \left(1 - \frac{n^2}{\alpha^2 x^2}\right) J_n^2\right]\right\}_0^a$$

$$= \frac{a^2}{2}\left\{[J_n'(\alpha a)]^2 + \left(1 - \frac{n^2}{\alpha^2 a^2}\right) J_n^2(\alpha a)\right\}. \qquad (8.5.13)$$

Again the integrated term vanishes at the lower limit $x = 0$.

If $h = 0$ in (8.5.12), i.e., α is a solution of $J_n'(\alpha a) = 0$ as in the case of a circular pond, then

$$\int_0^a x J_n^2(\alpha x) dx = \frac{a^2}{2}\left(1 - \frac{n^2}{\alpha^2 a^2}\right) J_n^2(\alpha a). \qquad (8.5.14)$$

On the other hand, if $J_n(\alpha a) = 0$ as in the case of a circular drum clamped along the rim, then

$$\int_0^a x J_n^2(\alpha x) dx = \frac{a^2}{2} [J_n'(\alpha a)]^2. \qquad (8.5.15)$$

It is now time to complete the initial-boundary-value problem in §8.1. Details will be given below only for the water-wave case.

8.5.3 Transient oscillations in a circular pond

In (8.5.7) k_{nm} must now be taken to be $k_{nm} = j'_{nm}/a$ and $\omega_{nm} = ck_{nm}$. To satisfy the initial condition for the displacement, we need

$$\eta(r, \theta, 0) = f(r, \theta)$$
$$= \sum_{n=0}^{\infty} \sum_{m=1}^{\infty} (A_{nm} \cos n\theta + C_{nm} \sin n\theta) J_n(k_{nm} r).$$

8.6 Hankel functions and wave propagation

Using the orthogonality of $(\sin n\theta, \cos n\theta)$ and $J_n(k_{nm}r)$, we get

$$\int_0^{2\pi}\int_0^a rf(r,\theta)\begin{pmatrix}\cos n\theta\\ \sin n\theta\end{pmatrix}J_n(k_{nm}r)\,d\theta dr$$

$$= \frac{2\pi}{\epsilon_n}\begin{pmatrix}A_{nm}\\ C_{nm}\end{pmatrix}\int_0^a rJ_n^2(k_{nm}r)\,dr$$

$$= \frac{2\pi}{\epsilon_n}\begin{pmatrix}A_{nm}\\ C_{nm}\end{pmatrix}\frac{a^2}{2}\left(1-\frac{n^2}{k_{nm}^2 a^2}\right)J_n^2(k_{nm}a), \qquad (8.5.16)$$

where ϵ_n is the Jacobi symbol. Therefore A_{nm} and C_{nm} are found. Similarly, from the initial condition on the surface velocity, we get

$$\eta_t(r,\theta,0) = g(r,\theta)$$

$$= \sum_{n=0}^{\infty}\sum_{m=1}^{\infty}\omega_{nm}(B_{nm}\cos n\theta + D_{nm}\sin n\theta)J_n(k_{nm}r)$$

with the result

$$\int_0^{2\pi}\int_0^a rg(r,\theta)\begin{pmatrix}\cos n\theta\\ \sin n\theta\end{pmatrix}J_n(k_{nm}r)\,d\theta dr$$

$$= \frac{2\pi}{\epsilon_n}\begin{pmatrix}B_{nm}\\ D_{nm}\end{pmatrix}\int_0^a rJ_n^2(k_{nm}\theta)\,dr$$

$$= \frac{2\pi}{\epsilon_n}\begin{pmatrix}B_{nm}\\ D_{nm}\end{pmatrix}\frac{a^2}{2}\left(1-\frac{n^2}{k_{nm}^2 a^2}\right)J_n^2(k_{nm}a). \qquad (8.5.17)$$

Thus B_{nm} and D_{nm} are also found, and the mathematical solution for the transient problem is complete. Further analysis can be pursued once f and g are specified.

8.6 Hankel functions and wave propagation

Recall that for trigonometric functions

$$\cos kx \pm i\sin kx = e^{\pm ikx}.$$

When multiplied by the time factor $e^{-i\omega t}$, the product $e^{ikx-i\omega t}$ represents a progressive wave propagating to the right, while the product $e^{-ikx-i\omega t}$ represents a progressive wave propagating to the left. Let us introduce a similar combination

$$H_n^{(1)}(x) = J_n(x) + iY_n(x), \qquad H_n^{(2)}(x) = J_n(x) - iY_n(x), \qquad (8.6.1)$$

which are called the Hankel function of the first and second kind, respectively, and are also independent solutions of the Bessel equation of order n. In view of (8.4.7) and (8.4.8) the Hankel functions behave for large x as

$$H_n^{(1)}(x) \sim \sqrt{\frac{2}{\pi x}} e^{i\left(x - \frac{\pi}{4} - \frac{n\pi}{2}\right)} \tag{8.6.2}$$

and

$$H_n^{(2)}(x) \sim \sqrt{\frac{2}{\pi x}} e^{-i\left(x - \frac{\pi}{4} - \frac{n\pi}{2}\right)}. \tag{8.6.3}$$

Changing x to kr and multiplying by the time factor $e^{-i\omega t}$, we have

$$H_n^{(1)}(kr) e^{-i\omega t} \sim \sqrt{\frac{2}{\pi kr}} e^{i\left(kr - \omega t - \frac{\pi}{4} - \frac{n\pi}{2}\right)}$$

$$H_n^{(2)}(kr) e^{-i\omega t} \sim \sqrt{\frac{2}{\pi kr}} e^{-i\left(kr + \omega t + \frac{\pi}{4} + \frac{n\pi}{2}\right)},$$

which signify radially outgoing and incoming waves, respectively.

It is for these asymptotic behaviors that Hankel functions play a prominent role in the theory of wave propagation, as demonstrated below.

8.6.1 Wave radiation from a circular cylinder

We now apply Hankel functions to the radiation of shallow water waves by a circular storage tank in an earthquake. Let the storage tank, standing on the horizontal seabed and extending above the sea surface, oscillate horizontally along a diameter, as depicted in Figure 8.4. The governing equation is still (8.1.1) for $r > a$. Recall the momentum equation in the radial direction

$$\frac{\partial \mathbf{q}}{\partial t} \cdot \mathbf{e}_r = -g \frac{\partial \zeta}{\partial r},$$

where $\mathbf{e}_r$ denotes a unit vector in the radial direction. Let the cylinder oscillate along its diameter $\theta = 0$ at the frequency ω and the velocity amplitude V, then ζ may be written in the form of (8.5.1)

$$\zeta = \text{Re}\left\{\eta(r, \theta) e^{-i\omega t}\right\} \tag{8.6.4}$$

with η being the complex spatial factor satisfying (8.5.2). The boundary condition on the cylinder surface is

$$\mathbf{q} \cdot \mathbf{e}_r = \text{Re}\left\{V e^{-i\omega t}\right\} \cos\theta, \quad \text{on} \quad r = a,$$

8.6 Hankel functions and wave propagation

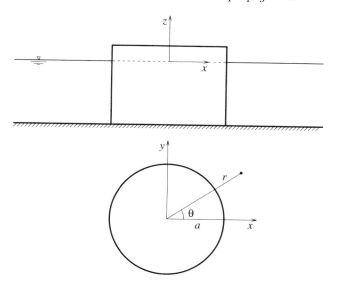

Fig. 8.4. An oscillating circular cylinder.

i.e.,

$$\frac{\partial \eta}{\partial r} = \frac{i\omega}{g} V \cos\theta \quad r = a. \tag{8.6.5}$$

Since disturbances are caused by the cylinder, waves, if any, should only propagate away from it. We therefore require that at $r \sim \infty$, η must represent an outgoing wave. This statement is the two-dimensional version of the radiation condition and is very important in the theory of sinusoidal wave propagation in an infinite plane. A justification of this condition will be given in Chapter 9. Because of the radiation condition the general solution can only include Hankel functions of the first kind

$$\eta = \sum_{n=0}^{\infty} (A_n \cos n\theta + B_n \sin n\theta) H_n^{(1)}(kr). \tag{8.6.6}$$

Clearly, if the time factor had been $e^{i\omega t}$, then Hankel functions of the second kind must be used instead.

Now apply the boundary condition at $r = a$

$$\frac{i\omega V}{g} \cos\theta = \sum_{n=0}^{\infty} k H_n^{(1)'}(kr) (A_n \cos n\theta + B_n \sin n\theta),$$

where the prime denotes the derivative with respect to the argument, i.e., $H_n^{(1)'}(z) = dH_n^{(1)}(z)/dz$. By the orthogonality of $\{\cos n\theta, \sin n\theta\}$ we obtain

$$A_n = 0, \quad n = 0, 2, 3, \ldots, \quad \text{and} \quad B_n = 0, \quad n = 0, 1, 2, 3, \ldots.$$

For $n = 1$

$$kA_1 H_1^{(1)'}(ka) = \frac{i\omega V}{g}$$

so that

$$A_1 = \frac{i\omega V}{gk} \frac{1}{H_1^{(1)'}(ka)}.$$

The final solution is

$$\zeta = \text{Re}\left(\eta e^{-i\omega t}\right) = \text{Re}\left\{\frac{i\omega V}{gk} \frac{H_1^{(1)}(kr)}{H_1^{(1)'}(ka)} e^{-i\omega t}\right\} \cos\theta. \tag{8.6.7}$$

We may now examine the physical implications of the solution. In the far field where $kr \gg 1$, the asymptotic formula for the Hankel function gives

$$\zeta \sim \text{Re}\left\{\frac{i\omega V}{gkH_1^{(1)'}(ka)} \sqrt{\frac{2}{\pi kr}} e^{i\left(kr - \omega t - \frac{3\pi}{4}\right)} \cos\theta\right\}. \tag{8.6.8}$$

The amplitude of the outgoing wave attenuates with the radial distance as $1/\sqrt{kr}$; it is the strongest along, and zero transverse to, the line of oscillation.

The hydrodynamic pressure everywhere in water is

$$p = \rho g \zeta = \rho g \text{Re}\left(\eta e^{-i\omega t}\right), \tag{8.6.9}$$

while the radial velocity of water is

$$q_r = \mathbf{q} \cdot \mathbf{e}_r = \text{Re}\left(\frac{g}{i\omega} \frac{\partial \eta}{\partial r} e^{-i\omega t}\right). \tag{8.6.10}$$

The radiated power in any θ, averaged over one wave period, is equal to the rate of work done by the fluid inside a circle r to the fluid outside

$$\frac{1}{r}\frac{dP}{d\theta} = \frac{\omega}{2\pi}\int_t^{t+2\pi/\omega} pq_r \, dt \stackrel{\text{def}}{=} \overline{pq_r}.$$

8.6 Hankel functions and wave propagation

With respect to a period the time-average of the product of two time-harmonic quantities, $a = \operatorname{Re} Ae^{-i\omega t}$ and $b = \operatorname{Re} Be^{-i\omega t}$, is

$$\overline{ab} = \frac{1}{2}\operatorname{Re}(AB^*) = \frac{1}{2}\operatorname{Re}(A^*B), \qquad (8.6.11)$$

where A^* denotes the complex conjugate of A. This identity can be verified as follows:

$$\overline{ab} = \frac{\omega}{2\pi}\int_0^{\frac{2\pi}{\omega}} dt\, ab$$

$$= \frac{\omega}{2\pi}\int_0^{\frac{2\pi}{\omega}} dt\, \frac{1}{4}\left(Ae^{-i\omega t} + A^*e^{i\omega t}\right)\left(Be^{-i\omega t} + B^*e^{i\omega t}\right)$$

$$= \frac{1}{4}(AB^* + A^*B) = \frac{1}{2}\operatorname{Re}(AB^*) = \frac{1}{2}\operatorname{Re}(A^*B).$$

Hence, the time-averaged energy flux rate across a large circular cylinder of unit depth is

$$\overline{pq_r} = \frac{1}{2}\rho g \operatorname{Re}\left(\eta^* \frac{\partial \eta}{\partial r}\frac{g}{i\omega}\right)$$

$$= \frac{1}{2}\rho g \operatorname{Re}\left\{\left[\left(\frac{i\omega V}{gkH_1^{(1)'}(ka)}\right)^*\sqrt{\frac{2}{\pi kr}}e^{-i(kr-\frac{3\pi}{4})}\cos\theta\right]\right.$$

$$\left.\cdot\left[\left(\frac{i\omega V}{gkH_1^{(1)'}(ka)}\right)\sqrt{\frac{2}{\pi kr}}\frac{g}{i\omega}ike^{i(kr-\frac{3\pi}{4})}\cos\theta\right]\right\}$$

$$= \frac{\rho g}{2}\left(\frac{\omega}{g}\right)^2 \frac{g}{\omega}\left|\frac{V}{kH_1^{(1)'}(ka)}\right|^2 \frac{2}{\pi r}\cos^2\theta$$

$$= \frac{\rho\omega}{2}\left|\frac{V}{kH_1^{(1)'}(ka)}\right|^2 \frac{2}{\pi r}\cos^2\theta,$$

which is a function of r and θ. Because of the factor $\cos^2\theta$, radiation is the strongest in the forward and backward directions $\theta = (0, \pi)$, and zero in the transverse directions $\theta = (\pi/2, 3\pi/2)$.

While the energy flux in any direction decays as $1/r$, the total energy flux rate across a large circle is

$$P = rh \int_0^{2\pi} \overline{pq_r} d\theta = \frac{\rho\omega h}{2} \left|\frac{V}{kH_1^{(1)'}(ka)}\right|^2 \frac{2}{\pi} \int_0^{2\pi} \cos^2\theta d\theta$$

$$= \omega\rho h \left|\frac{V}{kH_1^{(1)'}(ka)}\right|^2 = \rho \left|\frac{V}{kH_1^{(1)'}(ka)}\right|^2 (kh)\left(\frac{\omega}{k}\right). \quad (8.6.12)$$

Since for small z

$$H_1^{(1)}(z) = J_1(z) + iY_1(z) \sim -\frac{2i}{\pi z},$$

which is dominated by Y_1, we have, for $ka \ll 1$,

$$H_1^{(1)'}(ka) \sim \frac{2i}{\pi(ka)^2}.$$

Thus for a relatively small cylinder or long waves, $ka \ll 1$, the power radiated

$$P \sim \rho \left|\frac{\pi V}{2k}\right|^2 (ka)^4 h\omega$$

is proportional to $(ka)^4$ and is insignificant. For a large cylinder or short waves, $ka \gg 1$,

$$H_1^{(1)}(ka) \sim \sqrt{\frac{2}{\pi ka}} e^{i(ka-3\pi/4)}$$

$$H_1^{(1)'}(ka) \sim i\sqrt{\frac{2}{\pi ka}} e^{i(ka-3\pi/4)}$$

so that

$$P \sim \rho \left|\frac{V}{k\sqrt{\frac{2}{\pi ka}}}\right|^2 \sim \frac{\rho}{2}\left|\frac{V}{k}\right|^2 \pi kah\omega.$$

Thus the radiated power increases linearly with the size of the cylinder.

8.6.2 Scattering of plane waves by a circular cylinder

The study of the effects of a stationary object on an incident wave is called the scattering or diffraction theory. We illustrate this type of problem again for a circular cylinder standing vertically in a shallow sea. The

8.6 Hankel functions and wave propagation

incident wave η_i is assumed to be advancing along the positive x axis

$$\eta_i = Ae^{ikx}. \tag{8.6.13}$$

Let the total wave be expressed as the sum of incident and scattered η_s waves

$$\eta = \eta_i + \eta_s. \tag{8.6.14}$$

Since η_i satisfies the Helmholtz equation everywhere, while η satisfies it only outside the cylinder, then

$$\nabla^2 \eta_s + k^2 \eta_s = 0, \quad r > a. \tag{8.6.15}$$

On the cylinder $r = a$ the normal velocity vanishes so that

$$\frac{\partial \eta}{\partial r} = 0 \quad \text{or} \quad \frac{\partial \eta_s}{\partial r} = -\frac{\partial \eta_i}{\partial r}, \quad r = a. \tag{8.6.16}$$

As $r \to \infty$, η_s must be outgoing (the radiation condition).

In order to be outgoing at infinity, the scattered wave can only consist of Hankel functions of the first kind, i.e.,

$$\eta_s = \sum_{n=0}^{\infty} B_n H_n^{(1)}(kr) \cos n\theta.$$

To satisfy the boundary condition (8.6.16) on the cylinder, we must have

$$\sum_{n=0}^{\infty} \left(\epsilon_n i^n A k J_n'(ka) + B_n k H_n^{(1)'}(ka) \right) \cos n\theta = 0,$$

where the partial wave expansion (8.4.13) has been used. From every coefficient of $\cos n\theta$ we get

$$B_n = -A\epsilon_n i^n \frac{J_n'(ka)}{H_n^{(1)'}(ka)}.$$

The total wave is the sum of incident and scattered waves

$$\eta = A \sum_{n=0}^{\infty} \epsilon_n i^n \left[J_n(kr) - \frac{J_n'(ka)}{H_n^{(1)'}(ka)} H_n^{(1)}(kr) \right] \cos n\theta. \tag{8.6.17}$$

As a physical deduction, the spatial factor of the dynamic pressure on the cylinder is

$$p(a, \theta) = \rho g \eta(a, \theta) = \rho g A \sum_{n=0}^{\infty} \epsilon_n i^n \left[J_n(ka) - \frac{J_n'(ka)}{H_n^{(1)'}(ka)} H_n^{(1)}(ka) \right] \cos n\theta.$$

From the Wronskian identity (8.4.11), it can be shown that

$$J_n(z)H_n^{(1)'}(z) - H_n^{(1)}(z)J_n'(z) = \frac{2i}{\pi z}. \qquad (8.6.18)$$

The pressure on the cylinder reduces to

$$p(a,\theta) = \rho g A \sum_{n=0}^{\infty} 2\frac{\epsilon_n i^{n+1} \cos n\theta}{\pi k a H_n^{(1)'}(ka)}. \qquad (8.6.19)$$

The amplitude of the total horizontal force on the cylinder is in the direction of the x axis and given by

$$F = -ah \int_0^{2\pi} p(a,\theta) \cos\theta d\theta = \frac{4\rho g A a h}{k a H_1^{(1)'}(ka)}. \qquad (8.6.20)$$

For small ka the series solution (8.6.17) can be analyzed for any (r,θ) by using expansions of Bessel functions. For large ka the asymptotic analysis is considerably more advanced; interested readers are referred to Jones (1964) or Junger and Feit (1972). Numerical computation of the series is, however, straightforward.

8.7 Modified Bessel functions

Recall the relations between the trigonometric and exponential functions

$$\cos ix = \cosh x, \quad \sin ix = i\sinh x, \quad \cosh x \pm \sinh x = e^{\pm x}.$$

Similar relations exist among Bessel functions.

Consider the Bessel equation with the sign of $x^2 y$ reversed

$$x^2 y'' + xy' - (x^2 + \nu^2)y = 0. \qquad (8.7.1)$$

Let $ix = t$, then

$$t^2 y'' + ty' + (t^2 - \nu^2)y = 0, \qquad (8.7.2)$$

where primes denote derivatives with respect to t. The general solution is

$$y = \begin{cases} AJ_\nu(ix) + BJ_{-\nu}(ix), & \nu \neq \text{integer} \\ AJ_\nu(ix) + BY_\nu(ix), & \nu = \text{integer}. \end{cases}$$

8.7 Modified Bessel functions

From the series definition of $J_\nu(x)$

$$J_\nu(ix) = \sum_{p=0}^{\infty} \frac{(-1)^p i^{2p+\nu} \left(\frac{x}{2}\right)^{2p+\nu}}{p!\Gamma(p+\nu+1)} = i^\nu \sum_{p=0}^{\infty} \frac{\left(\frac{x}{2}\right)^{2p+\nu}}{p!\Gamma(p+\nu+1)}$$

we define the modified Bessel function of the first kind

$$I_\nu(x) = \sum_{p=0}^{\infty} \frac{\left(\frac{x}{2}\right)^{2p+\nu}}{p!\Gamma(p+\nu+1)} \tag{8.7.3}$$

so that

$$J_n(ix) = i^\nu I_\nu(x). \tag{8.7.4}$$

When $\nu \neq$ integer, the general solution of (8.7.1) is

$$y = A' I_\nu(x) + B' I_{-\nu}(x).$$

If, however, $\nu = n =$ integer, $I_n(x) = I_{-n}(x)$. Another independent solution to (8.7.1) is needed. Let us define the modified Bessel function of the second kind by

$$K_n(x) = \frac{\pi}{2} i^{n+1} \left(J_n(ix) + iY_n(ix)\right) = \frac{\pi}{2} i^{n+1} H_n^{(1)}(ix). \tag{8.7.5}$$

The general solution to (8.7.1) is then

$$y = A I_n(x) + B K_n(x). \tag{8.7.6}$$

For $n = 0, 1, 2$, the modified functions are plotted in Figure 8.5. For large x

$$I_\nu(x) \sim \frac{e^x}{\sqrt{2\pi x}}, \qquad K_\nu(x) \sim \frac{e^{-x}}{\sqrt{\frac{2}{\pi}x}}. \tag{8.7.7}$$

The following recursion formulas can be derived from their series expansions:

$$I_{p-1}(\alpha x) - I_{p+1}(\alpha x) = \frac{2p}{\alpha x} I_p(\alpha x) \tag{8.7.8}$$

and

$$K_{p-1}(\alpha x) - K_{p+1}(\alpha x) = -\frac{2p}{\alpha x} K_p(\alpha x). \tag{8.7.9}$$

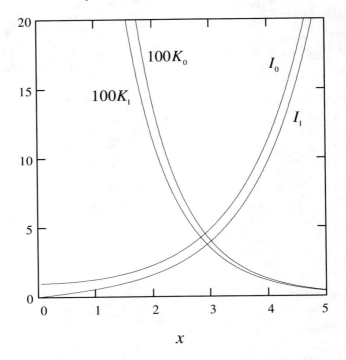

Fig. 8.5. Modified Bessel functions.

8.8 Bessel functions with complex argument

Consider the equation

$$x^2 y'' + xy' - (ix^2 + p^2)y = 0, \qquad (8.8.1)$$

which is (8.7.1) with x replaced by $(i)^{1/2}x$. Note that the first homogeneous solution, which is finite at $x = 0$, is

$$I_p(i^{1/2}x) = -iJ_p(i^{3/2}x)$$

in view of (8.7.4). The second independent solution, which is unbounded at $x = 0$, may be taken to $K_p(i^{1/2}x)$. It is customary to rename the real and imaginary parts of these solutions as follows:

$$J_p(i^{3/2}x) = \text{ber}_p(x) + i\text{bei}_p(x) \qquad (8.8.2)$$

and

$$K_p(i^{1/2}x) = i^p[\text{ker}_p(x) + i\text{kei}_p(x)]. \qquad (8.8.3)$$

8.9 Pipe flow through a vertical thermal gradient

The general homogeneous solution of (8.8.1) is, then,

$$y_p = aJ_p(i^{3/2}x) + bK_p(i^{1/2}x)$$
$$= [a\,\mathrm{ber}_p(x) + b\,\mathrm{ker}_p(x)] + i\,[a\,\mathrm{bei}_p(x) + b\,\mathrm{kei}_p(x)], \quad (8.8.4)$$

where a and b are two real constants. As a group the functions ber_p, bei_p, ker_p and kei_p are called the Kelvin functions of order p. For the special case of $p = 0$, the subscript is omitted, i.e.,

$$y_0 = [a\,\mathrm{ber}(x) + b\,\mathrm{ker}(x)] + i\,[a\,\mathrm{bei}(x) + ib\,\mathrm{kei}(x)]. \quad (8.8.5)$$

The series forms of Kelvin functions are directly found from the real and imaginary parts of $I_p(i^{3/2}x)$ and $K_p(i^{1/2}x)$. We only give results for $p = 0$ below:

$$\mathrm{ber}(x) = 1 - \frac{x^4}{2^2(2!)^2} + \frac{x^4}{2^8(4!)^2} - \cdots \quad (8.8.6)$$

$$\mathrm{bei}(x) = \frac{x^2}{2^2(1!)^2} - \frac{x^6}{2^6(3!)^2} + \frac{x^{10}}{2^{10}(5!)^2} - \cdots \quad (8.8.7)$$

$$\mathrm{ker}(x) = (\ln 2 - \gamma - \ln x)\mathrm{ber}(x) + \frac{\pi}{4}\mathrm{bei}(x) - \frac{(x/2)^4}{(2!)^2}\left(1 + \frac{1}{2}\right)$$
$$+ \frac{(x/2)^8}{(4!)^2}\left(1 + \frac{1}{2} + \frac{1}{3} + \frac{1}{4}\right) - \cdots \quad (8.8.8)$$

$$\mathrm{kei}(x) = (\ln 2 - \gamma - \ln x)\mathrm{bei}(x) + \frac{\pi}{4}\mathrm{ber}(x)$$
$$+ \left(\frac{x}{2}\right)^2 - \frac{(x/2)^6}{(3!)^2}\left(1 + \frac{1}{2} + \frac{1}{3}\right) + \cdots, \quad (8.8.9)$$

where $\gamma = 0.5773\ldots$ is the Euler constant.

An application of modified Bessel and Kelvin functions will be discussed next for an example of combined flow of heat and mass.

8.9 Pipe flow through a vertical thermal gradient

8.9.1 Formulation

Oil in a high-pressure reservoir can attain a temperature of $O(100°C)$. In Alaska, where oil reservoirs are buried beneath a permafrost, drilling wells must run thousands of meters vertically through cold soil in which there is an appreciable geothermal gradient. The insulation around the well casing is never perfect, hence heat is lost to the surrounding soil. What is the mutual effect of heat loss and flow in the well?

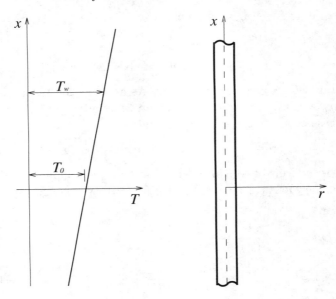

Fig. 8.6. Pipe flow through a vertical thermal gradient.

When the velocity is high enough, the flow in the well can be turbulent. Hence a realistic solution must be treated by incorporating empirical laws of turbulent flow and heat transfer; the physics is too complex to enter here. For low enough speeds the laminar flow model is appropriate, and a theory has been given by Morton (1960), who also gave an approximate analysis, to be solved by the method of perturbation later.

Referring to Figure 8.6, let us employ polar coordinates (r, θ, x) with the x axis coinciding with the well axis and pointing upward. Axial symmetry implies that convection and diffusion are independent of θ. For a deep well the end effects can be neglected throughout most of the length of the well, and the fluid velocity is unidirectional with only one component $\mathbf{u} = (u(r), 0, 0)$. The fluid temperature is $T = T_w - T'(r)$, where T_w and T' denote, respectively, the wall temperature and the temperature loss due to flow. Assuming a constant geothermal gradient τ in the surounding earth, we take

$$T_w = T_o + \tau \frac{x}{a}, \qquad (8.9.1)$$

where a is the well radius. For generality τ can be of either sign. Under

8.9 Pipe flow through a vertical thermal gradient

these assumptions the momentum equations in the oil are

$$0 = -\frac{\partial p}{\partial x} + \mu \nabla^2 u - \rho g \qquad (8.9.2)$$

in the x direction, and

$$0 = -\frac{\partial p}{\partial r} \qquad (8.9.3)$$

in the r direction. For slow enough flow viscous dissipation can be ignored. Heat energy is transferred by convection and diffusion. Conservation of energy requires

$$u \frac{\partial T}{\partial x} = k \nabla^2 T, \qquad (8.9.4)$$

where

$$\nabla^2 = \frac{\partial^2}{\partial r^2} + \frac{1}{r}\frac{\partial}{\partial r} + \frac{\partial^2}{\partial x^2}.$$

Denoting by ρ_w the density of fluid next to the well casing, we can write (8.9.2) as

$$0 = -\left(\frac{dp}{dx} + \rho_w g\right) + \mu \nabla^2 u + (\rho_w - \rho)g. \qquad (8.9.5)$$

If β denotes the thermal expansion coefficient, the equation of state reads

$$\rho = \rho_o(1 - \beta(T - T_o)), \qquad (8.9.6)$$

where ρ_o is the density at the reference temperature T_o. Thus a rise in $T - T_o$ leads to a reduction of fluid density. At the wall the fluid density is

$$\rho_w(x) = \rho_o(1 - \beta(T_w(x) - T_o))$$

so that

$$\rho_w - \rho = -\rho_o \beta (T_w - T).$$

Let

$$T' = T_w - T.$$

Equation (8.9.5) becomes

$$0 = -\left(\frac{1}{\rho_o}\frac{dp}{dx} + \frac{\rho_w}{\rho_o}g\right) + \frac{\mu}{\rho_o}\nabla^2 u - \beta g T'.$$

Because of the usually small thermal expansion coefficient β, the difference between ρ_w and ρ_o is negligible. We therefore replace the ratio ρ_w/ρ_o above by unity. This omission of temperature variation in every term except in buoyancy is called the Boussinesq approximation. The axial momentum equation now reads

$$\frac{\mu}{\rho_o}\left(\frac{d^2u}{dr^2} + \frac{1}{r}\frac{du}{dr}\right) = \left(\frac{1}{\rho_o}\frac{dp}{dx} + g\right) + \beta g T'. \qquad (8.9.7)$$

In view of (8.9.1), the law of energy conservation (8.9.4) becomes

$$\left(\frac{d^2T'}{dr^2} + \frac{1}{r}\frac{dT'}{dr}\right) = -\frac{\tau}{ka}u. \qquad (8.9.8)$$

The boundary conditions are

$$\frac{du}{dr} = \frac{dT'}{dr} = 0, \quad r = 0 \qquad (8.9.9)$$

along the axis because of symmetry, and

$$u = T' = 0, \quad r = a \qquad (8.9.10)$$

along the wall.

Let us introduce the following normalization:

$$U = \frac{ua}{k}, \quad \theta = \frac{T'}{\tau} = \frac{T'}{|\tau|\mathrm{sgn}\tau}, \quad R = \frac{r}{a}, \quad P = \frac{a^3}{k\mu}\left(\frac{dp}{dx} + g\right). \qquad (8.9.11)$$

The dimensionless governing equations are

$$\frac{d^2U}{dR^2} + \frac{1}{R}\frac{dU}{dR} = -P + \epsilon\theta\mathrm{sgn}\tau \qquad (8.9.12)$$

and

$$\frac{d^2\theta}{dR^2} + \frac{1}{R}\frac{d\theta}{dR} = -U \qquad (8.9.13)$$

in $0 < R < 1$, with the boundary conditions

$$\frac{dU}{dR} = \frac{d\theta}{dR} = 0, \quad R = 0 \qquad (8.9.14)$$

and

$$U = \theta = 0, \quad R = 1. \qquad (8.9.15)$$

P is the normalized pressure gradient driving the flow, and

$$\epsilon = \frac{\rho_o \beta g a^3}{k\mu}|\tau| \qquad (8.9.16)$$

is the Rayleigh number, which is a measure of the ratio of buoyancy force to viscous force. For a temperature gradient increasing (decreasing) in height, $\mathrm{sgn}\tau = 1$ (or -1). The negative sign corresponds to warm oil

8.9 Pipe flow through a vertical thermal gradient

flowing out of a well through permafrost. The linear boundary-value problem involves only ordinary differential equations and is solved in terms of Bessel and Kelvin functions as follows.

8.9.2 Solution for rising ambient temperature

By cross differentiation the temperature θ is first eliminated from (8.9.12) and (8.9.13) to give

$$\left(\frac{d^2}{dR^2} + \frac{1}{R}\frac{d}{dR}\right)^2 U + \epsilon \, \mathrm{sgn}\tau \, U = 0. \tag{8.9.17}$$

Let us restrict to the case of positive thermal gradient so that $\tau > 0$. The preceding equation may be written as

$$\left(\frac{d^2}{dR^2} + \frac{1}{R}\frac{d}{dR} + i\epsilon^{1/2}\right)\left(\frac{d^2}{dR^2} + \frac{1}{R}\frac{d}{dR} - i\epsilon^{1/2}\right) U = 0. \tag{8.9.18}$$

The homogeneous solutions of

$$\left(\frac{d^2}{dR^2} + \frac{1}{R}\frac{d}{dR} - i\epsilon^{1/2}\right) F = 0$$

are

$$J_0(i^{3/2}\epsilon^{1/4}R) = \mathrm{ber}(\epsilon^{1/4}R) + i\,\mathrm{bei}(\epsilon^{1/4}R)$$

and

$$K_0(i^{1/2}\epsilon^{1/4}R) = \mathrm{ker}(\epsilon^{1/4}R) + i\,\mathrm{kei}(\epsilon^{1/4}R),$$

while the homogeneous solutions of

$$\left(\frac{d^2}{dR^2} + \frac{1}{R}\frac{d}{dR} + i\epsilon^{1/2}\right) F^* = 0$$

are complex conjugates of F. The most general real solution of the real equation (8.9.18) is, therefore,

$$U = aJ_0(i^{3/2}\epsilon^{1/4}R) + bK_0(i^{1/2}\epsilon^{1/4}R) + *, \tag{8.9.19}$$

where a and b are complex constants, and $*$ represents the complex conjugate of all preceding terms, i.e., $f + * \equiv f + f^*$.

We now apply the boundary conditions required by symmetry on the axis $R = 0$

$$\frac{d\theta}{dR} = 0 \quad \text{or} \quad \frac{d}{dR}\left(\frac{d^2 U}{dR^2} + \frac{1}{R}\frac{dU}{dR}\right) = 0. \tag{8.9.20}$$

From the expansions of the Kelvin functions for small R, the derivatives of ker and kei are unbounded at the origin, hence the coefficients b and b^* must vanish, leaving

$$U = a\left[\text{ber}(\epsilon^{1/4}R) + i\,\text{bei}(\epsilon^{1/4}R)\right] + *. \tag{8.9.21}$$

The condition $U(1) = 0$ demands that

$$a\left[\text{ber}(\epsilon^{1/4}) + i\,\text{bei}(\epsilon^{1/4})\right] + * = 0,$$

i.e.,

$$(\text{Re}\,a)\text{ber}(\epsilon^{1/4}) - (\text{Im}\,a)\text{bei}(\epsilon^{1/4}) = 0. \tag{8.9.22}$$

In view of (8.9.12), vanishing of θ on the pipe wall implies

$$\left(\frac{d^2}{dR^2} + \frac{1}{R}\frac{d}{dR}\right)U = -P, \quad R = 1.$$

Rewriting (8.9.21) as

$$U = aJ_0(i^{3/2}\epsilon^{1/4}) + *,$$

we satisfy the above boundary condition by requiring

$$i\epsilon^{1/2}aJ_0(i^{3/2}\epsilon^{1/4}) + * = -P$$

or

$$(\text{Im}\,a)\text{ber}(\epsilon^{1/4}) + (\text{Re}\,a)\text{bei}(\epsilon^{1/4}) = \frac{P}{2}\epsilon^{-1/2}. \tag{8.9.23}$$

Equations (8.9.22) and (8.9.23) may be solved for the coefficients

$$2(\text{Re}\,a) = \frac{\epsilon^{-1/2}P\,\text{bei}(\epsilon^{1/4})}{\text{ber}^2(\epsilon^{1/4}) + \text{bei}^2(\epsilon^{1/4})}$$

and

$$2(\text{Im}\,a) = \frac{\epsilon^{-1/2}P\,\text{ber}(\epsilon^{1/4})}{\text{ber}^2(\epsilon^{1/4}) + \text{bei}^2(\epsilon^{1/4})}.$$

The final solution for the fluid velocity U is

$$\frac{\epsilon^{1/2}U}{P} = 2(\text{Re}\,a)\text{ber}(\epsilon^{1/4}R) - 2(\text{Im}\,a)\text{bei}(\epsilon^{1/4}R)$$

$$= \frac{\text{bei}(\epsilon^{1/4})\text{ber}(\epsilon^{1/4}R) - \text{ber}(\epsilon^{1/4})\text{bei}(\epsilon^{1/4}R)}{\text{ber}^2(\epsilon^{1/4}) + \text{bei}^2(\epsilon^{1/4})}. \tag{8.9.24}$$

From (8.9.12) the temperature θ is also found

$$\frac{\epsilon^{1/2}\theta}{P} = 1 - \frac{\text{ber}(\epsilon^{1/4})\text{ber}(\epsilon^{1/4}R) + \text{bei}(\epsilon^{1/4})\text{bei}(\epsilon^{1/4}R)}{\text{ber}^2(\epsilon^{1/4}) + \text{bei}^2(\epsilon^{1/4})}. \tag{8.9.25}$$

8.10 Differential equations reducible to Bessel form

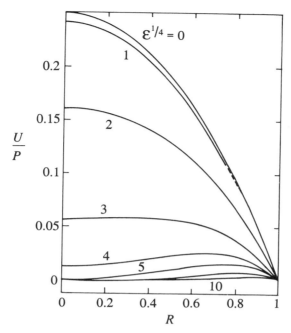

Fig. 8.7. Fluid velocity variation in a well through a positive vertical thermal gradient.

These results have been calculated by Morton (1960) for a wide range of ϵ. Figure 8.7 gives the velocity profile. The interesting feature that occurs at large Rayleigh number ϵ is that the flow is nearly stopped uniformly across the pipe except near the walls, and reversal of flow is possible near the center of the pipe. The case of negative geothermal gradient $\tau < 0$ is also discussed by Morton and left for the reader as an exercise.

The reader will also find McLachlan (1955) to be a rich fountain of practical information on the Kelvin and related functions, as well as their usages in problems involving ordinary differential equations.

8.10 Differential equations reducible to Bessel form

As a final topic of this chapter, we cite a recipe for converting a broad class of ordinary differential equations to the canonical form of (8.1.9) of the Bessel equation. The direct problem is this. Given a second-order ordinary differential equation, can one transform it into a Bessel equation?

A reverse way is to prescribe a transformation and determine the new equations which Bessel's equation must become. In other words, we begin with the presumption that $Y(X)$ satisfies Bessel's equation and seek the equation satisfied by $y(x)$ if

$$Y(X) = y(x)/g(x), \qquad X = f(x)$$

for given f and g. The new equation for y is found simply by substituting the preceding transforms into the standard equation for $Y(X)$. The solution to the new equation is, then,

$$y(x) = g(x)Y(f(x)).$$

A particularly useful transformation (Hildebrand, 1964) is

$$f = Cx^s, \qquad g = x^A e^{-Bx^r}.$$

Then $y(x)$ is governed by

$$x^2 y'' + x(a + 2bx^r)y' + [c + dx^{2s} - b(1-a-r)x^r + b^2 x^{2r}]y = 0. \quad (8.10.1)$$

The general solution has the following form:

$$y(x) = x^{\frac{1-a}{2}} \exp\left(-\frac{bx^r}{r}\right) Z_p\left(\frac{\sqrt{d}}{s} x^s\right), \quad (8.10.2)$$

where a, b, c, d are defined by

$$a = 1 - 2A, \quad b = rB, \quad c = A^2 - p^2 s^2, \quad d = s^2 C^2,$$

and Z_p is one of the Bessel functions of order p, with

$$p = \frac{1}{s}\sqrt{\left(\frac{1-a}{2}\right)^2 - c}. \quad (8.10.3)$$

Specifically,

$$Z_p = (J_p, J_{-p}) \quad \text{if} \quad p \neq \text{integer} \quad (8.10.4)$$
$$= (J_n, Y_n) \quad \text{if} \quad p = n = 0, 1, 2, 3, \ldots \quad (8.10.5)$$

for real $\sqrt{d}/s$, and

$$Z_p = (I_p, I_{-p}) \quad \text{if} \quad p \neq \text{integer} \quad (8.10.6)$$
$$= (I_n, K_n) \quad \text{if} \quad p = n = 0, 1, 2, 3, \ldots \quad (8.10.7)$$

for imaginary $\sqrt{d}/s$.

8.10 Differential equations reducible to Bessel form

Some important special cases are listed below:

$$x^2 y'' + xy' - (\beta^2 - \alpha x^s) y = 0, \quad y = Z_{\beta/s}\left(\frac{\sqrt{\alpha} x^s}{s}\right), \quad s \neq 0 \quad (8.10.8)$$

and

$$(x^n y')' + k x^m y = 0, \quad y = x^{ps} Z_p\left(\frac{\sqrt{k}}{s} x^s\right), \quad (8.10.9)$$

where

$$s = 1 + \frac{1}{2}(m - n), \quad p = \frac{1}{2s}(1 - n), \quad n \neq m + 2. \quad (8.10.10)$$

If $n = 0, m = 1$, then

$$y'' + kxy = 0, \quad (8.10.11)$$

which is the Airy equation. Since $s = 3/2, p = 1/3$, the solutions are

$$y = x^{1/2}\left\{ J_{\frac{1}{3}}\left(\frac{2\sqrt{k}}{3} x^{3/2}\right) \text{ and } J_{-\frac{1}{3}}\left(\frac{2\sqrt{k}}{3} x^{3/2}\right) \right\}. \quad (8.10.12)$$

More commonly, these solutions are replaced by Airy's functions defined below

$$Ai(-x) = \frac{x^{1/2}}{3}\left\{ J_{\frac{1}{3}}\left(\frac{2\sqrt{k}}{3} x^{3/2}\right) + J_{-\frac{1}{3}}\left(\frac{2\sqrt{k}}{3} x^{3/2}\right) \right\} \quad (8.10.13)$$

and

$$Bi(-x) = \frac{x^{1/2}}{3}\left\{ -J_{\frac{1}{3}}\left(\frac{2\sqrt{k}}{3} x^{3/2}\right) + J_{-\frac{1}{3}}\left(\frac{2\sqrt{k}}{3} x^{3/2}\right) \right\}, \quad (8.10.14)$$

which arise frequently in wave problems.

In this chapter, we have only introduced the bare essentials of the Bessel function theory. For a more detailed theoretical treatment on an elementary level, readers may wish to consult Hildebrand (1964) for his clear exposition and convenient collection of formulas. For other special functions (Kelvin, Airy, Struve, Lommel, etc.,) related to Bessel functions as well as numerous applications, McLachlan (1955) is highly recommended. For a truly comprehensive collection of formulas, Erdelyi (1953a) and Abramowitz and Stegun (1964) are indispensable.

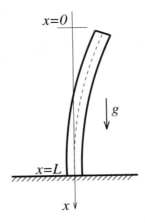

Fig. 8.8. Uniform column under its own weight.

Exercises

8.1 The bottom of a cylindrical water bucket of circular cross section is a soft and impervious membrane. If water is filled to the depth h measured from the bottom rim, what is the static profile of the membrane due to gravity? Use the linearized approximation only. Study the limit of small radius.

8.2 Referring to Figure 8.8, show that a uniform column deforms laterally under its own weight according to

$$\frac{d^2\theta}{dx^2} + k^2 x \theta = 0 \quad \text{with} \quad k^2 = AW/EI,$$

where x = vertical distance from the top, $\theta(x)$ = inclination of the column axis with respect to the original vertical axis, A = cross-sectional area, W = unit weight of column, E = Young's modulus of elasticity, and I = moment of inertia. Note that the displacement at the top must be finite.

Assume the column to be uniform in x and find the eigenmodes of buckling in terms of Bessel functions of 1/3. Use a computer to calculate the Bessel functions, the eigenvalues k^2 and the modal shapes for the first two modes.

8.3 Consider the buckling of a tapered column (McLachlan, 1955). Referring to Figure 8.9, let us consider a column of length L, clamped at the base, and loaded on top by the vertical force P. When buckling occurs the axis of the column is no longer

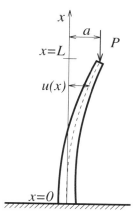

Fig. 8.9. Buckling of a tapered column.

vertical so that the top is displaced horizontally by an amount a. Let the moment of inertia of the column be $I = I(0)e^{-Kx/L}$ and the displacement away from the original vertical axis be u.

i) Show by balancing torques acting on the column that

$$EI(0)\frac{d^2u}{dx^2} + e^{Kx/L}P(u-a) = 0. \qquad (\text{E8.1})$$

ii) Show that the differential equation can be reduced to the homogeneous form

$$\frac{d^2w}{dx^2} + n^2 e^{2bx} w = 0 \qquad (\text{E8.2})$$

by introducing

$$w = u - a, \qquad 2b = \frac{K}{L}, \qquad n^2 = \frac{P}{EI(0)}.$$

iii) Show that the above equation can be transformed to the familiar Bessel equation by the following change of independent variable $\xi = ne^{bx}$, yielding

$$\xi^2 w'' + \xi w' + \left(\frac{\xi}{b}\right)^2 w = 0.$$

This transformation is not covered by §8.10.

iv) Show that the boundary conditions are homogeneous

$$\frac{dw}{d\xi} = 0, \quad x = 0; \quad w = 0, \quad x = L.$$

v) Verify the following eigenvalue condition

$$J_0(\phi\theta)Y_1(\theta) - Y_0(\phi\theta)J_1(\theta) = 0,$$

where

$$\theta = \frac{n}{b} = \frac{2Ln}{D} = \frac{2L}{K}\sqrt{\frac{P}{EI}}, \quad \phi = e^{K/2}.$$

Given ϕ, there can be infinitely many eigenvalues of θ, each corresponding to a buckling load P.

vi) For the degenerate case of a uniform column $I(0)/I(L) = 1$, show that the eigenvalue condition becomes

$$\sin\left(\theta(\phi-1) + \frac{\pi}{2}\right) = 0.$$

vii) Lastly, show from the smallest positive root of the preceding equation that the critical load for the buckling of a uniform column is

$$P_c = \frac{1}{4}\frac{\pi^2 E I_0}{L^2}, \tag{E8.3}$$

which was first obtained by Euler.

8.4   Solve the differential equation (E8.2)

$$\frac{d^2 x}{dt^2} + e^{-\epsilon t} x = 0$$

subject to the initial conditions

$$x(0) = 0 \quad \text{and} \quad \frac{dx}{dt}(0) = 1.$$

Show that

$$x = \frac{\pi}{\epsilon}\left[Y_0\left(\frac{2}{\epsilon}\right) J_0\left(\frac{2}{\epsilon}e^{-\epsilon t/2}\right) - J_0\left(\frac{2}{\epsilon}\right) Y_0\left(\frac{2}{\epsilon}e^{-\epsilon t/2}\right)\right].$$

Deduce approximations under the following circumstances:
(i): $\epsilon \ll 1$, all t; (ii): $\epsilon \ll 1$ and $\epsilon t \ll 1$, (iii): $\epsilon \ll 1$ and $\epsilon t = O(1)$, and (iv): $\epsilon \ll 1$ and $\epsilon e^{\epsilon t} \gg 1$ (Cheng and Wu, 1970).

8.5   Consider water waves in a shallow estuary of variable cross section. Let the channel section be rectangular with width $b(x)$ and depth $h(x)$ varying slowly along the x axis. The law of mass conservation is approximately

$$b\frac{\partial \zeta}{\partial t} = -\frac{\partial (ubh)}{\partial x}.$$

The linearized momentum equation is still

$$\frac{\partial u}{\partial t} = -g\frac{\partial \zeta}{\partial x}.$$

Eliminate u from these two equations and show that

$$\frac{\partial^2 \zeta}{\partial t^2} = \frac{1}{b}\frac{\partial}{\partial x}\left(bh\frac{\partial u}{\partial t}\right) = \frac{g}{b}\frac{\partial}{\partial x}\left(bh\frac{\partial \zeta}{\partial x}\right).$$

For simple harmonic waves $\zeta = \text{Re}\left(\eta(x)e^{-i\omega t}\right)$, find the spatial factor η for the following special cases:

Case (i): Estuary

$$h = \text{constant}, \quad b = \alpha x, \quad 0 < x < L.$$

At the open end $x = L$ the estuary is connected to a large ocean. Ignore radiation and let the wave amplitude at L be a constant A. Examine the condition for resonance.

Case (ii): Plane beach

$$h = sx, \quad 0 < x < \infty.$$

Examine the solution for large x and interpret the result as superposition of incident and reflected sinusoidal waves of varying amplitude and length.

Require that η be bounded at the shore.

8.6 Show that for $kr, ka \ll 1$ the first two terms ($n = 0, 1$) in the series of (8.6.17) give the leading-order result that satisfies Laplace's equation and corresponds to a steady inviscid flow past a circular cylinder.

8.7 A circular sill of radius a and height h_o rests on the bed of a shallow sea of constant depth h, with $h_o < h$. A plane wave of surface displacement $\zeta = \text{Re}\left(Ae^{(ikx-i\omega t)}\right)$ is incident on the sill. Find the diffracted waves on and around the sill. Assume the continuity of fluid pressure and horizontal flux $-h\,\partial\zeta/\partial r$ across the depth discontinuity (Longuet-Higgins, 1967).

8.8 A hollow elastic cylinder of length L is twisted by a distributed shear stress at one end. Show by considering the static equilibrium of a circular ring defined by $r, r+dr$ and $z, z+dz$ that the shear stress components are governed by

$$\frac{1}{r^2}\frac{\partial}{\partial r}\left(r^2 \tau_{r\theta}\right) + \frac{\partial \tau_{z\theta}}{\partial z} = 0.$$

In polar coordinates Hooke's law requires

$$\tau_{r\theta} = Gr\frac{\partial}{\partial r}\left(\frac{u}{r}\right), \qquad \tau_{z\theta} = G\frac{\partial u}{\partial z},$$

where $u(r, z)$ denotes the azimuthal displacement. Assume that the shear stresses are given at both ends of the cylinder

$$\tau_{r\theta}(r, 0) = f(r), \quad a < r < b$$

and

$$\tau_{r\theta}(r, L) = 0, \quad a < r < b.$$

On the inner surface the shear stress is prescribed

$$\tau_{z\theta}(a, z) = 0, \quad 0 < z < L.$$

The outer surface is held fixed. Find the stresses and displacement u. Suggestion: Write $u = r\phi(r, z)$ and solve for ϕ.

8.9 Solve the problem of pipe flow in a negative geothermal gradient (cf. §8.9), i.e., the ambient temperature decreases with height. This problem describes a model of oil flow through a well in a permafrost.

8.10 When surface waves propagate in a shallow sea, a thin turbulent boundary layer is formed near the seabed. Within this layer the fluid velocity diminishes downward from a finite value to zero at the seabed. Based on a semi-empirical model the turbulent shear stress is related to the horizontal velocity by

$$\tau = \alpha z \frac{\partial u}{\partial z},$$

where u denotes the horizontal velocity inside the boundary layer and αz is the eddy viscosity increasing with height z above the seabed. The conservation law of horizontal momentum reads

$$\frac{\partial u}{\partial t} = \frac{\partial U}{\partial t} + \frac{\partial \tau}{\partial z},$$

where $U(t)$ is the horizontal fluid velocity just outside the boundary layer. The bottom of the turbulent boundary layer is taken to be the top of sand grains, where the velocity vanishes

$$u = 0, \qquad z = z_o > 0.$$

At the top of the boundary layer, u approaches its ambient value

$$u = U, \qquad z \sim \infty.$$

Let the wave motion be simple harmonic

$$U = U_o \mathrm{Re}\, e^{-i\omega t}.$$

Verify the following solution due to Kajiura (1968):

$$\frac{u}{U_o} = \mathrm{Re}\left\{\left[1 - \frac{\ker(2\sqrt{\zeta}) - i\kei(2\sqrt{\zeta})}{\ker(2\sqrt{\zeta_o}) - i\kei(2\sqrt{\zeta_o})}\right] e^{-i\omega t}\right\},$$

where

$$\zeta = \frac{z}{\ell}, \quad \zeta_o = \frac{z_o}{\ell}, \quad \text{and} \quad \ell = \frac{\alpha}{\omega}.$$

Deduce the behavior for small and large ζ.

8.11   A long cylinder of circular cross section is heated from within by a uniformly distributed source

$$\frac{\partial T}{\partial t} = \frac{k}{r}\frac{\partial}{\partial r}\left(r\frac{\partial T}{\partial r}\right) + f, \quad r < a$$

and is surrounded by zero temperature. How does the temperature inside the cylinder change with space and time if the initial temperature is also zero?

9
Complex variables

Thus far we have only dealt with real variables; the use of complex representation for a sinusoidal function of time is just a matter of convenience involving only the real variable t and no new principles. To an analytical engineer the techniques of complex variables are essential because of their wide range of applications. Many two-dimensional potential theories in classical hydrodynamics, static electricity, steady diffusion, etc., can be directly solved by complex functions. The inverse Fourier and Laplace transforms are often most efficiently evaluated in a complex plane. In contrast to most methods of real variables where the mathematical details are tailored to suit the geometry of the boundaries, conformal mapping is a radically different tool whose effectiveness lies in altering the boundaries themselves.

In the following four chapters, we give a guided tour of the basic principles of complex functions, together with applications that range from the elementary to the slightly advanced. In the present chapter the basics of analytic functions and the rules of differential and integration are explained. In Chapter 10 these basics are applied to the techniques of Laplace transform. In Chapter 11 elements of conformal mapping are introduced with examples from hydrodynamics. One of the most beautiful applications of complex functions in continuum mechanics is the formulation and solution of certain mixed boundary-value problems. In Chapter 12 two examples from hydrodynamics and elasticity are examined and the Riemann–Hilbert technique is explained.

9.1 Complex numbers

We denote by $i = \sqrt{-1}$ the imaginary unit and $z = x + iy$ the complex number with x being the real part and y the imaginary part of z.

9.1 Complex numbers

Geometrically, z defines a point in the complex plane with x as the abscissa and y the ordinate, as shown in Figure 9.1. Thus z is just another symbol to denote the position of a point in the x, y plane or the vector from the origin to the point. As the same point can also be defined by polar coordinates r, θ with

$$x = r\cos\theta, \qquad y = r\sin\theta,$$

therefore,

$$z = r(\cos\theta + i\sin\theta). \tag{9.1.1}$$

Denoting the complex number in the parentheses by Q

$$Q(\theta) = \cos\theta + i\sin\theta,$$

we observe that $Q(\theta)$ satisfies the differential equation

$$\frac{dQ}{d\theta} = -\sin\theta + i\cos\theta = iQ$$

and the initial condition $Q(0) = 1$. The solution for Q is $e^{i\theta}$, hence

$$e^{i\theta} = \cos\theta + i\sin\theta,$$

which is well known. The polar form of the complex number z is, therefore,

$$z = re^{i\theta} \tag{9.1.2}$$

with $r = |z|$ being the magnitude of z and θ the argument. Since $e^{i\theta}$ is

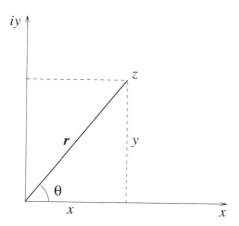

Fig. 9.1. The complex plane.

periodic in θ with period 2π, we often limit the range of θ to $0 \le \theta < 2\pi$ (or $-\pi \le \theta < \pi$). In this range, θ is said to take its principal value.

Let us list some elementary properties of complex numbers.

(i) The sum of two complex numbers z_1 and z_2 is another complex number whose real (imaginary) part is the sum of the real (imaginary) parts of z_1 and z_2

$$z_1 + z_2 = x_1 + iy_1 + x_2 + iy_2 = (x_1 + x_2) + i(y_1 + y_2).$$

(ii) The product of two complex numbers is

$$z_1 z_2 = (x_1 + iy_1)(x_2 + iy_2)$$
$$= (x_1 x_2 - y_1 y_2) + i(y_1 x_2 + x_1 y_2)$$

in cartesian form, and

$$z_1 z_2 = r_1 e^{i\theta_1} r_2 e^{i\theta_2} = r_1 r_2 e^{i(\theta_1 + \theta_2)}$$

in polar form. Thus in a product the magnitudes of the factors multiply and phases add. The mth power of a complex number is

$$z^m = r^m e^{im\theta} = r^m (\cos m\theta + i \sin m\theta).$$

(iii) Two complex numbers are equal if and only if their real and imaginary parts are separately equal, i.e.,

$$x_1 + iy_1 = x_2 + iy_2 \quad \text{implies} \quad x_1 = x_2, \quad y_1 = y_2.$$

(iv) The magnitude of a complex number is just the length of a vector with components x, y

$$|z| = \sqrt{x^2 + y^2} = r.$$

9.2 Complex functions

A function is a map. A real function $f(x)$ of the real variable x maps a point on the x axis onto a point on the y axis through the relation $y = f(x)$; see Figure 9.2a. If every x is mapped only to one point on the y axis, $f(x)$ is single-valued. Now consider a complex function $f(z)$ of z. Let $u(x, y)$ and $v(x, y)$ be the real and imaginary parts of f, i.e., $f(z) = u(x, y) + iv(x, y)$, and introduce a new complex plane with real and imaginary axes u and v. A point in the $w = u + iv$ plane is the image of a point in the z plane via the mapping $w = f(z)$; see Figure 9.2b.

9.2 Complex functions

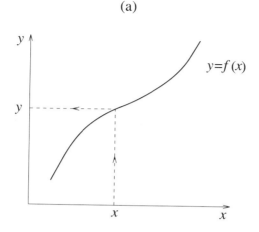

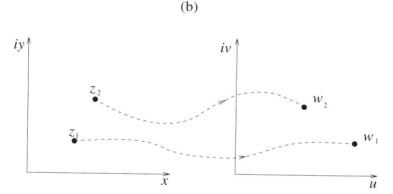

Fig. 9.2. Mapping by functions. (a) Real function of a real variable; (b) complex function of a complex variable.

Let us examine the geometry of some simple complex functions.
(i) $f(z) = Cz$, where C is a complex constant. In polar form

$$C = |C|e^{i\alpha}$$

and

$$f = (|C|r)e^{i(\theta+\alpha)}.$$

The function Cz maps a vector $z = re^{i\theta}$ onto the vector $w = (|C|r)e^{i(\theta+\alpha)}$

whose length is $|C|r$ and angle is $\theta + \alpha$. Thus the mapping has the effect of stretching (or shrinking) and rotation.

(ii) $f(z) = z^2$. In polar form we have
$$w^2 \equiv |w|e^{i\theta_w} = r^2 e^{2i\theta}.$$

In the complex plane of $w = u + iv$, the image of the complex vector z has the length r^2 and orientation 2θ. Since
$$w = u(x,y) + iv(x,y)$$
$$= (x+iy)^2 = (x^2 - y^2) + i2xy,$$
it follows that
$$u = x^2 - y^2, \qquad v = 2xy.$$

Thus hyperbolas with asymptotes $x = \pm y$ in the z plane map onto straight lines $u = $ constant in the w plane, while hyperbolas with asymptotes $x = 0$ and $y = 0$ map onto straight lines $v = $ constant in the w plane. Since
$$|w| = \sqrt{u^2 + v^2} = |z^2| = |r^2 e^{2i\theta}| = r^2,$$
a circle of radius r in the z plane maps onto a circle of radius r^2 in the w plane.

(iii) $f(z) = 1/z$. Since
$$w = u + iv = \frac{1}{x + iy} = \frac{x - iy}{(x+iy)(x-iy)} = \frac{x - iy}{x^2 + y^2},$$
the real and imaginary parts are
$$u = \frac{x}{x^2 + y^2} \quad \text{and} \quad v = \frac{-y}{x^2 + y^2},$$
which may be rearranged to
$$\left(x - \frac{1}{2u}\right)^2 + y^2 = \left(\frac{1}{2u}\right)^2$$
and
$$x^2 + \left(y + \frac{1}{2v}\right)^2 = \left(\frac{1}{2v}\right)^2.$$

A straight line $u = $ constant in the w plane is the image of a circle in the z plane. The circle must be centered at $x = 1/2u, y = 0$ and tangent to the y axis at the origin. Similarly, from the z plane, $1/z$ maps a circle with radius $1/2v$ and centered at $x = 0, y = -1/2v$ onto a straight line $v = $ constant in the w plane.

9.3 Branch cuts and Riemann surfaces

In real variables it is a familar fact that the square root of a positive number a has two values $\pm\sqrt{a}$. In complex variables we can circumvent the multivaluedness by the new concepts of branch cuts and Riemann surfaces. Let us explain the ideas through simple examples.

Example 1: $f(z) = z^{1/2}$. Since $e^{i\theta}$ is 2π periodic, i.e., $e^{i\theta} = e^{i(2n\pi+\theta)}$ with $n =$ integer, we have

$$f = r^{\frac{1}{2}} e^{i\left(n\pi + \frac{\theta}{2}\right)}.$$

Taking $n = 0, 1, 2, \ldots$ in succession, we get

$$f = \begin{cases} r^{1/2} e^{i\theta/2} & n = \text{ even} \\ -r^{1/2} e^{i\theta/2} & n = \text{ odd}. \end{cases}$$

Thus, $z^{1/2}$ is double-valued for any $z = re^{i\theta}$.

Geometrically, let us start with a point $z = re^{i\theta}$ in the z plane. After tracing one complete loop around the origin and returning to the same point, θ is increased by 2π. In the f plane, however, the image point starts at $r^{1/2} e^{i\theta/2}$ but ends at a different point $r^{1/2} e^{i(\pi + \theta/2)} = -r^{1/2} e^{i\theta/2}$. The same is true if z traces an odd number of loops around the origin. Only after z makes an even number of loops around the origin does the imaging point return to the same location in the f plane. Thus, the same point z corresponds to two different values of f. On the other hand, if the closed loop does not encircle the origin, then the value of θ is not changed from start to finish; f is single-valued. Clearly, the origin is a special point for $f(z) = z^{1/2}$.

We now define a *branch point* for any $f(z)$ as follows. A point z_o is a branch point of $f(z)$ if, when z traces a closed curve surrounding z_o, $f(z)$ does not return to its starting value.

Thus the origin is a branch point for $f(z) = z^{1/2}$. Note that for $f(z) = z^{1/2}$ the point at infinity is also a branch point. This fact can be seen by letting $z = 1/\zeta$, then $f(z(\zeta)) = \zeta^{-1/2}$, and $z = \infty$ corresponds to $\zeta = 0$. Now if ζ traces a closed loop around $\zeta = 0$, f also does not return to its original value, hence $\zeta = 0$ (or $z = \infty$) is another branch point of $z^{1/2}$.

To avoid nonuniqueness, a simple way is to cut the z plane along a line (any line) connecting the branch points, and crossing of the cut is disallowed. Then z is prohibited from tracing a complete loop around a branch point; no ambiguities will ever arise. The cut is called the *branch cut*, and the z plane split by the branch cut is called a *branch*. For the function $f(z) = z^{1/2}$ we may cut the z plane along the positive x axis

Fig. 9.3. Riemann surface for $f = z^{1/2}$.

and define the principal branch by limiting θ to the range $0 < \theta \leq 2\pi$. On this branch the function $z^{1/2}$ is single-valued.

Two matters are worthy of note: (i) We can also cut the z plane along the positive x axis but define the branch by the angular range $2\pi < \theta \leq 4\pi$. On this branch the function $z^{1/2}$ is single-valued but is opposite in sign from the principal branch defined previously by $0 < \theta \leq 2\pi$. (ii) The branch cut can be any curve connecting the branch points 0 and ∞, for example, the negative x axis. The principal branch is then defined by $-\pi \leq \theta < \pi$ on which $z^{1/2}$ is single-valued. The second branch is defined by $\pi \leq \theta < 3\pi$.

To unify the two possible choices of branch with the same branch cut, we now introduce the idea of Riemann surfaces.

Let us imagine that the original z plane is replaced by a surface with two sheets connected magically along the positive x axis, which has a two-way trap door. The upper sheet is defined by $0 \leq \theta < 2\pi$, while the lower sheet by $2\pi \leq \theta < 4\pi$. The two sheets are not only continuous along $\theta = 2\pi$, but the two lines $\theta = 0$ and 4π are perfectly joined, as sketched in Figure 9.3. We call the twin-sheet structure the Riemann surface, and each sheet is called a *branch*. The two-way escalator along the lines $\theta = 0, 2\pi, 4\pi$ coincide with the *branch cut*.

To see how f is single-valued when z changes, we regard $f = z^{1/2}$ as a mapping between the Riemann surface and the f plane. The structure of a Riemann surface is something like an underground garage.† As long as the point z is on the first Riemann sheet $0 \leq \theta < 2\pi$, i.e., you are on

---

† The underground garage metaphor is due to Professor Hung Cheng, Department of Mathematics, MIT.

the street level, your image is in the upper half of the f plane. Spiraling across the line at $\theta = 2\pi$, you must descend to the second Riemann sheet $2\pi \leq \theta < 4\pi$ (the underground level). Now your image is on the lower half of the f plane. After crossing the line $\theta = 4\pi$, you must resurface on the first sheet (street level) again. In this way, two images $\pm r^{1/2} e^{i\theta}$ on the f plane correspond uniquely to two different points on the Riemann surface. Double-valuedness is therefore avoided. The principal branch is just the top Riemann sheet.

Example 2: $f(z) = z^{1/3}$. In polar form

$$f = f(z) = r^{1/3} \exp\{i(2n\pi + \theta)/3\}.$$

For $n = 0, 1, 2$, $f(z)$ assumes three different values

$$r^{1/3} \left[\exp\left(\frac{i\theta}{3}\right), \exp\left(\frac{i\theta}{3} + \frac{2\pi i}{3}\right), \exp\left(\frac{i\theta}{3} + \frac{4\pi i}{3}\right) \right].$$

For $n = 3, 4, 5, \ldots$, the same three values are repeated. Therefore, $z^{1/3}$ is tripled-valued.

To avoid triple-valuedness, we may cut the z plane along $\theta = 0$ and choose the branch with $0 \leq \theta < 2\pi$. On this branch f is single-valued. Or, we may construct a Riemann surface with three sheets (an underground garage with three levels). With a cut along $\theta = 0$, the joining of the three sheets has the sideview shown in Figure 9.4. Starting from the first sheet, we spiral around the origin and descend to the second sheet through the cut along $\theta = 2\pi$. After another circle we descend to the third sheet through the cut. Finally, after 6π, we ascend to the first

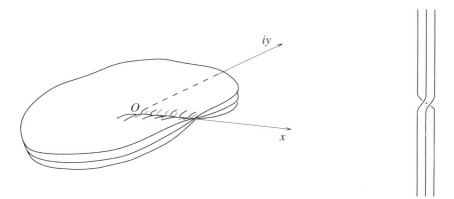

Fig. 9.4. Riemann surface for $f = z^{1/3}$.

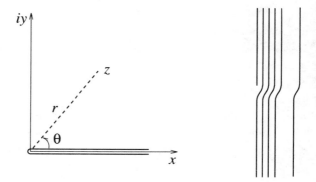

Fig. 9.5. Riemann surface for $\ln z$.

sheet again. In this way our three images in the w plane correspond to three different positions in the triple-decked Riemann surface, and the mapping is one-to-one. The principal branch corresponds to the top sheet of the Riemann surface.

Example 3: $f(z) = \ln z$. Since

$$w = f(z) = \ln[r\exp\{i(\theta + 2n\pi)\}] = \ln r + i(2n\pi + \theta),$$

$\ln z$ is infinitely many-valued for the same r, θ but different $n = 0, 1, 2, \ldots$. To avoid multivaluedness, we may cut the z plane along $\theta = 0$ and choose as the principal branch $0 < \theta \leq 2\pi$. Now the Riemann surface consists of infinitely many sheets, as shown in Figure 9.5. Again, the top sheet of the Riemann surface is the principal branch.

Example 4: $f = \sqrt{(z-a)(z-b)}$. There are two branch points $z = a, b$ in the finite plane. We can introduce either two semi-infinite cuts, as shown in Figure 9.6a, or one finite cut along $a < x < b, y = 0$, as shown in Figure 9.6b. Two Riemann sheets are connected through the cuts in a criss-crossing manner.

Let us see how f is evaluated when z is on the first sheet (i.e., first branch). Referring to Figure 9.6a, we define

$$z - a = r_1 e^{i\theta_1}, \qquad z - b = r_2 e^{i\theta_2}. \tag{9.3.1}$$

When on the first sheet, θ_1, θ_2 are measured counterclockwise from the positive x axis. To get the value of f imagine a rubber band pinned at the two branch points $z = a, b$. Attached to the middle of the band, a sharp pencil can be moved around to point at any z on the plane. The two

9.3 Branch cuts and Riemann surfaces

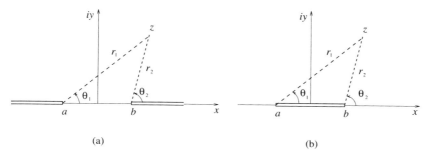

Fig. 9.6. Two possible branch cuts for $z = [(z-a)(z-b)]^{1/2}$.

segments of the rubber band then form two vectors $z-a$ and $z-b$. When the pencil point, or z, is anywhere along the positive x axis and on the upper edge of the right-hand cut, $\theta_1 = \theta_2 = 0$ so that $f = \sqrt{r_1 r_2} = |f|$ is real and positive. Moving the pencil to the left along the x axis and stopping somewhere between $a < x < b$, we get $\theta_1 = 0$ and $\theta_2 = \pi$ so that $f = \sqrt{r_1 r_2} e^{i\pi/2} = i\sqrt{r_1 r_2}$. When the pencil point z is on the upper edge of the left-hand cut, $\theta_1 = \theta_2 = \pi$ so that $f = |f|e^{i\pi} = -\sqrt{r_1 r_2}$. However, when z is on the lower edge of the left cut, $\theta_1 = -\pi, \theta_2 = \pi$ because the pencil cannot slide across the cut but must go around the branch point $z = a$ clockwise on the right. Thus $f = |f| = \sqrt{r_1 r_2}$ and assumes opposite signs on opposite sides of the same cut. Finally, let z lie on the lower edge of the right-hand cut, $\theta_1 = 0, \theta_2 = 2\pi$ so that $f = -|f| = -\sqrt{r_1 r_2}$.

Consider the second alternative of the finite cut along the real segment $a \leq x \leq b$. For the first branch (sheet), we introduce the rubber band and define the angles as before. If z is real and to the right of the cut, $f = |f| = \sqrt{r_1 r_2}$ since $\theta_1 = \theta_2 = 0$. If z is real and to the left of the cut, $f = -|f| = -\sqrt{r_1 r_2}$ since $\theta_1 = \theta_2 = \pi$. If z is on the upper (or lower) edge of the cut, $\theta_1 = 0$ while $\theta_2 = \pi$ (or $-\pi$), hence $f = i|f|$ (or $-i|f|$) since the cut must not be crossed. On this branch we have $f \to z$ as $|z| \to \infty$.

In most applications one is interested in one of the branches. It is necessary to declare the choice by first describing the location of the cut, defining how angles are measured, and specifying the range of angles. Thus for the present example, we may say that f is defined on the plane with the finite cut from a to b along the real axis and that the branch is chosen such that $f \to z$ as $|z| \to \infty$.

9.4 Analytic functions

A complex function $w(z)$ is said to be analytic if it is single-valued and differentiable. For $w(z) = u(x,y) + iv(x,y)$ to be analytic at the point $z = x + iy$, not only must $\partial u/\partial x$, $\partial u/\partial y$, $\partial v/\partial x$ and $\partial v/\partial y$ exist but they must obey the so-called Cauchy–Riemann conditions

$$\frac{\partial u}{\partial x} = \frac{\partial v}{\partial y}, \qquad \frac{\partial v}{\partial x} = -\frac{\partial u}{\partial y}. \qquad (9.4.1)$$

Conversely, if the real and imaginary parts u and v of a complex function w satisfy the Cauchy–Riemann conditions in some region of the z plane, then w is an analytic function of z in that region. Thus the Cauchy–Riemann conditions are the acid test for analyticity.

Let us prove the necessity of the Cauchy–Riemann conditions. The derivative of a complex function of z is

$$\frac{dw}{dz} = \lim_{dz \to 0} \frac{w(z+dz) - w(z)}{dz},$$

where $dz = dx + idy$ is a complex distance. The requirement that w is differentiable means that dw/dz must exist and be the same no matter how dz is chosen. Let us consider two special choices: (i) $dz = dx$ is real, and (ii) $dz = idy$ is imaginary. In the first

$$\frac{dw}{dz} = \lim_{dx \to 0} \frac{1}{dx} [u(x+dx,y) + iv(x+dx,y) - u(x,y) - iv(x,y)]$$

$$= \lim_{dx \to 0} \left\{ \frac{1}{dx}[u(x+dx,y) - u(x,y)] + \frac{i}{dx}[v(x+dx,y) - v(x,y)] \right\}$$

$$= \frac{\partial u}{\partial x} + i\frac{\partial v}{\partial x}.$$

In the second

$$\frac{dw}{dz} = \lim_{dy \to 0} \frac{1}{idy} [u(x,y+dy) + iv(x,y+dy) - u(x,y) - iv(x,y)]$$

$$= \lim_{dy \to 0} \left\{ \frac{1}{idy}[u(x,y+dx) - u(x,y)] + \frac{i}{idy}[v(x,y+dy) - v(x,y)] \right\}$$

$$= -i\frac{\partial u}{\partial y} + \frac{\partial v}{\partial y}.$$

The Taylor expansion theorem for real functions has been used in deriving both results above. For uniqueness the two results must be the same, hence the real and imaginary parts must be separately equal, implying that the Cauchy–Riemann conditions must be satisfied.

9.4 Analytic functions

Conversely, if the Cauchy–Riemann conditions are satisfied, the derivative of a complex function exists and is unique, hence f is analytic. The proof is as follows.

$$\frac{dw}{dz} = \lim_{dx,dy \to 0} \frac{1}{dx + idy} [u(x+dx, y+dy) + iv(x+dx, y+dy)$$
$$- u(x,y) - iv(x,y)]$$
$$= \lim_{dx,dy \to 0} \frac{1}{dx + idy} \left(\frac{\partial u}{\partial x} dx + \frac{\partial u}{\partial y} dy + i\frac{\partial v}{\partial x} dx + i\frac{\partial v}{\partial y} dy \right)$$
$$= \lim_{dx,dy \to 0} \frac{1}{dx + idy} \left(\frac{\partial u}{\partial x} dx - \frac{\partial v}{\partial x} dy + i\frac{\partial v}{\partial x} dx + i\frac{\partial u}{\partial x} dy \right)$$
$$= \lim_{dx,dy \to 0} \frac{1}{dx + idy} \left(\frac{\partial u}{\partial x} + i\frac{\partial v}{\partial x} \right)(dx + idy)$$
$$= \frac{\partial u}{\partial x} + i\frac{\partial v}{\partial x},$$

where the Cauchy–Riemann conditions have been used after the second equality. The derivative dw/dz is therefore unique for any choice of $dz = dx + idy$.

We now employ the Cauchy–Riemann conditions to check the analyticity of some elementary complex functions.

The function $w(z) = z^2$ can be written as

$$f = u + iv = (x^2 - y^2) + 2ixy$$

so that

$$u = x^2 - y^2, \qquad v = 2xy.$$

By differentiation we get

$$\frac{\partial u}{\partial x} = 2x = \frac{\partial v}{\partial y}, \qquad \frac{\partial u}{\partial y} = -2y = -\frac{\partial v}{\partial x}.$$

Clearly, the Cauchy–Riemann conditions are satisfied, and z^2 is analytic for all finite z.

The function $w(z) = 1/z$ is analytic everywhere except $z = 0$. Since

$$f = \frac{1}{x + iy} = \frac{x - iy}{x^2 + y^2},$$

we get

$$u = \frac{x}{x^2 + y}, \qquad v = \frac{-y}{x^2 + y^2}.$$

The Cauchy–Riemann conditions are satisfied since

$$\frac{\partial u}{\partial x} = \frac{\partial v}{\partial y} = \frac{y^2 - x^2}{(x^2 + y^2)^2}$$

$$\frac{\partial v}{\partial x} = -\frac{\partial u}{\partial y} = \frac{-2xy}{(x^2 + y^2)^2}.$$

The conjugate $z^* = x - iy$ and the magnitude $|z|$ of z are, however, not analytic. It is easy to check that the Cauchy–Riemann conditions are not satisfied.

Let us derive one immediate consequence of analyticity. If $w(z)$ is analytic, then its real and imaginary parts satisfy the Laplace equations. To see this we differentiate the Cauchy–Riemann conditions and get

$$\frac{\partial^2 u}{\partial x^2} = \frac{\partial^2 v}{\partial x \partial y} = \frac{\partial^2 v}{\partial y \partial x} = -\frac{\partial^2 u}{\partial y^2}$$

and

$$\frac{\partial^2 v}{\partial x^2} = -\frac{\partial^2 u}{\partial x \partial y} = -\frac{\partial^2 u}{\partial y \partial x} = -\frac{\partial^2 v}{\partial y^2},$$

hence

$$\frac{\partial^2 u}{\partial x^2} + \frac{\partial^2 u}{\partial y^2} = 0, \qquad \frac{\partial^2 v}{\partial x^2} + \frac{\partial^2 v}{\partial y^2} = 0.$$

This result suggests that analytic functions and two-dimensional Laplace equations are intimately related.

Any real function satisfying the Laplace equation is called a harmonic function. The real and imaginary parts of an analytic function are both harmonic and are called the *harmonic conjugates* of each other.

Note also that

$$\frac{\partial u}{\partial x}\frac{\partial v}{\partial x} + \frac{\partial u}{\partial y}\frac{\partial v}{\partial y} = \frac{\partial u}{\partial x}\left(-\frac{\partial u}{\partial y}\right) + \frac{\partial u}{\partial y}\left(\frac{\partial u}{\partial x}\right) = 0.$$

Since

$$\nabla u = \left(\frac{\partial u}{\partial x}, \frac{\partial u}{\partial y}\right), \qquad \nabla v = \left(\frac{\partial v}{\partial x}, \frac{\partial v}{\partial y}\right),$$

it follows that

$$\nabla u \cdot \nabla v = 0.$$

Thus, if u and v are harmonic conjugates, their contour lines are perpendicular to one another.

9.5 Plane seepage flows in porous media

To demonstrate one area of physical application we first show how the problem of seepage flow can be formulated in terms of analytic functions. Recall from Chapter 1 the empirical Darcy's law in two dimensions

$$u = \frac{\partial \phi}{\partial x}, \quad v = \frac{\partial \phi}{\partial y} \quad (9.5.1)$$

with the velocity potential related to the pore pressure by

$$\phi = -k\left(\frac{p}{\rho g} + y\right). \quad (9.5.2)$$

By mass conservation

$$\frac{\partial u}{\partial x} + \frac{\partial v}{\partial y} = 0, \quad (9.5.3)$$

which implies

$$\nabla^2 \phi = 0. \quad (9.5.4)$$

On the other hand (9.5.3) can be satisfied automatically if the velocity components (u, v) are expressed in terms of the *stream function* ψ according to

$$u = \frac{\partial \psi}{\partial y}, \quad v = -\frac{\partial \psi}{\partial x}. \quad (9.5.5)$$

Clearly, ϕ and ψ satisfy the Cauchy–Riemann conditions. Let

$$w(z) \stackrel{\text{def}}{=} \phi(x, y) + i\psi(x, y), \quad (9.5.6)$$

then the complex potential $w(z)$ must be analytic in the fluid.

As an example involving a water table, consider a sheet pile driven to the depth h below the surface of a saturated ground. On one side of the pile the water depth is H; on the other side water rises to the water table $y = \eta(x)$ below the ground surface (Figure 9.7). What is the flow pattern around the pile, and where is the water table (phreatic surface)?

Along the ground surface AB to the left of the pile $x < 0$, $y = 0$, the hydrostatic pressure is $p = \rho g H$, implying $\operatorname{Re} w = \phi = \rho g H$. The component $u = \partial \psi/\partial y$ vanishes along the wetted part of the pile, hence the stream function must be a constant. Without loss of generality, this constant is taken to be zero, i.e.,

$$\operatorname{Im} w = \psi = 0$$

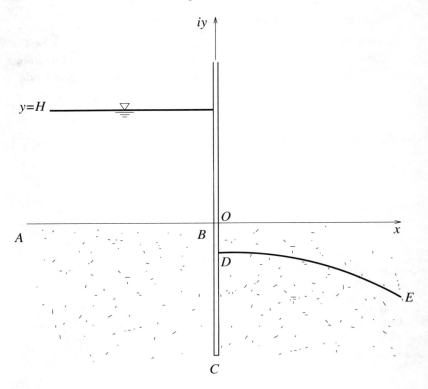

Fig. 9.7. A sheet pile through porous ground.

along BCD. Along the water table DE, the pressure is constant, which may also be taken to be zero

$$\operatorname{Re}(w - ikz) = \phi + ky = 0 \quad \text{on} \quad y = \eta(x).$$

Kinematically, the flow must also be tangential to the phreatic surface DE, which must be the same streamline as the pile

$$\operatorname{Im} w = \psi = 0 \quad \text{on} \quad y = \eta(x).$$

The last two boundary conditions are nonlinear because the height $\eta(x)$ of the water table is a part of the unknown solution.

An ingenious way to simplify the nonlinear mathematics is to introduce the Zhukovsky function θ, defined by

$$\theta(z) = \theta_1(x,y) + i\theta_2(x,y)$$
$$\stackrel{\text{def}}{=} w(z) - ikz = (\phi + ky) + i(\psi - kx). \quad (9.5.7)$$

Thus the real and imaginary parts of θ are

$$\theta_1 = \phi + ky \quad \text{and} \quad \theta_2 = \psi - kx.$$

In terms of θ_1 and θ_2 the boundary conditions become simply

$$\theta_1 = \operatorname{Re}\theta = \rho g H \quad \text{on } AB \tag{9.5.8}$$

$$\theta_2 = \operatorname{Im}\theta = 0 \quad \text{on } BCD \tag{9.5.9}$$

$$\theta_1 = \operatorname{Re}\theta = 0 \quad \text{on } DE. \tag{9.5.10}$$

Equations (9.5.7–9.5.10) constitute a boundary-value problem of the mixed type. An example of the use of $\theta(z)$ will be discussed in Chapter 11.

9.6 Plane flow of a perfect fluid

In a two-dimensional flow of inviscid and incompressible fluid with constant density, the velocity vector $\mathbf{q} = (u, v)$ must satisfy the continuity relation

$$\nabla \cdot \mathbf{q} = \frac{\partial u}{\partial x} + \frac{\partial v}{\partial y} = 0. \tag{9.6.1}$$

Momentum conservation requires that

$$\frac{\partial \mathbf{q}}{\partial t} + \mathbf{q} \cdot \nabla \mathbf{q} = -\frac{1}{\rho}\nabla p. \tag{9.6.2}$$

The quadratic term above may be rewritten as

$$\mathbf{q} \cdot \nabla \mathbf{q} = \frac{1}{2}\nabla(\mathbf{q} \cdot \mathbf{q}) - \mathbf{q} \times (\nabla \times \mathbf{q}).$$

Taking the curl of the momentum equation, we get

$$\frac{\partial}{\partial t}(\nabla \times \mathbf{q}) + \nabla \times \left[\frac{1}{2}\nabla(\mathbf{q} \cdot \mathbf{q}) - \mathbf{q} \times (\nabla \times \mathbf{q})\right]$$

$$= -\frac{1}{\rho}\nabla \times \nabla p. \tag{9.6.3}$$

The vector $\boldsymbol{\zeta} = \nabla \times \mathbf{q}$ is called *vorticity* in fluid mechanics and is related to the rate of rotation of a fluid element. Recall from vector analysis the following identities:

$$\nabla \cdot (\nabla \times \mathbf{a}) = 0 \tag{9.6.4}$$

$$\nabla \times (\mathbf{a} \times \mathbf{b}) = (\nabla \cdot \mathbf{b})\mathbf{a} - \mathbf{b}(\nabla \cdot \mathbf{a}) - \mathbf{a} \cdot \nabla \mathbf{b} + \mathbf{b} \cdot \nabla \mathbf{a}. \tag{9.6.5}$$

With the help of (9.6.5) we get

$$-\nabla \times (\mathbf{q} \times (\nabla \times \mathbf{q})) = -\nabla \times (\mathbf{q} \times \boldsymbol{\zeta})$$
$$= (\nabla \cdot \mathbf{q})\boldsymbol{\zeta} - (\nabla \cdot \boldsymbol{\zeta})\mathbf{q} - \boldsymbol{\zeta} \cdot \nabla \mathbf{q} + \mathbf{q} \cdot \nabla \boldsymbol{\zeta}.$$

In view of (9.6.1), (9.6.4) and the fact that $\boldsymbol{\zeta}$ is perpendicular to $\mathbf{q}$ for all two-dimensional flows, only the last term above does not vanish identically. Equation (9.6.3) then becomes

$$\frac{\partial \boldsymbol{\zeta}}{\partial t} + \mathbf{q} \cdot \nabla \boldsymbol{\zeta} = 0. \tag{9.6.6}$$

Thus, along the path of a fluid particle, vorticity remains constant in time. If $\boldsymbol{\zeta}$ is zero everywhere initially, it remains zero everywhere for all time, i.e.,

$$\frac{\partial u}{\partial y} - \frac{\partial u}{\partial x} = 0. \tag{9.6.7}$$

Such a flow is called *irrotational*. For an irrotational flow, (9.6.7) may replace (9.6.2) to form with (9.6.1) the set of governing equations for the two velocity components.

To satisfy continuity we again introduce the stream function ψ as in (9.5.5). On the other hand, irrotationality is also automatically satisfied if the velocity potential ϕ, defined by (9.5.1), is used. Clearly, ϕ and ψ obey the Cauchy–Riemann conditions, hence ϕ and ψ are harmonic conjugates and can be regarded as the real and imaginary parts of an analytic function

$$w(z) = \phi(x, y) + i\psi(x, y) \tag{9.6.8}$$

as in (9.5.6). Being analytic, the derivative of w can be taken in any direction, say along $y = $ constant,

$$\frac{dw}{dz} = \frac{\partial w}{\partial x} = \frac{\partial \phi}{\partial x} + i\frac{\partial \psi}{\partial x} = u - iv,$$

which is called the complex velocity (actually the complex conjugate of the velocity).

Another useful result for two- and three-dimensional irrotational flows can be derived by rewriting (9.6.2) as

$$\nabla \left(\frac{\partial \phi}{\partial t} + \frac{p}{\rho} + \frac{\mathbf{q} \cdot \mathbf{q}}{2} \right) = 0.$$

Upon integration we get

$$\frac{\partial \phi}{\partial t} + \frac{p}{\rho} + \frac{\mathbf{q} \cdot \mathbf{q}}{2} = C(t), \tag{9.6.9}$$

9.7 Simple irrotational flows

In simple cases it is convenient to find the potential flows represented by some given analytical functions. In this inverse approach we seek the problem of a solution.

(i) A uniform flow: $w = Uz = U(x+iy)$. The potential and stream functions are

$$\phi = Ux, \quad \psi = -Uy,$$

and the complex velocity is

$$\frac{dw}{dz} = u - iv = U,$$

hence $u = U, v = 0$. This flow is uniform in the x direction.

(ii) Point source or sink: $w = (m/2\pi)\ln z$ with real m. The complex velocity is

$$\frac{dw}{dz} = \frac{m}{2\pi z} = \frac{m}{2\pi r}e^{-i\theta}.$$

In polar form the complex velocity is $dw/dz = qe^{-i\lambda}$, where q is the speed and λ is the direction. Clearly,

$$q = \frac{m}{2\pi r}, \quad \lambda = \theta.$$

Thus the velocity is in the radial direction, and its magnitude is inversely proportional to r. This result corresponds to a source if $m > 0$ and a sink if $m < 0$. The total volume flux through a circle of radius r is $2\pi r q = m$. Thus m is the strength of the source of sink. Consider the complex potential itself

$$w = \phi + i\psi = \frac{m}{2\pi}\ln re^{i\theta} = \frac{m}{2\pi}\left(\ln r + \frac{im\theta}{2\pi}\right),$$

thus

$$\phi = \frac{m}{2\pi}\ln r, \quad \psi = \frac{m\theta}{2\pi}.$$

All equipotential lines are concentric circles, while all streamlines are radial, as shown in Figure 9.8.

$m > 0$

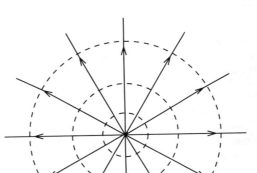

Fig. 9.8. Streamlines and equipotential lines for a source.

(iii) Point vortex: $w = (i\Gamma/2\pi) \ln z$, where Γ is a real constant. The complex velocity is

$$\frac{dw}{dz} = -\frac{i\Gamma}{2\pi z} = -\frac{i\Gamma}{2\pi r} e^{-i\theta} = \frac{\Gamma}{2\pi r} e^{-i(\theta+\pi/2)}.$$

In polar form

$$\frac{dw}{dz} = q\, e^{-i\lambda},$$

the velocity vector has the magnitude $q = \Gamma/2\pi r$ and is in the direction $\lambda = \theta + \pi/2$, i.e., 90 degrees counterclockwise from the position vector $z = r\, e^{i\theta}$, hence is tangent to the circle of radius r and counterclockwise for $\Gamma > 0$. Along the same circle the speed is constant. Along a larger circle the speed $\Gamma/2\pi r$ is lower.

The complex potential w is

$$w = \frac{i\Gamma}{2\pi} \ln z = \frac{i\Gamma}{2\pi} \ln e^{i\theta} = \frac{\Gamma}{2\pi}(-\theta + i\ln r).$$

Hence

$$\phi = -\frac{\Gamma}{2\pi}\theta, \qquad \psi = \frac{\Gamma}{2\pi} \ln r.$$

Thus the equipotential lines $\phi =$ constant are radial, and the streamlines are concentric circles.

(iv) **Doublet (dipole):** $w = -\mu/z$. The complex potential is

$$\phi + i\psi = -\frac{\mu}{x+iy} = -\frac{\mu(x-iy)}{x^2+y^2}$$

so that

$$\phi = -\frac{\mu x}{x^2+y^2}, \qquad \psi = \frac{\mu y}{x^2+y^2}.$$

Along a streamline, $\psi =$ constant; we get from the second equation

$$x^2 + \left(y - \frac{\mu}{2\psi}\right)^2 = \left(\frac{\mu}{2\psi}\right)^2.$$

Hence the streamlines are circles centered at $(x=0, y=\mu/2\psi)$ and tangent to the x axis at the origin, as shown in Figure 9.9. The direction of the streamlines can be found from the complex velocity

$$\frac{dw}{dz} = u - iv = \frac{\mu}{z^2} = \frac{\mu\left[(x^2-y^2) - 2ixy\right]}{(x^2+y^2)^2}.$$

Hence

$$u = \frac{\mu\left(x^2-y^2\right)}{(x^2+y^2)^2}, \qquad v = \frac{2\mu xy}{(x^2+y^2)^2}.$$

For $\mu > 0$, v is positive in the first and third quadrants and negative in the second and fourth quadrants. The directions of the streamlines are also shown in Figure 9.9. This flow is called a doublet; a similar solution in electromagnetism is called a dipole.

9.8 Cauchy's theorem

Consider a curve Γ from z_A to z_B in the complex plane. Let Γ be discretized into small segments by closely separated points $z_0, z_1, z_2, \ldots, z_k, z_{k+1} \ldots z_{N-1}$ and z_N where $z_0 = z_A$ and $z_N = z_B$, and let ζ_k denote any points between z_k and z_{k+1}. Then the line integral of $f(z)$ is defined to be

$$I = \lim_{n \to \infty} \sum_{k=1}^{n} (z_k - z_{k-1}) f(\zeta_k) = \int_{\Gamma} f(z)\, dz.$$

If the curve Γ is closed, the line integral is called a contour integral. Unless otherwise specified, the direction of integration along the closed

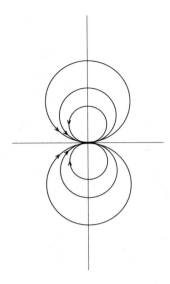

Fig. 9.9. Streamlines of a doublet.

contour is always counterclockwise so that the enclosed region is on the left.

Now we state the most celebrated theorem in the complex function theory.

Cauchy's theorem *If $f(z)$ is analytic and $f'(z)$ is continuous in the region A and along its boundary C, then the contour integral along C vanishes*

$$\oint_C f(z)\,dz = 0. \tag{9.8.1}$$

To prove this theorem, we note that

$$\oint_C f\,dz = \oint_C (u+iv)(dx+idy)$$
$$= \oint_C (u\,dx - v\,dy) + i\oint_C (v\,dx + u\,dy)$$
$$= \oint_C \left(u\frac{dx}{ds} - v\frac{dy}{ds}\right)ds + i\oint_C \left(v\frac{dx}{ds} + u\frac{dy}{ds}\right)ds,$$

9.8 Cauchy's theorem

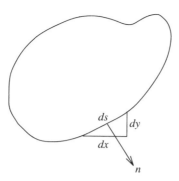

Fig. 9.10. A closed contour.

where ds is a line element along C. As shown in Figure 9.10, at any point along C the unit tangent vector is

$$\mathbf{t} = \left(\frac{dx}{ds}, \frac{dy}{ds}\right)$$

and the unit outward normal is

$$\mathbf{n} = \left(\frac{dy}{ds}, -\frac{dx}{ds}\right).$$

The last two contour integrals may be rewritten as

$$-\oint_C (v, u) \cdot \mathbf{n}\, ds + i \oint_C (u, -v) \cdot \mathbf{n}\, ds,$$

where (a, b) denotes a vector with components (a, b) in the (x, y) directions. By Gauss' theorem the last result may also be written as

$$\iint_A \nabla \cdot (v, u)\, dA + i \iint_A \nabla \cdot (u, -v)\, dA$$

$$= -\iint_A \left(\frac{\partial v}{\partial x} + \frac{\partial u}{\partial y}\right) dA + i \iint_A \left(\frac{\partial u}{\partial x} - \frac{\partial v}{\partial y}\right) dA.$$

Because f is analytic in A, u and v satisfy the Cauchy–Riemann conditions, hence both integrands vanish. This result proves Cauchy's theorem.

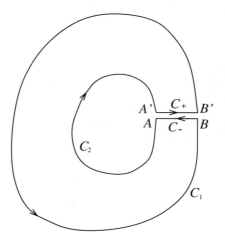

Fig. 9.11. Two closed contours.

The converse of Cauchy's theorem is called Morea's theorem, which we state below without proof:

Morea's theorem *If $f(z)$ is continuous and single-valued within a closed contour C and if*

$$\oint_\Gamma f(z)\,dz = 0$$

for every Γ inside C, then $f(z)$ is analytic inside C.

A proof can be found in Copson (1935, pp. 70–72).

Before illustrating the applications of the powerful Cauchy's theorem, let us check its truth via simple examples and a more familiar argument. Consider the contour integral of $f = z$ along a circle of radius r. Since $f(z) = re^{i\theta}$ and $dz = ire^{i\theta}\,d\theta$,

$$\oint f(z)\,dz = \int_0^{2\pi} d\theta\, ire^{i\theta} re^{i\theta} = ir^2 \int_0^{2\pi} e^{2i\theta}\,d\theta = 0.$$

The integral indeed vanishes for any circle.

Corollary 1 *If the closed contour C_1 is inside the closed contour C_2 and $f(z)$ is analytic between them, then*

$$\oint_{C_1} f(z)\,dz = \oint_{C_2} f(z)\,dz. \qquad (9.8.2)$$

Referring to Figure 9.11, let us cut C_1 and C_2 and connect them

9.8 Cauchy's theorem

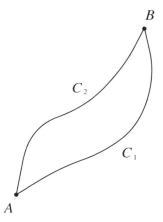

Fig. 9.12. Different paths connecting two points.

with two paths C_+ and C_-, which lie side by side and are taken to be straight for convenience. Let the closed contour C be composed of the curved path from A to A' along C_1 (counterclockwise), the straight path from A' to B', the curved path from B' to B along C_2 (clockwise) and the straight path C_+ from B back to A. Since $f(z)$ is analytic in the region enclosed by C_1, Cauchy's theorem applies

$$\int_{C_1} + \int_{C_+} + \int_{C_-} + \int_{C_2} f \, dz = 0.$$

Since C_+ and C_- are infinitesimally close and the directions of integration along them are opposite, the two-line integrals cancel each other. Changing the C_2 integral from clockwise to counterclockwise reverses its sign, hence (9.8.2) is proved.

Corollary 2 *In a region where $f(z)$ is analytic the line integral of $f(z)$ between two points A and B is independent of the path connecting points A and B, i.e.,*

$$\int_{C_1} f(z) \, dz = \int_{C_2} f(z) \, dz, \qquad (9.8.3)$$

where C_1 and C_2 are any two paths beginning at A and ending at B in the region of analyticity.

As shown in Figure 9.12, the contour beginning from A to B along C_1 and returning from B to A along C_2 is closed, so Cauchy's theorem requires that

$$\left[\int_A^B f(z)\,dz\right]_{C_1} + \left[\int_B^A f(z)\,dz\right]_{C_2} = 0.$$

Reversing the direction in the second integral changes the sign so that

$$\left[\int_A^B f(z)\,dz\right]_{C_1} - \left[\int_A^B f(z)\,dz\right]_{C_2} = 0,$$

which proves (9.8.3).

Both corollaries allow us to replace a given contour by a new one more convenient for the evaluation of integrals.

As an immediate application of Corollary 1, let us show that

$$I(1) \stackrel{\text{def}}{=} \oint_C \frac{dz}{z-a} = 2\pi i, \qquad (9.8.4)$$

where C is any closed contour enclosing a. By Corollary 2 we can replace the contour C by a circle C_ϵ centered at a and of some radius ϵ so that C_ϵ is inside C. Along C_ϵ, $z = a + \epsilon e^{i\theta}$ and $dz = i\epsilon e^{i\theta}\,d\theta$ so that

$$I(1) = \oint_{C_\epsilon} \frac{dz}{z-a} = \int_0^{2\pi} \frac{i\epsilon e^{i\theta}\,d\theta}{\epsilon e^{i\theta}}$$

$$= i\int_0^{2\pi} d\theta = 2\pi i.$$

For later use let us also show that

$$I(n) \stackrel{\text{def}}{=} \oint_C \frac{dz}{(z-a)^n} = 0 \qquad (9.8.5)$$

for any integer n not equal to unity. Again, replacing C by the circle C_ϵ,

$$I(n) = \oint_{C_\epsilon} \frac{dz}{(z-a)^n} = \int_0^{2\pi} \frac{i\epsilon e^{i\theta}\,d\theta}{\epsilon^n e^{in\theta}}$$

$$= i\epsilon^{n-1}\int_0^{2\pi} e^{i(1-n)\theta}\,d\theta = 0$$

because $e^{i(1-n)\theta}$ is periodic with the period 2π.

Let us test Corollary 2 by taking $f(z) = z$. Referring to Figure 9.13, we take two different paths from $z = 0$ to $z = a + ib$. Along the solid path, $z = x$, $dz = dx$ along the horizontal leg and $z = a + iy$, $dz = i\,dy$

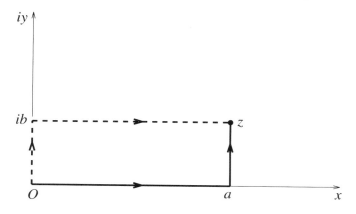

Fig. 9.13. Two different paths of integration.

along the vertical leg, hence

$$\int_0^z f(z)\,dz = \int_0^a x\,dx + \int_0^b (a+iy)i\,dy$$
$$= \frac{a^2}{2} + iab - \frac{b^2}{2} = \frac{1}{2}(a+ib)^2.$$

Along the dashed path, $z = iy$, $dz = i\,dy$ along the vertical leg and $z = x + ib$, $dz = dx$ along the horizontal leg, hence

$$\int_0^z f(z)\,dz = \int_0^b iy i\,dy + \int_0^a (x+ib)\,dx$$
$$= -\frac{b^2}{2} + \frac{a^2}{2} + iab = \frac{1}{2}(a+ib)^2.$$

The two paths indeed give the same result.

9.9 Cauchy's integral formula and inequality

Cauchy's integral formula If $f(z)$ is analytic in the area enclosed by the contour C, then

$$\frac{1}{2\pi i}\oint_C \frac{f(z)}{z-a}\,dz = f(a) \qquad (9.9.1)$$

if the point a is also inside C.

Proof The integral is singular at $z = a$. Let us surround the singularity by a circle C_ϵ of infinitesimal radius ϵ centered at a. The integrand is analytic in the region between C and C_ϵ, and Corollary 1 of Cauchy's theorem can be applied to give

$$\frac{1}{2\pi i} \oint_C \frac{f(z)}{z-a} \, dz = \frac{1}{2\pi i} \oint_{C_\epsilon} \frac{f(z)}{z-a} \, dz.$$

Now for the integral on the right we let $z = a + \epsilon e^{i\theta}$ so that $dz = i\epsilon e^{i\theta} d\theta$. Along the circle C_ϵ

$$f(z) = f\left(a + \epsilon e^{i\theta}\right).$$

By continuity $f(z)$ approaches $f(a)$ for diminishing ϵ. In the limit of $\epsilon = 0$ we replace $f(z)$ by $f(a)$ so that

$$\lim_{\epsilon \to 0} \frac{1}{2\pi i} \oint_{C_\epsilon} \frac{f(a)}{\epsilon e^{i\theta}} i\epsilon e^{i\theta} d\theta = \frac{f(a)}{2\pi i} \int_0^{2\pi} i d\theta = f(a).$$

Written in another way, (9.9.1) reads

$$f(z) = \frac{1}{2\pi i} \oint_C \frac{f(\zeta) \, d\zeta}{\zeta - z}, \tag{9.9.2}$$

which says that if f is given on a closed contour and is analytic inside, then $f(z)$ is known everywhere inside C.

From Cauchy's integral formula we can show that the derivative $f'(z)$ can be obtained by differentiating under the integral sign, i.e.,

$$f'(z) = \frac{1}{2\pi i} \oint_C \frac{f(\zeta)}{(\zeta - z)^2} \, d\zeta \tag{9.9.3}$$

or, more generally,

$$f^{(n)}(z) = \frac{n!}{2\pi i} \oint_C \frac{f(\zeta)}{(\zeta - z)^{n+1}} \, d\zeta. \tag{9.9.4}$$

The proof of (9.9.3) begins by forming the ratio

$$\frac{f(z + dz) - f(z)}{dz} = \frac{1}{2\pi i} \oint_C f(\zeta) \left(\frac{1}{\zeta - z - dz} - \frac{1}{\zeta - z} \right) d\zeta$$

$$= \frac{1}{2\pi i} \oint_C \frac{f(\zeta) \, d\zeta}{(\zeta - z)^2} + \frac{dz}{2\pi i} \oint_C \frac{f(\zeta) \, d\zeta}{(\zeta - z)^2 (\zeta - z - dz)}.$$

As $dz \to 0$ the left-hand side becomes $f'(z)$. Let the shortest distance from C to z be ℓ, then $|\zeta - z| \geq \ell$ for every $z \in C$ and $1/|\zeta - z| \leq 1/\ell$. By analyticity, $f(\zeta)$ is bounded by some finite constant M for all ζ on C. Hence the second integral on the right-hand side is bounded by

$$\frac{dz}{2\pi} \frac{ML}{\ell^3},$$

where L is the length of C. Clearly, this upper bound vanishes in the limit of $dz \to 0$, and (9.9.3) is proven.

By repeating the argument for $n = 2, 3, \ldots$ and by induction, (9.9.4) can be proven.

Equation (9.9.4) can be used to derive the following inequality.

Cauchy's inequality Let $f(z)$ be analytic and regular within a circle C of radius R centered at ℓ, then

$$|f^{(n)}(a)| \leq \frac{Mn!}{R^n}, \qquad (9.9.5)$$

where M is the upper bound of $|f(z)|$ on C.

The inequality follows from (9.9.4)

$$|f^{(n)}(a)| \leq \frac{Mn!(2\pi R)}{2\pi R^{n+1}} = \frac{Mn!}{R^n}.$$

9.10 Liouville's theorem

An immediate consequence of Cauchy's inequality is:

Liouville's theorem If $f(z)$ is analytic everywhere in z including infinity, then $f(z)$ is a constant.

To prove this theorem consider the derivative $f'(z)$. Since $f(z)$ is analytic everywhere, it must be bounded, i.e., $|f(z)| < M$, where M is a finite number. By Cauchy's inequality

$$f'(z) \leq \frac{M}{R}$$

for any R. Now let R increase without bound. The right-hand side must vanish, implying that $f'(z) = 0$ for all z. Therefore, $f(z)$ must be a constant for all z.

9.11 Singularities

The seemingly innocuous Liouville's theorem implies that an interesting complex function should not be analytic everywhere but must be singular somewhere.

A singular point of $f(z)$ is a point where $f(z)$ is not analytic. For example, $f = \ln z, z^{-1}, z^{-2}$ and $z^{-1/2}$ are singular at $z = 0$, while $f = z, z^2, z^{1/2}$ and $\ln z$ are singular at $z \sim \infty$. There are three types of singularities: (i) branch points, (ii) poles and (iii) essential singularities. A branch point is a point around which $f(z)$ is not unique, as has been discussed in §9.3.

If $f(z)$ is singular at $z = z_o$ near which $f(z)$ is dominated by

$$f \cong \frac{a_{-n}}{(z - z_o)^n},$$

then $f(z)$ has an nth-order pole at $z = z_o$. In particular, if $n = 1, f(z)$ has a simple pole at $z = z_o$. For example,

$$\frac{1}{z^2 + 1} = \frac{1}{(z + i)(z - i)}$$

has two simple poles, one at i and one at $-i$.

At an essential singularity, $f(z)$ can take on any value. For example,

$$f = e^{1/z} = \exp\left(\frac{1}{r}e^{i\theta}\right) = \exp\left[\frac{1}{r}(\cos\theta - i\sin\theta)\right]$$

so that

$$|f| = \exp\left(\frac{\cos\theta}{r}\right), \quad \arg f = -\frac{\sin\theta}{r}.$$

For any $-\pi/2 < \theta < \pi/2$, $\cos\theta > 0$, and $|f|$ increases without bound if $r \to 0$. However, for any $\pi/2 < \theta < 3\pi/2$, $\cos\theta < 0$ so that $|f| \to 0$ as $r \to 0$. If $z \to 0$ along the curve $r = a\cos\theta$, which is a circle centered at $x = a/2$ and tangent to the y axis, $|f| = e^{1/a}$ can be any positive number depending on the sign and magnitude of a. On such a circle the argument of f is $-(\tan\theta)/a$, which varies from 0 to $-\infty$ as θ changes from 0 to $\pi/2$. Therefore, as z approaches zero along the circle, $\theta \to \pi/2$, and $|f|$ is unbounded. The argument changes infinitely fast, and f can take on any complex value infinite times. The essential singularity is, therefore, the most undefinable of them all.

9.12 Evaluation of integrals by Cauchy's theorems

One of the mathematical applications of Cauchy's theorem is to facilitate the explicit evaluation of integrals along a real line. The typical procedure is to (i) change the real integration variable to the complex variable, (ii) find the singularities of the integrand in the complex z plane, (iii) connect the original path of integration with an additional path to form a closed contour, (iv) apply Cauchy's integral formula to evaluate the integral along the closed contour, (v) find the integral along the additional path and, finally, (vi) subtract from the results of (v) from (iv) to get the original integral. Some simple examples are given in this section.

Example 1: We first consider an integral with an even integrand

$$I = \int_0^\infty \frac{dx}{1+x^2}. \tag{9.12.1}$$

Because of the evenness of the integrand, I can be rewritten as

$$I = \frac{1}{2} \int_{-\infty}^{\infty} \frac{dx}{1+x^2} = \frac{1}{2} \int_{-\infty}^{\infty} \frac{dz}{z^2+1}.$$

In the last integral the real variable x has been replaced by the complex variable z. This replacement does not change the integral since the path

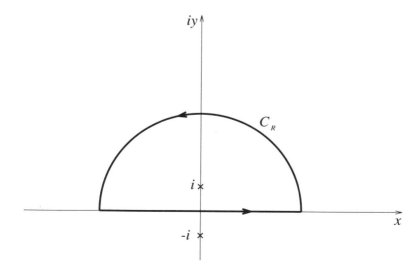

Fig. 9.14. Contour for Example 1.

of integration is still the entire real axis. Now in the complex plane the integrand has two simple poles at $z = \pm i$. Let us introduce a semicircular arc C_R of large radius R in the upper half plane to form a closed contour with the real axis, as shown in Figure 9.14. Since only the pole at $z = i$ is within the contour, the Cauchy integral theorem gives

$$\frac{1}{2} \oint_C \frac{dz}{z^2+1} = \frac{1}{2} \left(\int_{-\infty}^{\infty} + \int_{C_R} \right) \frac{dz}{z^2+1}$$

$$= \frac{\pi i}{2\pi i} \oint_C \frac{dz}{(z+i)(z-i)} = \pi i \left(\frac{1}{z+i} \right)_{z=i} = \frac{\pi}{2}.$$

With $z = Re^{i\theta}$, the integral along the semicircle C_∞ is

$$\frac{1}{2} \int_0^\pi \frac{Rie^{i\theta} \, d\theta}{1 + R^2 e^{2i\theta}} \sim \frac{i}{2R} \int_0^\pi e^{-i\theta} \, d\theta,$$

which vanishes at the limit of $R \to \infty$. It follows that

$$I = \frac{\pi}{2}.$$

In this example C_R can also be chosen to be in the lower half plane.

Example 2: We now consider an integral with an odd integrand

$$I = \int_0^\infty \frac{dx}{1+x^3} = \int_0^\infty \frac{dz}{1+z^3}. \tag{9.12.2}$$

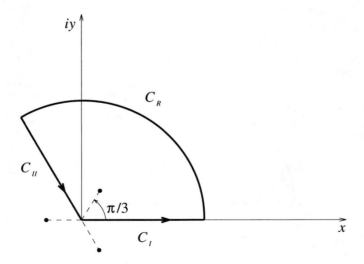

Fig. 9.15. Contour for Example 2.

9.12 Evaluation of integrals by Cauchy's theorems

Referring to Figure 9.15, we form a closed contour C by adding a circular arc C_R and a ray C_{II} inclined at $2\pi/3$ from the real axis C_I. The poles of the integrand are at the zeros of $z^3 + 1 = 0$ or

$$z = e^{i\pi/3}, \quad e^{i\pi}, \quad e^{5i\pi/3}.$$

Only the first pole lies inside C, hence,

$$\oint_C \frac{dz}{1+z^3} = \oint_C \frac{dz}{\left(z - e^{i\pi/3}\right)\left(z - e^{i\pi}\right)\left(z - e^{5i\pi/3}\right)}$$

$$= \frac{2\pi i}{\left(e^{i\pi/3} - e^{i\pi}\right)\left(e^{i\pi/3} - e^{5i\pi/3}\right)} = \frac{2\pi i}{3} e^{-2i\pi/3}.$$

The integral along C_R can be shown to vanish in the limit of $R \to \infty$ because the integrand diminishes as R^{-3} when $R \to \infty$. The integral along the inclined path C_{II} can be rearranged by the following change of variables $z = re^{2i\pi/3}$ and $dz = dre^{2i\pi/3}$

$$\int_{C_{II}} \frac{dz}{1+z^3} = \int_\infty^0 \frac{dr\, e^{2i\frac{\pi}{3}}}{1+r^3} = -e^{2i\frac{\pi}{3}} \int_0^\infty \frac{dr}{1+r^3} = -Ie^{2i\frac{\pi}{3}},$$

hence

$$I\left(1 - e^{2i\frac{\pi}{3}}\right) = Ie^{\frac{\pi i}{3}} \left(-2i \sin\frac{\pi}{3}\right) = \frac{2\pi i}{3} e^{-2i\frac{\pi}{3}}.$$

Finally, the answer is

$$I = \frac{-\frac{\pi}{3} e^{-\pi i}}{\sin\frac{\pi}{3}} = \frac{2\pi}{3\sqrt{3}}.$$

Example 3: Let us consider the Fourier integral

$$I = 2 \int_0^\infty \frac{\sin x \, dx}{x\left(a^2 + x^2\right)}. \tag{9.12.3}$$

At $x = 0$ the integral has a removable singularity, namely, there is only an apparent singularity since $\sin x \sim x + O(x^3)$ also vanishes at $x = 0$ and $\sin x / x \to 1$ is finite. Let us rewrite I as

$$I = \int_{-\infty}^\infty \frac{\sin x \, dx}{x\left(a^2 + x^2\right)} = \frac{1}{2i} \int_{-\infty}^\infty \frac{e^{ix} - e^{-ix}}{x\left(a^2 + x^2\right)} dx$$

$$= \frac{1}{2i} \int_{-\infty}^\infty \frac{e^{iz} - e^{-iz}}{z\left(a^2 + z^2\right)} dz$$

$$= \frac{1}{2i} \left(I_+ - I_-\right),$$

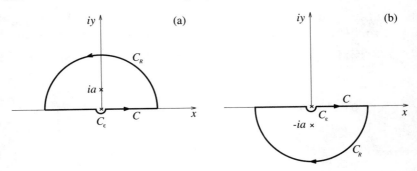

Fig. 9.16. Contours for Example 3.

where

$$I_\pm = \int_{-\infty}^{\infty} \frac{e^{\pm iz}}{z(a^2 + z^2)}\,dz.$$

Consider first I_+ along the closed contour, as shown in Figure 9.16a, with a large semicircle C_R in the upper half plane. Because $z = 0$ is a pole, we indent the contour along a small semicircle below the origin. Since the indented semicircle contributes equally to I_+ and I_-, it will not affect $I = (I_+ - I_-)/2i$. Two poles ($z = 0, z = ia$) are inside C. The integral along the closed contour is, therefore,

$$\bar{I}_+ = \oint \frac{e^{iz}\,dz}{z(z^2 + a^2)} = \oint \frac{e^{iz}\,dz}{z(z + ia)(z - ia)}$$

$$= 2\pi i \left(\frac{e^{i(ia)}}{(ia)(2ia)} + \frac{1}{(ia)(-ia)} \right)$$

$$= 2\pi i \left(\frac{1}{a^2} - \frac{e^{-a}}{2a^2} \right).$$

Now the contribution from the semicircle is

$$\lim_{R\to\infty} \left| \int_0^\pi \frac{e^{iRe^{i\theta}} iRe^{i\theta}\,d\theta}{R^3 e^{3i\theta}} \right|$$

$$= \lim_{R\to\infty} \frac{1}{R^2} \int_0^\pi \left| \frac{e^{iR\cos\theta} e^{-R\sin\theta}}{e^{2i\theta}} \right| d\theta$$

$$\leq \lim_{R\to\infty} \frac{1}{R^2} \int_0^\pi e^{-R\sin\theta}\,d\theta$$

$$= \lim_{R\to\infty} \frac{2}{R^2} \int_0^{\pi/2} e^{-R\sin\theta}\,d\theta.$$

9.12 Evaluation of integrals by Cauchy's theorems

The integral is clearly convergent for $0 \leq \theta \leq \pi$, where $\sin\theta \geq 0$. Indeed, convergence was the motivation behind the choice of this semicircle in the upper half for the factor e^{iz}. Within the domain of integration $0 \leq \theta \leq \frac{\pi}{2}$,

$$\frac{2\theta}{\pi} \leq \sin\theta < \theta,$$

therefore the above integral is bounded by

$$\lim_{R\to\infty} \frac{2}{R^2} \int_0^{\frac{\pi}{2}} e^{-2R\theta/\pi} \, d\theta = \lim_{R\to\infty} \frac{2}{R^2} \frac{1}{-\frac{2}{\pi}R} \left[e^{-2R\theta/\pi} \right]_0^{\frac{\pi}{2}}$$
$$= \lim_{R\to\infty} O(R^{-3}) = 0.$$

It follows that $\overline{I}_+$ is contributed by the indented path alone

$$\frac{1}{2i}\overline{I}_+ = \frac{1}{2i} I_+ = \frac{1}{2i} \int_{-\infty}^{\infty} \frac{e^{iz}}{z(a^2+z^2)} dz$$
$$= \pi \left(\frac{1}{a^2} - \frac{e^{-a}}{2a^2} \right).$$

Now we calculate I_-. Because of the factor e^{-iz}, we now choose a closed contour by adding a large semicircle in the lower half plane to the indented path, as shown in Figure 9.16b. The direction of integration is clockwise, and only one pole at $z = -ia$ is enclosed; the integral is

$$\frac{\overline{I}_-}{2i} = -\frac{1}{2i} \oint \frac{e^{-iz}}{z(a^2+z^2)} dz = -\frac{2\pi i}{2i} \frac{e^{-i(-ia)}}{-ia(-2ia)} = \frac{\pi e^{-a}}{2a^2}.$$

The final result is

$$I = \frac{1}{2}(I_+ - I_-) = \frac{\pi}{a^2}(1 - e^{-a}).$$

It is easy to show that the result for I is unchanged if the indentation is above the origin, because the singularity arises in I_+ and I_- only as a consequence of splitting.

Example 4: We now illustrate an integrand which must involve a branch cut when extended to the complex plane

$$I = \int_0^\infty \frac{\sqrt{x}\, dx}{1+x^2}. \tag{9.12.4}$$

Consider instead a contour integral in the z plane along C that wraps around the branch cut along the positive real axis, as shown in Figure 9.17

$$\overline{I} = \oint_C \frac{\sqrt{z}\, dz}{1+z^2}.$$

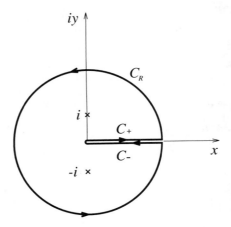

Fig. 9.17. Contour for Example 4.

The square root $\sqrt{z}$ is rendered single-valued by choosing the branch such that

$$\sqrt{z} = \sqrt{r}\, e^{i\theta/2} \quad \text{with} \quad 0 \le \theta < 2\pi.$$

In this closed contour there are two poles $z = \pm i$. The sum of residues is

$$\overline{I} = \oint \frac{\sqrt{z}\, dz}{(z+i)(z-i)} = 2\pi i \left(\frac{\sqrt{i}}{2i} + \frac{\sqrt{-i}}{-2i} \right).$$

By our choice of the branch we have

$$\sqrt{i} = \left(e^{i\pi/2}\right)^{1/2} = e^{i\pi/4} = \frac{1+i}{\sqrt{2}}$$

$$\sqrt{-i} = \left(e^{3\pi i/2}\right)^{1/2} = e^{3i\pi/4} = \frac{-1+i}{\sqrt{2}},$$

hence,

$$\overline{I} = \frac{2\pi i}{2i} \left(\frac{1+i}{\sqrt{2}} - \frac{-1+i}{\sqrt{2}} \right) = \sqrt{2}\, \pi.$$

Now the contribution from the infinite circle is zero because

$$\lim_{R \to \infty} \int_0^{2\pi} \frac{R^{1/2} R}{R^2}\, d\theta = O\left(\frac{1}{R^{1/2}}\right) \to 0.$$

Along C_+, $\sqrt{z} = \sqrt{x}$, while along C_-

$$\sqrt{z} = \left(xe^{2i\pi}\right)^{1/2} = -\sqrt{x},$$

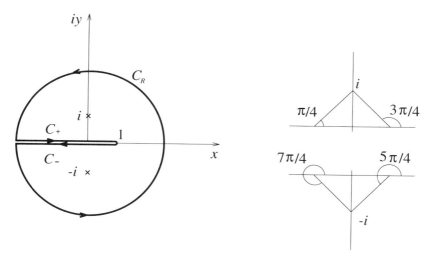

Fig. 9.18. Contour for Example 5.

hence the integrals along C_+ and C_- are equal and

$$2\int_0^\infty \frac{\sqrt{x}\,dx}{1+x^2} = \sqrt{2}\,\pi \quad \text{or} \quad I = \frac{\pi}{\sqrt{2}}.$$

Example 5: Let us illustrate one more case with a square root in the integrand

$$I = \int_{-1}^1 \frac{dx}{\sqrt{1-x^2}(1+x^2)}. \tag{9.12.5}$$

Consider the following contour integral

$$\bar{I} = \oint_C \frac{dz}{\sqrt{z^2-1}(1+z^2)},$$

where the closed contour C is shown in Figure 9.18. As discussed in §9.3, the square root $\sqrt{z^2-1}$ is defined by introducing a cut along $-1 < x < 1$ and by defining

$$\sqrt{z^2-1} = r_1 r_2 e^{i(\theta_1+\theta_2)/2}$$

with θ_1 and θ_2 measured from the positive x axis. Clearly, from the poles $z = \pm i$ we get

$$\bar{I} = 2\pi i \left[\frac{1}{\sqrt{i^2-1}(2i)} + \frac{1}{\sqrt{(-i)^2-1}(-2i)} \right].$$

As in §9.3 we find $\sqrt{i^2-1} = 2i$. Similarly, the second square root at $z = -i$ is found to be $\sqrt{(-i)^2-1} = 2e^{i3\pi/2} = -2i$. It follows that

$$\bar{I} = 2\pi i \left(\frac{1}{(2i)(2i)} + \frac{1}{(-2i)(-2i)} \right) = -\pi i.$$

Now

$$\bar{I} = \left(\int_{C_R} + \int_{C_+} + \int_{C_-} \right) \frac{dz}{\sqrt{z^2-1}(1+z^2)}.$$

The integral along C_R vanishes in the limit of $R \to \infty$ since the integrand dies out as R^{-3}. The two integrals along C_+ and C_- cancel except along the opposite edges of the branch cut, where

$$\sqrt{z^2-1} = \pm i\sqrt{1-x^2}.$$

Hence we get

$$\bar{I} = 2i \int_{-1}^{1} \frac{dx}{\sqrt{1-x^2}(1+x^2)} = 2iI = -\pi i$$

and, finally,

$$I = -\frac{\pi}{2}.$$

9.13 Jordan's lemma

For evaluating integrals in the Fourier transforms or their inverses, one often passes from the real axis to the complex plane and chooses a closed path by adding a great semicircular arc C_R. The following lemma gives the conditions under which the contribution from C_R vanishes in the limit of $R \to \infty$ and is useful in deciding how to choose C_R.

Jordan's lemma If $f(z)$ is analytic in $\operatorname{Im} z \geq 0$ except at poles and $|f(z)| \to 0$ on the semicircular arc C_R in the upper half plane as $R \to \infty$, then for $m > 0$

$$\lim_{R \to \infty} \int_{C_R} f(z) e^{imz} \, dz = 0. \tag{9.13.1}$$

The proof is along the same lines explained in Example 4 above. Let

$$|f(z)| < \delta$$

with $\delta \to 0$ as $R \to \infty$. Using the fact that

$$|e^{imz}| = |e^{im(R\cos\theta + iR\sin\theta)}| = e^{-mR\sin\theta}$$

9.14 Forced harmonic waves and the radiation condition

for $0 < \theta < \frac{\pi}{2}$, we have

$$|I| = \lim_{R \to \infty} \left| \int_{C_R} f(z)e^{imz} \, dz \right| \leq \left| \int_0^\pi \delta i R e^{i\theta} e^{-mR\sin\theta} \, d\theta \right|$$

$$= 2\delta R \int_0^{\pi/2} e^{-mR\sin\theta} \, d\theta.$$

Since

$$\frac{2\theta}{\pi} \leq \sin\theta < \theta \quad \text{in} \quad 0 \leq \theta \leq \frac{\pi}{2},$$

it follows that

$$|I| \leq \lim_{R \to \infty} 2\delta R \int_0^{\pi/2} e^{-\frac{2mR}{\pi}\theta} \, d\theta$$

$$= \lim_{R \to 0} 2\delta R \frac{\pi}{2mR} e^{-\frac{2mR}{\pi}\theta} \bigg|_0^{\pi/2}$$

$$= \lim_{R \to 0} O(\delta) \to 0,$$

which proves Jordan's lemma.

If $m < 0$ and $f(z)$ is analytic in $\text{Im}\, z \leq 0$ except at poles, and $|f(z)| \to 0$ as $|z| \to \infty$ in the lower half plane, then Jordan's lemma (9.13.1) holds along a semicircle C_R in the lower half plane.

For applications in the Laplace transforms, to be discussed in the next chapter, we need Jordan's lemma in the following form.

If $f(z)$ is analytic in $\text{Re}\, z \leq 0$ (or $\text{Re}\, z \geq 0$) except at poles and $|f(z)| \to 0$ on the semicircle C_R in the left (or right) half plane as $R \to \infty$, then for $m > 0$ (or $m < 0$)

$$\lim_{R \to \infty} \int_{C_R} f(z)e^{mz} \, dz = 0. \tag{9.13.2}$$

9.14 Forced harmonic waves and the radiation condition

As an application of complex variables to the Fourier transform we examine the Green function of a harmonically excited string with a small damping

$$\frac{1}{c^2}\frac{\partial^2 u}{\partial t^2} = \frac{\partial^2 u}{\partial x^2} - \alpha \frac{\partial u}{\partial t} + \delta(x)e^{-i\omega t} \tag{9.14.1}$$

for $-\infty < x < \infty$, where the real part of the final result is to be taken. Let the excitation begin at $t \sim -\infty$. For any finite t the response is

expected to be simple harmonic with the same frequency. Therefore we seek a solution
$$u(x,t) = \overline{u}(x)e^{-i\omega t}, \qquad (9.14.2)$$
where $\overline{u}(x)$ satisfies
$$\frac{\partial^2 \overline{u}}{\partial x^2} + \left(\frac{\omega^2}{c^2} + i\alpha\omega\right)\overline{u} = -\delta(x). \qquad (9.14.3)$$

Let
$$\frac{\omega}{c} = K \quad \text{and} \quad 2\epsilon = \alpha c \ll 1,$$
then
$$\frac{\partial^2 \overline{u}}{\partial x^2} + \left(K^2 + 2i\epsilon K\right)\overline{u} = -\delta(x).$$

The exponential Fourier transform of $\overline{u}$ is found to be
$$U(k) = \frac{1}{k^2 - (K^2 + 2i\epsilon K)}.$$

The inverse transform is
$$\overline{u}(x) = \frac{1}{2\pi}\int_{-\infty}^{\infty} \frac{e^{ikx}\,dk}{k^2 - (K^2 + 2i\epsilon K)}. \qquad (9.14.4)$$

Let us write
$$K^2 + 2i\epsilon K = (K + i\epsilon)^2 + \epsilon^2 \equiv K_o^2 + \epsilon^2,$$
where $K_o = K + i\epsilon$. Ignoring $O(\epsilon^2)$, we may write (9.14.4) as
$$\overline{u}(x) = \frac{1}{2\pi}\int_{-\infty}^{\infty} \frac{e^{ikx}\,dk}{k^2 - K_o^2}. \qquad (9.14.5)$$

To evaluate the integral in (9.14.5) we pass to the complex k plane. There are two poles in the integrand: one at $k = K_o = K + i\epsilon$ slightly above the real k axis, and one at $k = -K_o = -K - i\epsilon$ slightly below, as shown in Figure 9.19.

For $x > 0$ we choose a half circle C_R in the upper half plane to form a closed path C with the real axis. By Jordan's lemma this integral along C_R vanishes, hence only the pole at $k = -K_o$ matters, and by Cauchy's integral formula
$$\overline{u}(x) = \frac{1}{2\pi}\oint_C \frac{e^{ikx}\,dk}{(k - K_o)(k + K_o)}$$
$$= -\frac{ie^{-iK_o x}}{2K_o} \simeq \frac{i}{2}\frac{e^{iKx}}{K}e^{-\epsilon x}, \quad x > 0.$$

9.14 Forced harmonic waves and the radiation condition

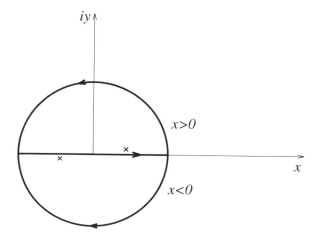

Fig. 9.19. Path of Fourier integral for damped radiation.

For $x < 0$ one closes the contour by a half circle in the lower half plane. After a similar application of the Cauchy integral formula and Jordan's lemma the result is

$$\bar{u} \cong \frac{i}{2}\frac{e^{-iKx}}{K}e^{\epsilon x}, \quad x < 0.$$

Thus damping causes the displacement to decay exponentially. Together with the time factor,

$$u(x,t) \cong \frac{i}{2K}e^{iK(|x|-ct)}e^{-\epsilon|x|}.$$

The solution represents a slightly attenuated outgoing wave.

In the limit of zero damping the solution is

$$u(x,t) = \frac{i}{2K}e^{iK(|x|-ct)}, \qquad (9.14.6)$$

which represents an unattenuated outgoing wave.

In many wave radiation problems we often want simple harmonic responses in a system without damping. Formal application of the Fourier transform usually leads to an inverse Fourier integral whose integrand has real poles, hence the integral is undefined. In the present example if damping is ignored from the outset the inverse transform would be

$$u(x) = \frac{1}{2\pi}\int_{-\infty}^{\infty}\frac{e^{ikx}\,dk}{k^2 - K^2}$$

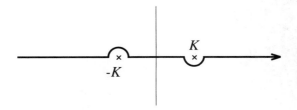

Fig. 9.20. Indented contour Γ for undamped radiation.

in which the integrand is singular at $k = \pm K$. To define the Fourier integral we must require the so-called *radiation condition* that waves due to local disturbances be outgoing at infinity. This requirement can be satisfied only if the path of the Fourier integral is indented above the pole at $-K$ and below the pole at K, as shown by Γ in Figure 9.20, as can be readily verified by Cauchy's integral formula. Let us show that this indented contour can also be reached by starting from weak damping $0 < \epsilon \ll 1$ and then letting $\epsilon \to 0$.

In taking the limit of zero damping the pole at $K_o = K + i\epsilon$ approaches the real k axis from above while the pole at $-K_o = -K - i\epsilon$ approaches the real k axis from below. When $\epsilon = 0$ the poles lie on the real axis. In order not to change the value of the line integral we must not allow the poles to cross the contour. Therefore, as the poles become real, the real axis must yield by indenting above the pole at $-K$ and below the pole at K. These indentations can be of any form, but semicircles are convenient; the resulting path is Γ, as shown in Figure 9.20. Thus for undamped waves the inverse Fourier transform (9.14.5) must be interpreted as the line integral along Γ in the complex k plane

$$\bar{u}(x) = \frac{1}{2\pi} \int_\Gamma \frac{e^{ikx}\,dk}{k^2 - K^2}. \tag{9.14.7}$$

9.15 Taylor and Laurent series, extended Liouville theorem and Cauchy's residue theorem

Taylor expansion theorem *If $f(z)$ is analytic inside and on C, which is a circle centered at $z = a$, then $f(z)$ can be expanded as an ascending power series about a:*

$$f(z) = \sum_{n=0}^{\infty} \frac{f^{(n)}(a)}{n!}(z-a)^n, \tag{9.15.1}$$

9.15 Taylor and Laurent series

where $f^{(n)}(a)$ is given by (9.9.4)

$$f^{(n)}(a) \equiv \left(\frac{d^n f}{dz^n}\right)_{z=a} = \frac{n!}{2\pi i}\oint_C \frac{f(\zeta)\,d\zeta}{(\zeta-a)^{n+1}}. \tag{9.15.2}$$

To prove Taylor's series theorem we start from Cauchy's formula

$$f(z) = \frac{1}{2\pi i}\oint_C \frac{f(\zeta)\,d\zeta}{\zeta - z} \tag{9.15.3}$$

with z in C. Let us first rewrite the integrand in (9.15.3)

$$\frac{1}{\zeta - z} = \frac{1}{(\zeta - a) - (z - a)} = \frac{1}{\zeta - a}\frac{1}{1 - \frac{z-a}{\zeta-a}}.$$

Since

$$\left|\frac{z-a}{\zeta-a}\right| < 1,$$

the last expression may be expressed as an absolutely convergent series

$$\frac{1}{\zeta - a} = \frac{1}{\zeta - a}\left\{1 + \frac{z-a}{(\zeta-a)} + \left(\frac{z-a}{\zeta-a}\right)^2 + \cdots + \left(\frac{z-a}{\zeta-a}\right)^n + \cdots\right\}.$$

Hence,

$$f(z) = \frac{1}{2\pi i}\oint_C \frac{f(\zeta)\,d\zeta}{\zeta - a}\left\{1 + \frac{z-a}{\zeta-a} + \cdots + \left(\frac{z-a}{\zeta-a}\right)^n + \cdots\right\}$$

$$= \frac{1}{2\pi i}\left\{\oint_C \frac{f(\zeta)\,d\zeta}{\zeta - a} + (z-a)\oint_C \frac{f(\zeta)\,d\zeta}{(\zeta-a)^2}\right.$$

$$\left. + \cdots + (z-a)^n \oint_C \frac{f(\zeta)\,d\zeta}{(\zeta-a)^{n+1}} + \cdots\right\}.$$

Making use of (9.15.2), we get

$$f(z) = \sum_{n=0}^{\infty} \frac{f^{(n)}(a)}{n!}(z-a)^n,$$

which is (9.15.1). Thus, a function analytic near a can be expanded as a Taylor series in power of $z - a$. The series coefficients are proportional to the derivatives evaluated at a.

Recall that analyticity is defined by the Cauchy–Riemann conditions, which involve only first derivatives. Now the Taylor series theorem assures that an analytic function has derivatives of all orders, with $f^{(n)}$ given by (9.15.2). An analytic function is, therefore, infinitely differentiable.

With Taylor's expansion theorem and Cauchy's inequality, we may now deduce:

Extended Liouville's theorem If $f(z)$ is analytic for all finite z and behaves as
$$f(z) = O(z^m)$$
when $|z| \to \infty$, then $f(z)$ is a polynomial of order no greater than m.

Since $f(z)$ is analytic for all finite z, it can be expanded as a Taylor series
$$f(z) = \sum_{n=0}^{\infty} \frac{f^{(n)}(0)}{n!} z^n = \sum_{n=0}^{\infty} a_n z^n.$$

From Cauchy's inequality (9.9.5)
$$a_n \leq \frac{M}{r^n},$$
where M is the upper bound of $|f(z)|$ on a circle of any radius r. We now let r be the radius of a large circle centered at the origin. By assumption $M = B r^m$, where B is some constant. It follows that
$$a_n \leq O\left(\frac{1}{r^{n-m}}\right)$$
for large r. Let $r \to \infty$, then a_n must vanish for all $n > m$, implying that $f(z)$ is a polynomial of degree no greater than m.

This theorem will be applied in Chapter 12.

If $f(z)$ has a pole at a, the Taylor expansion theorem is not valid and we need instead the following:

Laurent expansion theorem Let the region D be bounded by two concentric circles C_1 and C_2 centered at a, as shown in Figure 9.21. If $f(z)$ is analytic in D as well as on C_1 and C_2, then
$$f(z) = \sum_{n=0}^{\infty} a_n (z-a)^n + \sum_{n=1}^{\infty} b_n (z-a)^{-n} \tag{9.15.4}$$
with the coefficients given by
$$a_n = \frac{1}{2\pi i} \oint_{C_1} \frac{f(\zeta)\, d\zeta}{(\zeta - a)^{n+1}}, \quad b_n = \frac{1}{2\pi i} \oint_{C_2} \frac{f(\zeta)\, d\zeta}{(\zeta - a)^{1-n}}. \tag{9.15.5}$$

Let C_ϵ be the small circle with an infinitesimal radius centered at z, and let D' denote the region D excluding the disc formed by C_ϵ. Then

9.15 Taylor and Laurent series

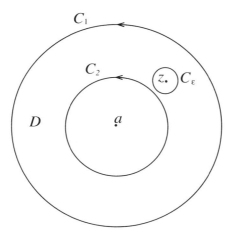

Fig. 9.21. Concentric circles around a.

$f(\zeta)/(\zeta - z)$ is regular within D' and on its boundaries C_1, C_2 and C_ϵ. By Cauchy's integral theorem we have

$$\oint_{C_1} \frac{f(\zeta)\,d\zeta}{\zeta - z} = \oint_{C_2} \frac{f(\zeta)\,d\zeta}{\zeta - z} + \oint_{C_\epsilon} \frac{f(\zeta)\,d\zeta}{\zeta - z},$$

where all contour integrals are counterclockwise. Clearly,

$$\oint_{C_\epsilon} \frac{f(\zeta)\,d\zeta}{\zeta - z} = 2\pi i f(z)$$

according to Cauchy's integral formula, hence

$$f(z) = \oint_{C_1} \frac{f(\zeta)\,d\zeta}{\zeta - z} - \oint_{C_2} \frac{f(\zeta)\,d\zeta}{\zeta - z}.$$

Now for ζ on C_1

$$\left|\frac{z-a}{\zeta-a}\right| < 1,$$

therefore $1/(\zeta - z)$ may be expanded as an absolutely convergent series

$$\frac{1}{\zeta - z} = \frac{1}{\zeta - a - (z-a)} = \frac{1}{(\zeta - a)\left(1 - \frac{z-a}{\zeta - a}\right)}$$

$$= \frac{1}{\zeta - a} + \frac{z-a}{(\zeta - a)^2} + \cdots + \frac{(z-a)^n}{(\zeta - a)^{n+1}} + \cdots.$$

On the other hand, for ζ on C_2

$$\left|\frac{\zeta - a}{z - a}\right| < 1,$$

$1/(\zeta - z)$ can be expanded as

$$\frac{1}{\zeta - z} = \frac{1}{\zeta - a - (z - a)} = -\frac{1}{(z - a)\left(1 - \frac{\zeta - a}{z - a}\right)}$$

$$= \frac{1}{z - a} + \frac{\zeta - a}{(z - a)^2} + \cdots + \frac{(\zeta - a)^{n-1}}{(z - a)^n} + \cdots.$$

The expansion theorem of Laurent then follows by applying Cauchy's integral formula.

Corollary: Cauchy's residue theorem *If $f(z)$ can be expanded as a Laurent series around a, which lies inside a closed curve C, then*

$$\oint_C f(z)\,dz = 2\pi i b_1,$$

where b_1 is the coefficient of the term $1/(z - a)$ and is called the residue.

This corollary is the immediate consequence of Cauchy's integral formula and (9.8.5).

9.16 More on contour integration

In many practical problems the integrand of a contour integral is the ratio of two regular functions, and the poles are the zeros of the denominator, i.e.,

$$I = \oint_C \frac{f(z)}{g(z)}\,dz,$$

where $g(a) = 0$ with a lying inside the contour C.

Assume first that

$$g'(a) \neq 0, \qquad f(a) \neq 0.$$

We may expand both f and g in the Taylor series to get

$$\frac{f(z)}{g(z)} = \frac{f(a) + f'(a)(z-a) + \cdots}{(z-a)g'(a) + \frac{1}{2}(z-a)^2 g''(a) + \cdots}$$

$$= \frac{f(a)}{(z-a)g'(a)} \frac{1 + \frac{f'(a)}{f(a)}(z-a) + \cdots}{1 + \frac{z-a}{2} \frac{g''(a)}{g'(a)} + \cdots}$$

$$= \frac{f(a)}{g'(a)} \frac{1}{z-a} \left\{ 1 + \left[\frac{f'(a)}{f(a)} - \frac{1}{2} \frac{g''(a)}{g'(a)} \right] (z-a) + \cdots \right\}.$$

It follows by the residue theorem that

$$\oint_C \frac{f(z)}{g(z)} dz = 2\pi i \frac{f(a)}{g'(a)}. \tag{9.16.1}$$

If $g(a) = g'(a) = 0$ but $g''(a) \neq 0$, then

$$g(z) = \frac{g''(a)}{2}(z-a)^2 + \frac{g'''}{6}(z-a)^3 + \cdots,$$

and the integrand is expanded as a Laurent series

$$\frac{f(z)}{g(z)} = \frac{f(a) + f'(a)(z-a) + \frac{f''(a)}{2}(z-a)^2 + \cdots}{\frac{g''(a)}{2}(z-a)^2 + \frac{g'''(a)}{6}(z-a)^3 + \cdots}$$

$$= \frac{f(a) \left[1 + \frac{f'(a)}{f(a)}(z-a) + \frac{f''(a)}{2f(a)}(z-a)^2 + \cdots \right]}{\frac{g''(a)}{2}(z-a)^2 \left[1 + \frac{1}{3}\frac{g'''(a)}{g''(a)}(z-a) + \cdots \right]}$$

$$= \frac{f(a)}{\frac{g''(a)}{2}(z-a)^2} \left\{ 1 + \left[\frac{f'(a)}{f(a)} - \frac{1}{3}\frac{g'''(a)}{g''(a)} \right] (z-a) + O(z-a)^2 \right\}.$$

The residue comes from the second term in the Laurent series

$$I = 2\pi i \frac{f(a)}{\frac{g''(a)}{2}} \left[\frac{f'(a)}{f(a)} - \frac{1}{3}\frac{g'''(a)}{g''(a)} \right]$$

$$= \frac{4\pi i}{g''(a)} \left[f'(a) - \frac{1}{3}\frac{f(a)g'''(a)}{g''(a)} \right]. \tag{9.16.2}$$

In particular, if $f(a) = 0$ but $f'(a) \neq 0$

$$I = \frac{4\pi i}{g''(a)} f'(a). \tag{9.16.3}$$

Similarly, if $g(a) = g'(a) = g''(a) = 0$ but $g'''(a) \neq 0$, one should expand $f(z)$ near a up to $f''(a)(z-a)^2$ and get the Laurent expansion of the ratio f/g. The residue is then obtained from the term proportional to $1/(z-a)$. The important thing is not to forget to expand $f(z)$.

Exercises

9.1 For the complex function
$$f(z) = \sqrt{(z^2+1)(z-2)}.$$
Choose two straight-line branch cuts with one along $-i$ to i and the other along 2 to ∞. Define the branch and give the expressions along two opposite sides of each branch cut.

9.2 The function
$$f(z) = \sqrt{(z^2-1)(z^2-9)}$$
has branch points at $z = \pm 1, \pm 3$. Define a branch with cuts along $-\infty < x < -3$, $-1 < x < 1$ and $\infty > x > 3$. What is f if z approaches each cut from above and below?

9.3 Show that the complex potential
$$w(z) = -Q \ln\left(\sinh\frac{\pi z}{2a}\right)$$
satisfies the boundary condition
$$\operatorname{Im} w = \text{constant}, \qquad -\infty < x < \infty, \ y = \pm a.$$
What are the constants along $y = \pm a$? Regard $\operatorname{Im} w = $ constant as a streamline of an irrotational flow. Sketch several streamlines by either numerical computation or analytical reasoning. Describe the physical picture.

9.4 The following complex function arises from mixed boundary-value problems in plane elasticity, to be discussed in Chapter 12,
$$X(z) = \frac{1}{\sqrt{z^2-1}}\left(\frac{z+1}{z-1}\right)^{i\mu},$$
where μ is real and positive. Define the branch cut to be along $-1 < \operatorname{Re} z < 1$ and show that
$$X^+(x) + \kappa X^-(x) = 0, \qquad -1 < x < 1,$$
where
$$\kappa = e^{2\pi\mu}$$
and
$$X^\pm(x) = \lim_{y\to\pm 0} X(z).$$

9.5 Show that
$$\int_0^\infty \frac{\ln x \, dx}{(x^2+1)^2} = \frac{\pi}{4}.$$

9.6 Evaluate the integral
$$\int_{-1}^1 \frac{dx}{\sqrt{1-x^2}(x^3+8)}.$$

9.7 Show that
$$\int_{-\infty}^\infty \frac{dx\, e^{-x^2}}{x-ia} = (\text{sgn } a) i\pi \, e^{a^2}(1 - \text{erf}|a|).$$

Deduce this result by shifting the integration path to a line from $-\infty + ia$ to $\infty + ia$ with a semicircular indentation above (below) $z = ia$ if $a < 0$ ($a > 0$). Separate the result into a principal-valued integral and an integral along the small semicircle, and evaluate the former by differentiating first with respect to a.

9.8 Verify the following equality:
$$\int_{-\infty}^\infty \frac{\sin \omega x \sinh x}{\cosh^2 x} dx = \frac{\pi \omega}{\cosh \frac{\pi \omega}{2}},$$

which appears in the theory of dynamical systems (Holmes, 1979). Replace x by the complex variable z and consider the contour integral along the rectangle with the lower side along the real axis from $-L$ to L, the upper side along a line parallel to the real axis from $-L + i\pi$ to $L + i\pi$, the right side from L to $L + i\pi$ and the left side from $-L$ to $-L + i\pi$. Calculate the residue from the pole at $z = i\pi/2$ enclosed by the rectangle and take the limit of $L \to \infty$.

9.9 Evaluate the following integrals that arise from mixed boundary-value problems of two-dimensional elasticity, to be discussed in Chapter 12. Show that
$$I(n) = \frac{1}{2\pi i} \int_{-1}^1 x^n X^+(x) \, dx, \qquad n = 0, 1, 2, 3, \ldots,$$

where $X^+(x)$ has been defined in Exercise 9.4.

First show that $I(n)$ is equal to the contour integral
$$I(n) = \frac{\kappa}{2\pi(\kappa+1)} \oint_C \zeta^n X(\zeta) \, d\zeta,$$

where C is a counterclockwise contour encircling the branch cut from -1 to 1. Next, replace C by another great circle C_∞ and show that for $|z| \gg 1$, $X(z)$ can be expanded in Laurent series

$$X(z) = \frac{1}{z} + \frac{2i\mu}{z^2} + \frac{1-4\mu^2}{2z^3} + \frac{i\mu}{3}\frac{5-4\mu^2}{z^4} + \cdots.$$

Finally, use the residue theorem to find $I(n)$ for $n = 0, 1, 2,$ and 3.

9.10 The governing equation for the lateral displacement of a taut spring surrounded by an elastic medium and forced by a sinusoidal load concentrated at the origin is

$$-m\omega^2 V - T\frac{d^2V}{dx^2} + KV = A\delta(x), \qquad |x| < \infty.$$

Show that there is a cut-off frequency ω_c such that the response is localized if $\omega < \omega_c$ and involves propagating waves if $\omega > \omega_c$. Use exponential Fourier transform to solve for Green's function. For $\omega > \omega_c$ indent the path of Fourier integral of the inverse transform so as to satisfy the radiation condition.

9.11 Referring to (7.8.5) in the example of §7.8, use the symmetry of the integrand in α and rewrite the Fourier sine integral as an exponential Fourier integral from $-\infty$ to ∞. Identify all the poles on the imaginary axis in the α plane, then use Jordan's lemma and the residue theorem to change the Fourier integral to an infinite series. Verify the series solution by separation of variables.

9.12 Due to gravity, disturbances on or below the free surface of water induce waves on the surface. For a two-dimensional irrotational flow in water of depth h the velocity potential satisfies

$$\nabla^2 \Phi = 0, \qquad -h < y < 0, \ |x| < \infty$$

in the fluid. If the wave amplitude is small compared to the wavelength, the free-surface displacement $\zeta(x,t)$ is approximately related to Φ by

$$\frac{\partial \zeta}{\partial t} = \frac{\partial \Phi}{\partial y}, \qquad y = 0.$$

If the atmospheric pressure is known to be $p_a(x,t)$, then Bernoulli's equation on the free surface implies, approximately,

$$g\zeta + \frac{\partial \Phi}{\partial t} = -\frac{p_a}{\rho}, \qquad y = 0.$$

On the horizontal seabed the vertical velocity vanishes

$$\frac{\partial \Phi}{\partial y} = 0, \quad y = -h.$$

Consider the special forcing

$$p_a = \mathrm{Re}\left\{P\delta(x)e^{-i\omega t}\right\}$$

and let

$$\Phi = \mathrm{Re}\left[\phi(x,y)e^{-i\omega t}\right], \qquad \zeta = \mathrm{Re}\left[\eta(x)e^{-i\omega t}\right].$$

Solve ϕ and η by exponential Fourier transform. Indent the path of the inverse integral so that propagating waves at infinity are outgoing.

9.13 A protrusion on an otherwise horizontal river bed induces stationary surface waves on a running stream. Assume the streaming velocity to be constant U and the protrusion to be two-dimensional. In the coordinates system fixed with the river bed, the disturbance potential is governed by

$$\nabla^2 \phi = 0, \qquad |x| < \infty, \ -h + b(x) < y < 0,$$

where $b(x)$ is nonzero only in a finite region and $b/h \ll 1$. On the free surface $y = 0$, the boundary conditions are

$$U\frac{\partial \eta}{\partial x} = \frac{\partial \phi}{\partial y}$$

and

$$g\eta + U\frac{\partial \phi}{\partial x} = 0.$$

The atmospheric pressure has been assumed to vanish. On the river bed the normal velocity must vanish

$$\frac{\partial \phi}{\partial y} = U\frac{db}{dx}, \quad y = -h(x).$$

Solve the problem by exponential Fourier transform. In the inverse transform integral identify all poles and indent the path properly so that waves appear only downstream of the protrusion.

10
Laplace transform and initial value problems

The Laplace transform is another important member of the family of integral transforms. It is most often applied to time in differential equations describing transient phenomena and can reduce the number of independent variables by one. The formal procedure is similar to that of Fourier transforms. By its definition, however, the inverse Laplace transform must be evaluated in the complex plane. Contour integration therefore plays a pivotal role in this chapter, as will be demonstrated through several examples.

10.1 The Laplace transform

As a mathematical motivation we observe that there are many functions such as $x, x^2, \ldots$ for which the Fourier transform integrals, as defined in Chapter 7, are not convergent. But this difficulty can be avoided if we consider another integrand $g(x)$

$$g(x) = H(x)e^{-cx}f(x) = \begin{cases} e^{-cx}f(x), & x > 0 \\ 0, & x < 0 \end{cases}$$

with c being real and positive and $H(x)$ being the Heaviside step function

$$H(x) = \begin{cases} 1, & x > 0 \\ 0, & x < 0. \end{cases}$$

The Fourier transform of $g(x)$ is

$$G(\lambda) = \int_{-\infty}^{\infty} e^{-i\lambda x} g(x)\, dx = \int_{-\infty}^{\infty} e^{-cx} H(x) f(x) e^{-i\lambda x}\, dx$$

$$= \int_{0}^{\infty} e^{-i\lambda x} e^{-cx} f(x)\, dx = \int_{0}^{\infty} e^{-sx} f(x)\, dx$$

10.1 The Laplace transform

with $s = c + i\lambda$. The constant c is chosen to be large enough, say $c > a$, so that the last integral exists. The value of a depends on the behavior of $f(x)$ at large x. The inverse transform is

$$H(x)f(x)e^{-cx} = \frac{1}{2\pi}\int_{-\infty}^{\infty} e^{-i\lambda x} G(\lambda)\,d\lambda$$

$$f(x)H(X) = \frac{1}{2\pi}\int_{-\infty}^{\infty} e^{(c+i\lambda)x} G(\lambda)\,d\lambda.$$

Changing λ to $-i(s-c)$ so that $d\lambda = -ids$, we get

$$f(x)H(x) = -\frac{1}{2\pi i}\int_{c+i\infty}^{c-i\infty} e^{sx} G[-i(s-c)]\,ds.$$

Let us define

$$\overline{F}(s) = G[-i(s-c)]$$

and

$$F(x) = \begin{cases} f(x)H(x), & x > 0 \\ 0, & x < 0, \end{cases}$$

then

$$\overline{F}(s) = \int_0^{\infty} e^{-sx} F(x)\,dx$$

$$F(x) = \frac{1}{2\pi i}\int_{c-i\infty}^{c+i\infty} e^{sx} \overline{F}(s)\,ds,$$

where the path of integration with respect to s is a vertical line parallel to and on the right of the imaginary axis in the complex s plane. $\overline{F}(s)$ is called the Laplace transform of $F(x)$, while $F(x)$ is the inverse Laplace transform of $\overline{F}(s)$. The integral operators of the Laplace and the inverse Laplace transforms are sometimes denoted by $\mathcal{L}$ and $\mathcal{L}^{-1}$, respectively.

In most physical applications the variable x stands for time. Therefore we shall rewrite the preceding pair of formulas as

$$\overline{F}(s) \equiv \mathcal{L}(F(t)) \equiv \int_0^{\infty} e^{-st} F(t)\,dt \qquad (10.1.1)$$

$$F(t) \equiv \mathcal{L}^{-1}(\overline{F}(s)) \equiv \frac{1}{2\pi i}\int_{c-i\infty}^{c+i\infty} e^{st}\overline{F}(s)\,ds. \qquad (10.1.2)$$

For brevity we shall occasionally denote the vertical path of the inversion integral by Γ.

Let us illustrate by an elementary example the role of contour integration in evaluating the inverse Laplace transform. Consider

$$f(t) = \begin{cases} \cos\omega t, & t > 0 \\ 0, & t < 0. \end{cases}$$

The Laplace transform is

$$\overline{f}(s) = \int_0^\infty e^{-st} \cos\omega t = \frac{1}{2}\int_0^\infty e^{-st}\left(e^{i\omega t} + e^{-i\omega t}\right)dt$$

$$= \frac{1}{2}\int_0^\infty \left(e^{-(s-i\omega)t} + e^{-(s+i\omega)t}\right)dt$$

$$= \frac{1}{2}\left(\frac{1}{s-i\omega} + \frac{1}{s+i\omega}\right) = \frac{s}{s^2+\omega^2}.$$

Let us check that the inverse Laplace transform of $\overline{f}(s)$ is indeed the original $f(t)$. By definition

$$f(t) = \frac{1}{2\pi i}\int_{c-i\infty}^{c+i\infty} \frac{e^{st}\,ds}{s^2+\omega^2}. \tag{10.1.3}$$

There are two simple poles at $s = \pm i\omega$ in the complex s plane. For $t < 0$ we close the contour by a large semicircle on the right of Γ, as shown in Figure 10.1. Since no poles are inside the contour, the contour integral vanishes. Since $t < 0$ and $(s^2+\omega^2)^{-1} \to 0$ as $s \to \infty$, Jordan's lemma

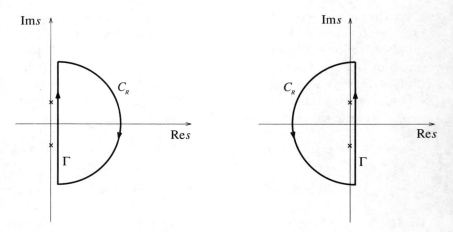

Fig. 10.1. Contours for (10.1.3).

10.2 Derivatives and the convolution theorem

assures us that the integral along the semicircle vanishes in the limit of $R \to \infty$. Therefore,
$$f(t) = 0, \quad t < 0.$$

For $t > 0$ we close the contour on the left (Figure 10.1). Since there are two poles inside the contour, Cauchy's integral formula gives

$$\frac{1}{2\pi i}\left(\int_\Gamma + \int_{C_R}\right) \frac{e^{st} s\, ds}{s^2 + \omega^2} = \frac{1}{2\pi i} \oint \frac{e^{st} s\, ds}{s^2 + \omega^2}$$

$$= \frac{1}{2\pi i} \oint \frac{e^{st}}{2}\left(\frac{1}{s - i\omega} + \frac{1}{s + i\omega}\right) ds$$

$$= \frac{1}{2}\left(e^{i\omega t} + e^{-i\omega t}\right) = \cos \omega t.$$

Again, Jordan's lemma assures that the integral along the semicircle C_R vanishes, hence
$$f(t) = \cos \omega t, \quad t > 0,$$
which is precisely the original function.

10.2 Derivatives and the convolution theorem

As in the Fourier transform theory, the Laplace transform of a derivative is obtained by partial integration

$$\overline{\frac{df}{dt}} = \int_0^\infty \frac{df}{dt} e^{-st}\, dt = \left[fe^{-st}\right]_0^\infty + s \int_0^\infty f e^{-st}\, dt$$
$$= s\overline{f} - f(0), \tag{10.2.1}$$

$$\overline{\frac{d^2 f}{dt^2}} = \int_0^\infty \frac{d^2 f}{dt^2} e^{-st}\, dt = \left[\frac{df}{dt} e^{-st}\right]_0^\infty + s \int_0^\infty \frac{df}{dt} e^{-st}\, dt$$
$$= -\frac{df(0)}{dt} + s\left[fe^{-st}|_0^\infty - (-s)\int_0^\infty fe^{-st}\, dt\right]$$
$$= s^2 \overline{f}(s) - sf(0) - \frac{df(0)}{dt}. \tag{10.2.2}$$

By induction one can prove that

$$\overline{\frac{d^n f}{dt^n}} = s^n \overline{f}(s) - s^{n-1} f(0) - \cdots - \frac{d^{n-1} f(0)}{dt^{n-1}}. \tag{10.2.3}$$

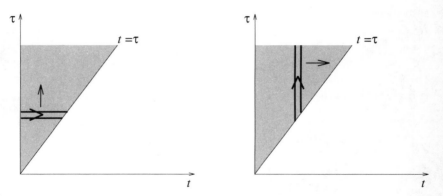

Fig. 10.2. Alternative ways of integration.

There is also a convolution theorem in the Laplace transform theory:

Convolution theorem *If the Laplace transforms of $f(t)$ and $g(t)$ are $\overline{f}(s)$ and $\overline{g}(s)$, respectively, then the inverse transform of $\overline{f}(s)\overline{g}(s)$ is the convolution integral*

$$\mathcal{L}^{-1}[\overline{f}(s)\overline{g}(s)] = \int_0^t f(t-\tau)g(\tau)\,d\tau. \tag{10.2.4}$$

This theorem is useful for finding the inverse transform of some $\overline{F}(s)$ if we can write $\overline{F}(s) = \overline{f}(s)\overline{g}(s)$ and the inverse of the two factors $\overline{f}$ and $\overline{g}$ are known.

To prove the convolution theorem we show that the Laplace transform of the convolution integral is equal to $\overline{f}\overline{g}$, i.e.,

$$\mathcal{L}\{\mathcal{L}^{-1}(\overline{f}\overline{g})\} = \int_0^\infty dt \int_0^t e^{-st} f(t-\tau)g(\tau)\,d\tau = \overline{f}\overline{g}.$$

As shown in Figure 10.2, the integration is over the shaded wedge in the t vs. τ plane. The same area can be covered by reversing the integration order first in t and then in τ. Therefore, the area integral above is equal to

$$\int_0^\infty g(\tau)d\tau \int_\tau^\infty e^{-st} f(t-\tau)\,dt$$

$$= \int_0^\infty e^{-s\tau} g(\tau)d\tau \int_\tau^\infty e^{-s(t-\tau)} f(t-\tau)\,d(t-\tau)$$

$$= \int_0^\infty e^{-s\tau} g(\tau)d\tau \int_0^\infty e^{-s\tau'} f(\tau')\,d\tau'$$

$$= \overline{f}(s)\,\overline{g}(s).$$

The inverse transform of this result is the convolution theorem.

We now devote the remaining sections to examples of applications to ordinary and partial differential equations.

10.3 Coupled pendula

In elementary applications the routine of the Laplace transform can be illustrated by a problem involving just ordinary differential equations. Let two pendula of equal length L and mass m be connected by a spring of elastic constant k. If the horizontal displacements are $x_1(t)$ and $x_2(t)$, both of which are infinitesimal, Newton's law gives

$$m\ddot{x}_1 = -\frac{mg}{L}x_1 + k(x_2 - x_1) \qquad (10.3.1)$$

$$m\ddot{x}_2 = -\frac{mg}{L}x_2 + k(x_1 - x_2). \qquad (10.3.2)$$

The initial conditions are assumed to be

$$x_1(0) = x_2(0) = 0 \quad \text{and} \quad \dot{x}_1(0) = V, \quad \dot{x}_2(0) = 0. \qquad (10.3.3)$$

Applying the Laplace transform to (10.3.1) and (10.3.2) and invoking the initial condition, we get

$$m(s^2\overline{x}_1 - V) = -\frac{mg}{L}\overline{x}_1 + k(\overline{x}_2 - \overline{x}_1)$$

$$ms^2\overline{x}_2 = -\frac{mg}{L}\overline{x}_2 + k(\overline{x}_1 - \overline{x}_2).$$

Thus the Laplace transform reduces ordinary differential equations to algebraic equations. This advantage is typical of all *time-invariant systems*, i.e., systems described by differential equations with time-independent coefficients. The solutions are

$$\overline{x}_1(s) = \frac{V\left(s^2 + \frac{g}{L} + \frac{k}{m}\right)}{\left(s^2 + \frac{g}{L}\right)\left(s^2 + \frac{g}{L} + \frac{2k}{m}\right)}$$

and

$$\overline{x}_2 = \frac{\frac{Vk}{m}}{\left(s^2 + \frac{g}{L}\right)\left(s^2 + \frac{g}{L} + \frac{2k}{m}\right)}.$$

We only demonstrate the inverse transform for $\overline{x}_1$

$$x_1(t) = \frac{V}{2\pi i}\int_{c-i\infty}^{c+i\infty} \frac{e^{st}\left(s^2 + \frac{g}{L} + \frac{k}{m}\right)}{\left(s^2 + \omega_1^2\right)\left(s^2 + \omega_2^2\right)}\,ds,$$

where

$$\omega_1 = \sqrt{\frac{g}{L}}, \quad \omega_2 = \sqrt{\frac{g}{L} + \frac{2k}{m}}.$$

There are four poles on the imaginary axis

$$s = \pm i\omega_1, \quad s = \pm i\omega_2.$$

For $t < 0$ we introduce a closed contour by adding a semicircle on the right of the integration path Γ. Cauchy's integral formula and Jordan's lemma give the expected result

$$x_1(t) = 0, \quad t < 0.$$

For $t > 0$ we introduce a closed contour by adding a large semicircle on the left of the integration path (Figure 10.3). By Jordan's lemma the contour integral is equal to the inversion integral, and the residues from all four poles inside the contour give

$$x_1(t) = \frac{V}{2}\left(\frac{\sin\omega_1 t}{\omega_1} + \frac{\sin\omega_2 t}{\omega_2}\right).$$

Thus the first mass oscillates at two frequencies: one is characteristic of the pendulum when the spring is absent, and the other is affected by the spring and the masses.

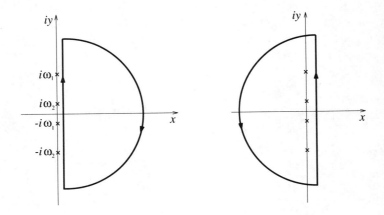

Fig. 10.3. Inversion contours for coupled pendula.

10.4 One-dimensional diffusion in a strip

We now return to partial differential equations and illustrate the use of the Laplace transform in a problem that can also be solved by separation of variables. Consider the concentration $C(x,t)$ of a diffusive substance governed by

$$\frac{\partial C}{\partial t} = k\frac{\partial^2 C}{\partial x^2}, \quad 0 < x < L, \quad t > 0, \tag{10.4.1}$$

subject to the boundary conditions

$$\frac{\partial C(0,t)}{\partial x} = 0, \quad C(L,t) = C_o \tag{10.4.2}$$

and the initial condition

$$C(x,0) = 0, \quad 0 < x < L. \tag{10.4.3}$$

Taking the Laplace transform of (10.4.1) with respect to t and using the initial condition, we get from the left-hand side

$$\overline{\frac{\partial C}{\partial t}} = s\overline{C}.$$

From the right-hand side we get

$$\overline{\frac{\partial^2 C}{\partial x^2}} = \frac{\partial^2 \overline{C}}{\partial x^2}$$

since the sequence of the time and space integrations can be interchanged. Therefore,

$$\frac{\partial^2 \overline{C}}{\partial x^2} - \frac{s}{k}\overline{C} = 0, \quad 0 < x < L.$$

The Laplace transforms of the boundary conditions are

$$\frac{\partial \overline{C}(0,s)}{\partial x} = 0$$

at the left end and

$$\overline{C}(L,s) = \int_0^\infty e^{-st} C(L,t)\, dt = \frac{C_o}{s}$$

at the right end. Thus the partial differential equation is reduced to an ordinary differential equation, and the initial-boundary-value problem to a simple boundary-value problem. The solution is

$$\frac{\overline{C}}{C_o} = \frac{1}{s}\frac{\cosh\sqrt{\frac{s}{k}}x}{\cosh\sqrt{\frac{s}{k}}L}.$$

The inverse transform is

$$\frac{C}{C_o} = \frac{1}{2\pi i}\int_{c-i\infty}^{c+i\infty}\frac{e^{st}\cosh\sqrt{\frac{s}{k}}x}{s\cosh\sqrt{\frac{s}{k}}L}\,ds.$$

There are simple poles at the origin

$$s = 0$$

and at

$$\sqrt{\frac{s}{k}}L = \left(n+\frac{1}{2}\right)\pi i$$

or

$$s = -s_n = -\frac{k}{4}\left[\left(n+\frac{1}{2}\right)\frac{\pi}{L}\right]^2, \quad n = 0,1,2,\ldots$$

on the negative real axis. Observe that $\sqrt{s}$ has a branch point at $s = 0$. However, by Taylor expansion

$$\cosh\sqrt{\frac{s}{k}}x = 1 + \frac{1}{2}\left(\frac{s}{k}x^2\right) + \cdots + \frac{1}{2n!}\left(\frac{s}{k}\right)^n x^{2n} + \cdots$$

is single-valued near $s = 0$. For $t < 0$ we close the contour by a semicircle in the right half plane, as shown in Figure 10.4. The inversion integral along the closed contour is zero because there is no singularity within. Use of Jordan's lemma yields the expected result

$$\frac{C}{C_o} = 0, \quad t < 0.$$

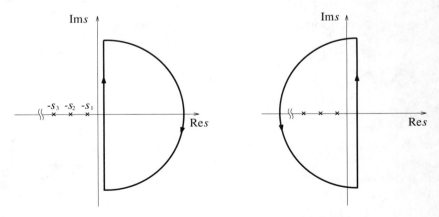

Fig. 10.4. Contour for one-dimensional diffusion.

For $t > 0$ we close the contour by a semicircle on the left plane. The line integral along the semicircle is again zero by Jordan's lemma. Now we must calculate the residues from the poles.

From the pole at $s = 0$ the residue is unity. From the pole at $s = -s_n$ the residue is

$$\frac{e^{-s_n t} \cosh\left[i\left(n + \frac{1}{2}\right)\frac{\pi x}{L}\right]}{\left(s\frac{\partial}{\partial s}\cosh\sqrt{\frac{s}{k}}L\right)_{-s_n}}$$

according to (9.15.5). The denominator is

$$\left(s\frac{\partial}{\partial s}\cosh\sqrt{\frac{s}{k}}L\right)_{s=-s_n} = \left(\frac{1}{2}\sqrt{\frac{s}{k}}L\sinh\sqrt{\frac{s}{k}}L\right)_{s=-s_n}$$

$$= \frac{1}{2}i\left(n+\frac{1}{2}\right)\pi\left[i\sin\left(n+\frac{1}{2}\right)\pi\right] = -\frac{1}{2}\left(n+\frac{1}{2}\right)\pi(-1)^n$$

since $\sinh i\alpha = i\sin\alpha$. The final result is the sum of all residues

$$\frac{C}{C_o} = 1 - \frac{4}{\pi}\sum_{n=0}^{\infty}\frac{(-1)^n}{(2n+1)}\cos\left[\left(n+\frac{1}{2}\right)\frac{\pi x}{L}\right]\cdot$$

$$\exp\left\{-kt\left[\left(n+\frac{1}{2}\right)\frac{\pi}{L}\right]^2\right\}. \qquad (10.4.4)$$

The reader may check the above result by separation of variables.

10.5 A string-oscillator system

As shown in Figure 10.5 an infinitely long taut string is attached at the origin to a mass m that slides in a vertical slot and is constrained by

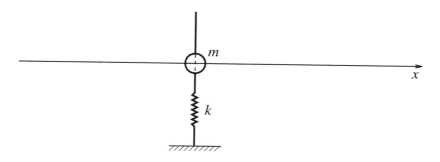

Fig. 10.5. A taut string tied to an oscillator.

a spring of constant k. If the mass is given an initial velocity U, what happens to the string and to the mass?†

The vertical displacement $V(x,t)$ of the string displacement is obviously even in x, and it is only necessary to limit our attention to $0 < x < \infty$. The governing equation for V is

$$\frac{\partial^2 V}{\partial t^2} = \frac{1}{c^2}\frac{\partial^2 V}{\partial x^2}, \quad 0 < x < \infty, \quad t > 0, \qquad (10.5.1)$$

where $c^2 = T/\rho$. The initial conditions are

$$V(x,0) = \frac{\partial V}{\partial t}(x,0) = 0, \quad x > 0, \qquad (10.5.2)$$

and the boundary condition at infinity is

$$V(\infty, t) \to 0, \quad t < \infty. \qquad (10.5.3)$$

Balancing the inertia of the mass with the spring force and the force from the string on both sides, we get for $V(0,t)$

$$m\left(\frac{\partial^2 V}{\partial t^2} + \omega^2 V\right)_{x=0} = -2T\left(\frac{\partial V}{\partial x}\right)_{x=0+}, \qquad (10.5.4)$$

where $\omega = \sqrt{\frac{k}{m}}$. Since the displacement of the mass is the same as the displacement of the string at $x = 0$, the initial conditions for the mass are

$$V(0,0) = 0, \quad \frac{\partial V(0,0)}{\partial t} = U. \qquad (10.5.5)$$

Now let us take the Laplace transform of all the governing equations and apply the initial conditions. From (10.5.1) and (10.5.2)

$$\frac{d^2 \overline{V}}{dx^2} - s^2 \overline{V} = 0.$$

The transform of (10.5.4) prescribes the boundary condition for $\overline{V}(x,s)$ at $x = 0$

$$m\left(s^2 + \omega^2\right)\overline{V} - 2T\frac{d\overline{V}}{dx} = mU, \quad x = 0. \qquad (10.5.6)$$

Clearly,

$$\overline{V}(\infty, s) = 0.$$

The solution for $\overline{V}$ is

$$\overline{V}(x,s) = \frac{Ue^{-s\frac{x}{c}}}{(s+\sigma)^2 + (\omega^2 - \sigma^2)}$$

† This example is taken from Carslaw and Jaeger (1963), p. 154.

10.5 A string-oscillator system

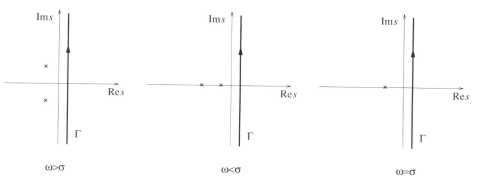

Fig. 10.6. Inversion path and poles.

with
$$\sigma^2 = \frac{T}{mc}.$$

By inverse transform the string displacement for $x > 0$ is

$$V(x,t) = \frac{U}{2\pi i} \int_{c-i\infty}^{c+i\infty} \frac{e^{s(t-\frac{x}{c})} ds}{(s+\sigma)^2 + (\omega^2 - \sigma^2)}. \qquad (10.5.7)$$

The integrand has simple poles at

$$s = -\sigma \pm i\sqrt{\omega^2 - \sigma^2}.$$

We now distinguish three cases.

Case (i): $\omega > \sigma$. The poles are complex conjugates, as shown in Figure 10.6. For $x > ct$, i.e., for an observer who is far to the right of the wave front, we close the contour on the right half plane and get, expectedly,

$$V(x,t) = 0, \quad x > ct.$$

For $x < ct$ we close the contour on the left half plane and use Cauchy's integral formula to find

$$\begin{aligned} V(x,t) &= \frac{U}{2\pi i} \oint \frac{e^{s(t-\frac{x}{c})} ds}{\left(s+\sigma+i\sqrt{\omega^2-\sigma^2}\right)\left(s+\sigma-i\sqrt{\omega^2-\sigma^2}\right)} \\ &= \frac{U \exp\left[\left(-\sigma - i\sqrt{\omega^2-\sigma^2}\right)\left(t-\frac{x}{c}\right)\right]}{-\sigma - i\sqrt{\omega^2-\sigma^2} + \sigma - i\sqrt{\omega^2-\sigma^2}} \\ &\quad + \frac{U \exp\left[\left(-\sigma + i\sqrt{\omega^2-\sigma^2}\right)\left(t-\frac{x}{c}\right)\right]}{-\sigma + i\sqrt{\omega^2-\sigma^2} + \sigma + i\sqrt{\omega^2-\sigma^2}} \\ &= \frac{U e^{-\sigma(t-\frac{x}{c})}}{2\sqrt{\omega^2-\sigma^2}} \sin\left[\sqrt{\omega^2-\sigma^2}\left(t-\frac{x}{c}\right)\right]. \qquad (10.5.8) \end{aligned}$$

There is a wave front advancing outward at the speed c. Behind the front the string oscillates at the frequency $\sqrt{\omega^2 - \sigma^2}$ and wavelength $2\pi c/\sqrt{\omega^2 - \sigma^2}$, while the wave amplitude attenuates exponentially behind the front.

Case (ii): $\omega < \sigma$. There is again no response if $x > ct$. For $x < ct$ Cauchy's integral formula gives

$$V(x,t) = \frac{U}{2\pi i} \oint \frac{e^{s(t-\frac{x}{c})} ds}{\left(s + \sigma + \sqrt{\sigma^2 - \omega^2}\right)\left(s + \sigma - \sqrt{\sigma^2 - \omega^2}\right)}$$

$$= \frac{Ue^{-\sigma(t-x/c)}}{2\sqrt{\sigma^2 - \omega^2}} \left[e^{\sqrt{\sigma^2-\omega^2}(t-x/c)} - e^{-\sqrt{\sigma^2-\omega^2}(t-x/c)}\right]$$

$$= \frac{Ue^{-\sigma\left(t-\frac{x}{c}\right)}}{\sqrt{\sigma^2 - \omega^2}} \sinh\left[\sqrt{\sigma^2 - \omega^2}\,(t - x/c)\right]. \tag{10.5.9}$$

Thus while the wave front advances at the speed c, there is no oscillation behind the attenuating front.

Case (iii): $\omega = \sigma$. For $x < ct$ one must cope with the second-order pole at $s = -\sigma$

$$V(x,t) = \frac{U}{2\pi i} \oint \frac{e^{s\left(t-\frac{x}{c}\right)} ds}{(s+\sigma)^2}$$

$$= \frac{U}{2\pi i} e^{-\sigma\left(t-\frac{x}{c}\right)} \oint \frac{e^{(s+\sigma)\left(t-\frac{x}{c}\right)}}{(s+\sigma)^2} ds.$$

The contour can be shrunk to a small circle around the pole by Cauchy's theorem, then the exponential term can be expanded for small $(s + \sigma)$

$$V(x,t) = \frac{Ue^{-\sigma\left(t-\frac{x}{c}\right)}}{2\pi i} \oint \frac{ds}{(s+\sigma)^2} \sum_{n=0}^{\infty} \frac{1}{n!}(s+\sigma)^n \left(t - \frac{x}{c}\right)^n. \tag{10.5.10}$$

The integrand is a Laurent series whose residue term gives, finally,

$$V(x,t) = U\left(t - \frac{x}{c}\right) e^{-\sigma\left(t-\frac{x}{c}\right)}.$$

10.6 Diffusion by sudden heating at the boundary

In Chapter 7 we demonstrated the use of Fourier transform in solving initial-boundary-value problems in a semi-infinite domain. Let us now apply the Laplace transform to a problem already solved in §7.7. A semi-infinite solid with zero temperature is brought to sudden contact with a heat reservoir at $t = 0$. How is heat diffused to the rest of the solid?

10.6 Diffusion by sudden heating at the boundary

The governing equations for the temperature $T(x,t)$ are:

$$\frac{\partial T}{\partial t} = k\frac{\partial^2 T}{\partial x^2}, \quad x, t > 0 \tag{10.6.1}$$

$$T(x,0) = 0 \tag{10.6.2}$$

$$T(0,t) = T_o \tag{10.6.3}$$

$$T(\infty,t) = 0. \tag{10.6.4}$$

Taking the Laplace transform of the differential equation with respect to t and making use of the initial condition, we get

$$\frac{d^2\overline{T}}{dx^2} - \frac{s}{k}\overline{T} = 0, \quad x > 0.$$

The Laplace transform of the solution should also diminish to zero at $x \to \infty$, hence the solution is

$$\overline{T} = Ae^{-\sqrt{s/k}\,x}.$$

The coefficient A is determined by the Laplace transform of the boundary condition at $x = 0$

$$\overline{T}(0,s) = \frac{T_o}{s}.$$

It follows that

$$\overline{T} = \frac{T_o}{s}e^{-\sqrt{s/k}\,x}$$

and the inverse transform is

$$T(x,t) = \frac{T_o}{2\pi i}\int_{c-i\infty}^{c+i\infty} e^{st-\sqrt{s/k}\,x}\,\frac{ds}{s},$$

where $c > 0$.

The origin in the complex plane of s is both a simple pole and a branch point. We first define a branch cut along the negative real axis. For $t < 0$ a large semicircular contour can be introduced in the right half plane of s. By Jordan's lemma it is easy to show that $T(x,t) = 0$, as expected. For $t > 0$ we choose a closed contour by a large semicircle C_R in the left half plane connected to two straight paths C_+ and C_- above and below the cut along the negative real axis and to a circle C_ϵ of infinitesimal

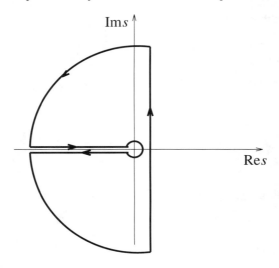

Fig. 10.7. Contour for the heat diffusion problem when $t > 0$.

radius ϵ surrounding the pole, as shown in Figure 10.7. Since there is no singularity inside the contour, Cauchy's integral formula applies

$$\left\{\int_{c-iR}^{c+iR} + \int_{C_R} + \int_{C_+} + \int_{C_\epsilon} + \int_{C_-}\right\} e^{st - \sqrt{s/k}\,x}\, \frac{ds}{s} = 0.$$

By Jordan's lemma, the integral along C_R vanishes in the limit of $R \to \infty$. Along C_ϵ we let $s = \epsilon e^{i\theta}$ so that

$$\int_{C_\epsilon} e^{st - \sqrt{s/k}\,x}\, \frac{ds}{s} = \lim_{\epsilon \to 0} \int_{-\pi}^{\pi} \frac{i\epsilon e^{i\theta}\, d\theta}{e^{i\theta}} = 2\pi i.$$

Along C_+ and C_-

$$s = ue^{i\pi},\, ue^{-i\pi} \quad \text{and} \quad \sqrt{s} = i\sqrt{u},\, -i\sqrt{u},$$

respectively, where u is real and positive. Hence the two corresponding integrals combine to give

$$\int_0^\infty e^{-ut}\left(-e^{i\sqrt{u/k}\,x} + e^{-i\sqrt{u/k}\,x}\right)\frac{du}{u} = -2i \int_0^\infty e^{-ut} \sin(\sqrt{u/k}\,x)\, \frac{du}{u}.$$

With $u = v^2$ the last integral can be rewritten as

$$2I \equiv 2 \int_0^\infty e^{-v^2 t} \sin\left(\frac{xv}{\sqrt{k}}\right) \frac{dv}{v}.$$

To facilitate the calculation we differentiate the integrand as follows:
$$\frac{dI}{d(x/\sqrt{k})} = J \equiv \int_0^\infty e^{-v^2 t} \cos\left(\frac{xv}{\sqrt{k}}\right) dv.$$

The integral J can be reduced to probability integrals by completing the squares
$$\begin{aligned} J &= \frac{1}{2}\int_0^\infty e^{-v^2 t}\left(e^{iv\sqrt{x/k}} + e^{-iv\sqrt{x/k}}\right) dv \\ &= \frac{1}{2}\int_{-\infty}^\infty e^{-x^2/4kt}\left\{\exp\left[-\left(v\sqrt{t} + \frac{ix}{2\sqrt{kt}}\right)^2\right]\right\} dv \\ &= \frac{1}{2\sqrt{t}}e^{-x^2/4kt}\int_{-\infty}^\infty e^{-\sigma^2} d\sigma \\ &= \frac{1}{2}\sqrt{\frac{\pi}{t}}e^{-x^2/4kt}. \end{aligned}$$

The integral I is, therefore,
$$I = \frac{1}{2}\sqrt{\frac{\pi}{t}}\int_0^{x/\sqrt{k}} e^{-\zeta^2/4t}\, d\zeta = \sqrt{\pi}\int_0^{x/2\sqrt{kt}} e^{-\xi^2}\, d\xi = \frac{\pi}{2}\mathrm{erf}\left(\frac{x}{2\sqrt{kt}}\right),$$

where use is made of the fact that $I(0) = 0$. Finally, the solution is
$$T(x,t) = T_o\left[1 - \mathrm{erf}\left(\frac{x}{2\sqrt{kt}}\right)\right], \qquad (10.6.5)$$

which is the same as (7.7.10).

Physically, the isothermals are given by the curves $x^2 = 2kt$, which are parabolas in the space-time diagram. At any time the temperature decreases monotonically from T_o to zero within a boundary layer whose thickness expands as $O(2\sqrt{kt})$.

10.7 Sound diffraction near the edge of a shadow – the parabolic approximation

Referring to Figure 10.8, we consider the two-dimensional diffraction of plane sound wave incident on a rigid barrier. Let the sound pressure be simple harmonic in time: $p(x, y, t) = \mathrm{Re}\left(P(x,y)e^{-i\omega t}\right)$, then P satisfies the Helmholtz equation
$$\nabla^2 P + k^2 P = 0, \quad k = \omega/C \qquad (10.7.1)$$

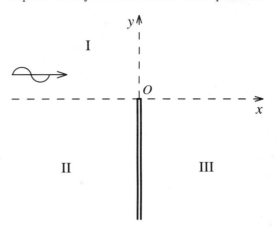

Fig. 10.8. Diffraction by a thin barrier.

and the boundary condition on the reflective barrier along the negative y axis

$$\frac{\partial P}{\partial x} = 0, \quad x = \pm 0, \quad -\infty < y < 0. \tag{10.7.2}$$

The exact solution, by A. Sommerfeld, of this problem is a landmark in mathematical physics. Here we shall discuss instead an appproximation that captures the essence of the physics and can be solved by the Laplace transform.

Heuristically, one can divide the x, y plane into three zones: the unobstructed zone I ($-\infty < x < \infty, y > 0$), the reflection zone II ($x < 0, y < 0$) and the shadow III ($x > 0, y < 0$). In each zone the pressure is crudely approximated by

$$P(x, y) = \begin{cases} P_o e^{ikx} & \text{in } I \\ P_o \left(e^{ikx} + e^{-ikx}\right) & \text{in } II \\ 0 & \text{in } III, \end{cases} \tag{10.7.3}$$

where P_o is a real constant. This approximation is inadequate since it is discontinuous across the x axis. Since a discontinuity is a poor representation of the derivative in the transverse direction, a remedy is to introduce a boundary layer in which the variation in the (y) direction is much more important than that in the longitudinal (x) direction. Consider in particular the neighborhood of the shadow edge (the positive x axis) and assume that the wave is essentially propagating forward

10.7 Sound diffraction near a shadow edge

except that its amplitude must be varying spatially

$$P(x,y) = A(x,y)e^{ikx}. \tag{10.7.4}$$

Substituting this assumption into Helmholtz's equation, we get

$$2ik\frac{\partial A}{\partial x} + \frac{\partial^2 A}{\partial x^2} + \frac{\partial^2 A}{\partial y^2} = 0.$$

As long as $kx \gg 1$, the ratio of the first two terms is

$$\frac{2ik\frac{\partial A}{\partial x}}{\frac{\partial^2 A}{\partial x^2}} = O(kx) \gg 1, \tag{10.7.5}$$

which suggests that

$$2ik\frac{\partial A}{\partial x} + \frac{\partial^2 A}{\partial y^2} \cong 0, \quad x > 0. \tag{10.7.6}$$

In order to match the amplitude away from the edge, we impose the boundary conditions that

$$A \to \begin{cases} P_o, & y \to \infty, \quad x > 0 \\ 0, & y \to -\infty, \quad x > 0. \end{cases} \tag{10.7.7}$$

Since the approximate equation is now parabolic, an initial condition along $x = 0$ can be imposed

$$A = \begin{cases} P_o, & y > 0 \\ 0, & y < 0. \end{cases} \tag{10.7.8}$$

This type of approximation, due to Fock (1960), is called the *parabolic approximation*.

Let us split A into real and imaginary parts

$$A = a(x,y) + ib(x,y)$$

so that (10.7.6) gives

$$2k\frac{\partial a}{\partial x} + \frac{\partial^2 b}{\partial y^2} = 0 \tag{10.7.9}$$

$$-2k\frac{\partial b}{\partial x} + \frac{\partial^2 a}{\partial y^2} = 0. \tag{10.7.10}$$

Eliminating b by cross differentiation, we get a real equation

$$4k^2\frac{\partial^2 a}{\partial x^2} + \frac{\partial^4 a}{\partial y^4} = 0, \tag{10.7.11}$$

which is the equation for the vibration of a beam with negligible rotatory inertia. It is easy to check that b satisfies the same equation. The initial condition (10.7.8) becomes

$$a(0, y) = P_o H(y) \tag{10.7.12}$$

$$b(0, y) = 0. \tag{10.7.13}$$

Because of (10.7.9), (10.7.13) implies

$$\frac{\partial a}{\partial x}(0, y) = 0. \tag{10.7.14}$$

A minor convenience can be gained by introducing

$$M = \frac{\partial a}{\partial y} \tag{10.7.15}$$

so that M satisfies (10.7.11) and the initial conditions

$$M(0, y) = P_o \delta(y) \tag{10.7.16}$$

and

$$\frac{\partial M}{\partial x}(0, y) = 0. \tag{10.7.17}$$

Let the Laplace transform be defined by

$$\overline{M}(s, y) = \int_0^\infty e^{-sx} M(x, y) \, dx.$$

The transform of (10.7.13) is

$$\frac{d^4 \overline{M}}{dy^4} + 4k^2 s^2 \overline{M} = 4k^2 s P_o \delta(y), \quad -\infty < y < \infty.$$

The solution of this equation can be inferred from $G(x)$ in (6.5.14).

$$\overline{M} = \frac{P_o}{2} \sqrt{\frac{k}{s}} e^{-\sqrt{ks}|y|} [\cos(\sqrt{ks}|y|) + \sin(\sqrt{ks}|y|)]. \tag{10.7.18}$$

The inverse transform can be evaluated as in the last section and is left as an exercise. Only the result is cited below:

$$M(x, y) = \frac{P_o}{2} \sqrt{\frac{k}{\pi x}} \left[\cos\left(\frac{ky^2}{2x}\right) + \sin\left(\frac{ky^2}{2x}\right) \right]. \tag{10.7.19}$$

10.7 Sound diffraction near a shadow edge

By integrating this result, we obtain

$$a = \int_{-\infty}^{y} M\,dy = \frac{P_o}{2}\sqrt{\frac{k}{\pi x}} \int_{-\infty}^{y} \left[\cos\left(\frac{ky^2}{2x}\right) + \sin\left(\frac{ky^2}{2x}\right)\right] dy$$

$$= \frac{P_o}{2} \int_{-\infty}^{y\sqrt{\frac{k}{\pi x}}} \left[\cos\left(\frac{\pi z^2}{2}\right) + \sin\left(\frac{\pi z^2}{2}\right)\right] dz.$$

Defining the Fresnel cosine and sine integrals by

$$C(z) = \int_0^z \cos\frac{\pi z^2}{2}\,dz, \qquad S(z) = \int_0^z \sin\frac{\pi z^2}{2}\,dz \qquad (10.7.20)$$

and using the fact that

$$C(\infty) = S(\infty) = \frac{1}{2}, \qquad (10.7.21)$$

we get

$$a(x,y) = \frac{P_o}{2}\left[1 + C\left(y\sqrt{\frac{k}{\pi x}}\right) + S\left(y\sqrt{\frac{k}{\pi x}}\right)\right]. \qquad (10.7.22)$$

The imaginary part b can be found straightforwardly by integrating (10.7.10). The result is

$$b(x,y) = \frac{P_o}{2}\left[-C\left(y\sqrt{\frac{k}{\pi x}}\right) + S\left(y\sqrt{\frac{k}{\pi x}}\right)\right], \qquad (10.7.23)$$

which may be combined with (10.7.22) to yield

$$P = a + ib = \frac{P_o}{\sqrt{2}}e^{-i\pi/4}\left\{\frac{1}{2} + C\left(y\sqrt{\frac{k}{\pi x}}\right) + i\left[\frac{1}{2} + S\left(y\sqrt{\frac{k}{\pi x}}\right)\right]\right\}. \qquad (10.7.24)$$

Using tabulated values of Fresnel integrals (Abramowitz and Stegun, 1964), we can compute the intensity ratio

$$\frac{|P|^2}{P_o^2} = \frac{1}{2}\left\{\left[\frac{1}{2} + C\left(y\sqrt{\frac{k}{\pi x}}\right)\right]^2 + \left[\frac{1}{2} + S\left(y\sqrt{\frac{k}{\pi x}}\right)\right]^2\right\}, \qquad (10.7.25)$$

which decays monotonically into the shadow as $y\sqrt{k/\pi x} \to -\infty$ but fluctuates about the ultimate limit of unity outside the shadow, as plotted in Figure 10.9. Contours of equal $|P|^2$ in the x,y plane are parabolas. In the similar problem of light diffraction these fluctuations cause alternatingly dark and bright bands, which are called the diffraction bands.

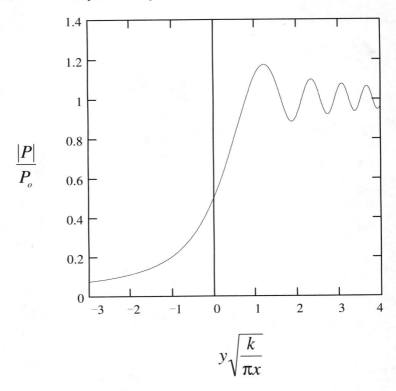

Fig. 10.9. Variation of wave intensity across the shadow boundary.

10.8 *Temperature in a layer of accumulating snow

At $t = 0$ fresh snow of temperature T_o starts to fall on a thick layer of pre-existing snow. The rate of accumulation is V m/sec. If the initial temperature in the old snow is uniform $T(x,0) = 0$, what is the temperature variation in the snow field at any $t > 0$?†

Let (x', t') denote the coordinates in a stationary frame of reference with x' being measured downward from the initial surface of the old layer. The temperature $T(x', t')$ field obeys the heat equation

$$\frac{\partial T}{\partial t'} = k\frac{\partial^2 T}{\partial x'^2}, \quad t' > 0 \qquad (10.8.1)$$

for all $x' > -Vt'$. In this frame of reference the top of the fresh snow is moving at the speed $-V$. Let us change to a moving coordinate system

† From Carrier and Pearson, 1976, p. 26.

10.8 Temperature in a layer of accumulating snow

fixed on the top surface of the fresh snow by the following transformation

$$x = x' + Vt', \quad t = t'. \tag{10.8.2}$$

For any function $T = T(x(x',t'), t(x',t'))$, the derivatives with respect to both coordinate systems are related by the chain rule

$$\frac{\partial T}{\partial t'} = \frac{\partial T}{\partial t}\frac{\partial t}{\partial t'} + \frac{\partial T}{\partial x}\frac{\partial x}{\partial t'} = \frac{\partial T}{\partial t} + V\frac{\partial T}{\partial x}$$

$$\frac{\partial T}{\partial x'} = \frac{\partial T}{\partial t}\frac{\partial t}{\partial x'} + \frac{\partial T}{\partial x}\frac{\partial x}{\partial x'} = \frac{\partial T}{\partial x}.$$

Consequently, in the moving coordinate system the heat equation reads

$$\frac{\partial T}{\partial t} + V\frac{\partial T}{\partial x} = k\frac{\partial^2 T}{\partial x^2}, \quad x > 0, t > 0. \tag{10.8.3}$$

The boundary conditions in the moving coordinates are

$$T(0,t) = T_o \tag{10.8.4}$$

and

$$T(\infty, t) = 0. \tag{10.8.5}$$

The initial condition is

$$T(x,0) = 0, \quad 0 < x < \infty. \tag{10.8.6}$$

The Laplace transform of (10.8.3) with respect to t gives

$$\frac{\partial^2 \overline{T}}{\partial x^2} - \frac{V}{k}\frac{\partial \overline{T}}{\partial x} - \frac{s}{k}\overline{T} = 0, \quad x > 0,$$

subject to the boundary conditions

$$\overline{T}(0, s) = T_o/s, \quad \overline{T}(\infty, s) = 0.$$

The solution for $\overline{T}(x, s)$ is

$$\overline{T}(x,s) = \frac{T_o}{s} \exp\left[\left(\frac{V}{2k} - \sqrt{\frac{s}{k} + \frac{V^2}{4k^2}}\right)x\right],$$

whose inverse transform is, formally,

$$\frac{T(x,t)}{T_o} = \frac{e^{\frac{V}{2k}x}}{2\pi i}\int_{c-i\infty}^{c+i\infty} \frac{e^{st}}{s} e^{-\sqrt{\frac{s}{k} + \frac{V^2}{4k^2}}\,x}$$

$$= \frac{e^{\frac{V}{2k}x - \frac{V^2}{4k}t}}{2\pi i}\int_{b-i\infty}^{b+i\infty} \frac{e^{pt}e^{-\sqrt{\frac{p}{k}}\,x}}{p - \frac{V^2}{4k}}\,dp \tag{10.8.7}$$

after the obvious transformation

$$s + \frac{V^2}{4k} = p, \qquad b = c + \frac{V^2}{4k}.$$

The inverse transform above can be found in mathematical handbooks (e.g., Erdelyi, 1953b, p. 246). It is instructive, however, to go through the analysis as an exercise in contour integration. The integrand has a pole at $p = V^2/4k$ and a branch point at $p = 0$. To make $\sqrt{p}$ single-valued we cut the complex p plane along the negative real axis and define $p = \rho e^{i\theta}$ with $-\pi < \theta \leq \pi$.

For $t > 0$ an infinitely large semicircle is introduced to connect Γ to the edges of the branch cut and form a closed contour, as shown in Figure 10.10. By Jordan's lemma there is no contribution from the semicircle so that

$$\frac{1}{2\pi i} \oint \frac{e^{pt} e^{-\sqrt{\frac{p}{k}} x}}{p - \frac{V^2}{4k}} dp$$

$$+ \frac{1}{2\pi i} \left\{ \int_{b-i\infty}^{b+i\infty} + \int_{C_+} + \int_{C_-} \right\} \frac{e^{pt} e^{-\sqrt{\frac{p}{k}} x}}{p - \frac{V^2}{4k}} dp.$$

The contour integral is equal to its residue at $p = V^2/4k$

$$e^{\frac{V^2 t}{4k} - \frac{Vx}{2k}},$$

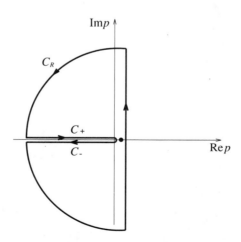

Fig. 10.10. Integration path in the p plane.

10.8 Temperature in a layer of accumulating snow

hence from (10.8.7)

$$\frac{T(x,t)}{T_o} = 1 - e^{\frac{Vx}{2k} - \frac{V^2 t}{4k}} J, \tag{10.8.8}$$

where J represents the sum of line integrals along $C_\pm$. Let us take $p = \rho e^{\pm i\pi}$ along $C_\pm$ so that

$$\frac{1}{2\pi i} \int_{C_+} = -\frac{1}{2\pi i} \int_\infty^0 \frac{e^{-\rho t} e^{-i\sqrt{\frac{\rho}{k}} x}}{-\left(\rho + \frac{V^2}{4k}\right)} d\rho = -\frac{1}{2\pi i} \int_0^\infty \frac{e^{-\rho t} e^{-i\sqrt{\frac{\rho}{k}} x}}{\rho + \frac{V^2}{4k}} d\rho$$

$$\frac{1}{2\pi i} \int_{C_-} = -\frac{1}{2\pi i} \int_0^\infty \frac{e^{-\rho t} e^{i\sqrt{\frac{\rho}{k}} x}}{-\left(\rho + \frac{V^2}{4k}\right)} d\rho = \frac{1}{2\pi i} \int_0^\infty \frac{e^{-\rho t} e^{i\sqrt{\frac{\rho}{k}} x}}{\rho + \frac{V^2}{4k}} d\rho,$$

whose sum is

$$J = \frac{1}{2\pi i} \left\{ \int_{C_+} + \int_{C_-} \right\} = \frac{1}{\pi} \int_0^\infty \frac{e^{-\rho t}}{\rho + \frac{V^2}{4k}} \sin \sqrt{\frac{\rho}{k}} x \, d\rho.$$

Let $\sqrt{\rho} = u^2$ in the integrand so that $\rho = u^2$ and $d\rho = 2u\,du$. The integral J becomes

$$J = \frac{1}{\pi} \int_0^\infty \frac{2u}{u^2 + \frac{V^2}{4k}} e^{-u^2 t} \sin \frac{ux}{\sqrt{k}} du$$

$$= \frac{1}{\pi} \int_{-\infty}^\infty \frac{u e^{-u^2 t}}{u^2 + \frac{V^2}{4k}} \sin \frac{ux}{\sqrt{k}} du$$

because of the evenness of the integrand. Since $u \cos ux/\sqrt{k}$ is odd in u, we can replace $u \sin ux/\sqrt{k}$ by $iu e^{-iux/\sqrt{k}}$ without affecting the integral so that

$$J = -\frac{1}{\pi i} \int_{-\infty}^\infty \frac{u}{u^2 + \frac{V^2}{4k}} e^{-u^2 t} e^{-\frac{ixu}{\sqrt{k}}} du$$

$$= -\frac{1}{2\pi i} \int_{-\infty}^\infty du\, e^{-\left(u^2 t + \frac{iux}{\sqrt{k}}\right)} \left(\frac{1}{u + iV/2\sqrt{k}} + \frac{1}{u - iV/2\sqrt{k}} \right).$$

Completing the squares by letting

$$u^2 t + \frac{iux}{\sqrt{k}} = \sigma^2 + \frac{x^2}{4kt} \quad \text{with} \quad \sigma = u\sqrt{t} + \frac{ix}{2\sqrt{kt}},$$

we obtain

$$J = -\frac{\exp\left(\frac{-x^2}{4kt}\right)}{2\pi i} \int_{-\infty+\frac{ix}{2\sqrt{kt}}}^{\infty+\frac{ix}{2\sqrt{kt}}} d\sigma\, e^{-\sigma^2} \left[\frac{1}{\sigma - \frac{i(x-Vt)}{2\sqrt{kt}}} + \frac{1}{\sigma - \frac{i(x+Vt)}{2\sqrt{kt}}}\right]$$

$$= -\frac{\exp\left(\frac{-x^2}{4kt}\right)}{2\pi i} \int_{-\infty+i\xi}^{\infty+i\xi} d\sigma\, e^{-\sigma^2} \left(\frac{1}{\sigma - i\xi_-} + \frac{1}{\sigma - i\xi_+}\right), \qquad (10.8.9)$$

where

$$\xi = \frac{x}{2\sqrt{kt}}, \qquad \xi_\pm = \frac{x \pm Vt}{2\sqrt{kt}}$$

so that $\xi_- < \xi < \xi_+$. In the complex σ plane the path of integration is a horizontal line above the real axis at the height $\xi = x/2\sqrt{kt}$. There are two imaginary poles in the integrand: $i\xi_+$ above the path and $i\xi_-$ below, as shown in Figure 10.11.

Now the integral

$$I = \int_{-\infty+i\xi}^{\infty+i\xi} \frac{d\sigma\, e^{-\sigma^2}}{\sigma - i\gamma}$$

has been evaluated in Exercise 9.7 with the result

$$I = -i\pi e^{\gamma^2}(1 + \mathrm{erf}\,\gamma)$$
$$= i\pi e^{\gamma^2}(-2 + \mathrm{erfc}\,\gamma) \quad \text{if} \quad \xi > \gamma$$

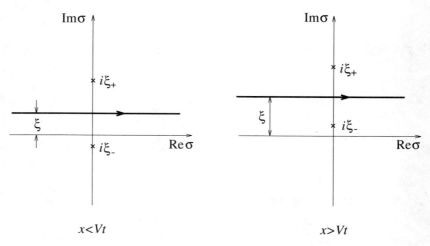

x<Vt   x>Vt

Fig. 10.11. Path of integration in the σ plane.

10.8 Temperature in a layer of accumulating snow

and
$$I = i\pi e^{\gamma^2}(1 - \text{erf}\gamma) = i\pi e^{\gamma^2}\text{erfc}\gamma \quad \text{if} \quad \xi < \gamma.$$

The final result for the temperature is, therefore,
$$\frac{T(x,t)}{T_o} = \frac{1}{2}\text{erfc}\left(\frac{x - Vt}{2\sqrt{kt}}\right) + \frac{1}{2}e^{\frac{Vx}{k}}\text{erfc}\left(\frac{x + Vt}{2\sqrt{kt}}\right). \tag{10.8.10}$$

As a check of the boundary condition at the top surface of the fresh snow where $x = 0$, we have
$$\frac{T(0,t)}{T_o} = \frac{1}{2}\left[1 - \text{erf}\left(\frac{-Vt}{2\sqrt{kt}}\right) + 1 - \text{erf}\left(\frac{Vt}{2\sqrt{kt}}\right)\right] = 1$$

because the error function is odd in its argument. Thus the boundary condition is verified. At the initial surface of the old snow $x = Vt$, the temperature is
$$\frac{T(Vt,t)}{T_o} = \frac{1}{2}\left\{1 + e^{\frac{V^2 t}{k}}\text{erfc}\left(\frac{V\sqrt{t}}{\sqrt{k}}\right)\right\}.$$

For large t the complementary error function approaches zero exponentially fast
$$\text{erfc}\left(\frac{V\sqrt{t}}{\sqrt{k}}\right) \sim \frac{e^{-\frac{V^2 t}{k}}}{V\sqrt{\frac{\pi t}{k}}}\left[1 + O\left(\frac{1}{t}\right)\right]$$

so that
$$\frac{T(Vt,t)}{T_o} \cong \frac{1}{2}\left(1 + \frac{1}{V}\sqrt{\frac{k}{\pi t}} + \cdots\right),$$

which has the limit of 1/2 as it should.

In the examples discussed in this chapter the inverse transforms can be explicitly evaluated. In many practical problems, alas, this is often not the case, and approximate analysis of the complex integral is necessary to extract information of physical interest. Various asymptotic techniques are available if either the variable t or some parameter is very large or very small. A comprehensive treatment of these more advanced techniques can be found in a number of fine texts such as Carlsaw and Jaeger (1963), Erdelyi (1953b), Carrier, Krook and Pearson (1966) and Nayfeh (1981). The last three references give excellent accounts of the steepest descent method, which is a showcase for the power of the complex function theory.

Exercises

10.1 Initially, the temperature in a semi-infinite medium is zero everywhere ($x > 0$). At the end $x = 0$ heat is radiated to an environment of temperature T_o according to the following law:

$$\frac{\partial T}{\partial x} = a(T - T_o).$$

Show that the temperature variation is

$$\frac{T}{T_o} = \text{erfc}\frac{x}{2\sqrt{kt}} - e^{ax+a^2kt}\text{erfc}\left(\frac{x}{2\sqrt{kt}} + a\sqrt{kt}\right).$$

10.2 Find the flow of heat in a semi-infinite rod $x > 0$ with the end at $x = 0$ kept at a prescribed temperature $T(0,t) = F(t)$ for $t > 0$, with (i) $F(t) = T_o \sin \omega t$, and (ii), $F(t) = T_o$, $0 < t < \tau$; $F(t) = 0, t > \tau$, where T_o and τ are constants. The initial temperature is zero everywhere.

10.3 An elastic bar of length L and cross-sectional area A is fixed at the end $x = 0$. Starting from $t = 0$, a constant force F is suddenly applied at the other end. Find the displacement $u(x,t)$ by the Laplace transform and check the result by separation of variables.

10.4 Consider Smoluchowski's model of the coagulation of suspended aerosols. A stationary spherical particle is in a cloud of like particles of the same radius. Let the number density $n(r,t)$ of particles obey the diffusion law in spherical polar coordinates

$$\frac{\partial n}{\partial t} = \frac{D}{r^2}\frac{\partial}{\partial r}\left(r^2 \frac{\partial n}{\partial r}\right), \quad r > a,$$

where D is the Brownian diffusivity due to random bombardment of air molecules, and a is the diameter of the sphere, i.e., the shortest distance between the centers of two particles. Deduce this equation directly by considering a control volume made of a spherical shell of radius r and thickness dr. If once two particles collide they do not separate again; a boundary condition is $n(a,t) = 0$. At infinity we must have $n(\infty,t) = n_o$. The initial condition is $n(r,0) = n_o$ for $r > a$. Solve for $n(r,t)$ by the Laplace transform and then calculate the rate of particle loss due to coagulation at $r = R$. (The governing equation can be reduced to the standard form for x,t. Try to solve it without resorting to such a transformation.)

10.5 Two semi-infinite solids of different materials are brought into contact along $x = 0$ at $t = 0$. Prior to contact the temperature is T_1 in one ($x < 0$) and T_2 in the other ($x > 0$). Find the subsequent temperature everywhere.

10.6 A tautly stretched string embedded in an elastic surrounding is governed by the following Klein-Gordon equation:

$$\frac{\partial^2 U}{\partial t^2} - c^2 \frac{\partial^2 U}{\partial x^2} + k^2 U = 0.$$

Find the fundamental solution G satisfying the equation

$$\frac{\partial^2 G}{\partial t^2} - c^2 \frac{\partial^2 G}{\partial x^2} + k^2 G = \delta(x - x')\delta(t - t')$$

for $-\infty < x < \infty$. The initial conditions are

$$G(x, t_o; x', t') = \frac{\partial G(x, t_o; x't')}{\partial t} = 0$$

for $t_o < t'$. Use the Laplace transform and the tabulated result

$$\frac{1}{2\pi i} \int_{c-i\infty}^{c+\infty} e^{-a\sqrt{s^2+b^2}} \frac{e^{st}\, ds}{\sqrt{s^2 + b^2}} = \begin{cases} 0, & 0 < t < a \\ J_0(b\sqrt{t^2 - a^2}), & t > a \end{cases}$$

to show that

$$G = \frac{1}{2c} J_0 \left(\frac{k}{c} \sqrt{c^2(t - t')^2 - (x - x')^2} \right) \quad \text{if} \quad |x - x'| < |t - t'|,$$

and G is equal to zero otherwise.

10.7 A wide river of uniform depth and slope has a rigid bottom for $x < 0$ but is laden with fine sediments for $x > 0$. The sediments are dislodged by turbulent fluctuations in water and are transported by the mean flow. In the simplest model, the flow velocity is constant and particles are of one size so that the sediment concentration satisfies the convective diffusion equation

$$U \frac{\partial C}{\partial x} - w \frac{\partial C}{\partial y} = D \frac{\partial^2 C}{\partial y^2}, \quad x > 0.$$

Here, w is the free-fall velocity of a particle in water and D the eddy diffusivity. Assume for simplicity that

$$C = C_o, \quad y = 0$$

$$wC + D \frac{\partial C}{\partial y} = 0, \quad y = h$$

with the initial condition that $C(0, y) = 0$. Find $C(x, y)$ by the Laplace transform.

10.8 A long beam of uniform cross section is at rest initially. Starting from $t = 0$, a constant torque M is applied at the end $x = 0$. Show that the bending moment elsewhere is

$$\frac{M}{2}\left[1 - C\left(\frac{x}{a\sqrt{2\pi t}}\right) - S\left(\frac{x}{a\sqrt{2\pi t}}\right)\right],$$

where $a^2 = \sqrt{EI/\rho A}$ with $E =$ Young's modulus, $I =$ moment of inertia of the cross section, $\rho =$ density per unit length, and $A =$ cross-sectional area. C and S are Fresnel integrals.

10.9 Perform the inverse transform of (10.7.18) to verify (10.7.19).

10.10 Referring to §10.7, let the reflected wave $P_o e^{-ikx}$ be discontinuous across the negative x axis (the border between regions I and II). Find by the parabolic approximation an improved solution describing the smooth transition in the boundary layer near the border.

10.11 Recall from §9.14 that the propagation of undamped sinusoidal waves requires the radiation condition. An alternative is to consider an initial-value problem with zero initial data, then let $t \to \infty$. Solve by the Laplace transform the following problem:

$$\frac{1}{c^2}\frac{\partial^2 u}{\partial t^2} = \frac{\partial^2 u}{\partial x^2} + \delta(x)e^{-i\omega t} \quad |x| < \infty, \quad t > 0$$

subject to

$$u(x, 0) = \frac{\partial u}{\partial t}(x, 0) = 0, \quad |x| < \infty.$$

Show that the physical solution is

$$\operatorname{Re} u(x, t) = \operatorname{Re}\left\{\frac{ic}{\omega}H\left(t - \frac{|x|}{c}\right)e^{i\omega\left(\frac{|x|}{c} - t\right)}\right\}.$$

11
Conformal mapping and hydrodynamics

11.1 What is conformal mapping?

In §9.2 mapping of a curve in z plane to another plane of ζ has been examined when the two are related by some specific analytic functions $\zeta = f(z)$. A general property of all analytic functions is that the mapping is *conformal*, meaning that a configuration in the z plane is locally similar to its image in the mapped plane. Specifically, an infinitesimal wedge near a point in the z plane is similar to its image near the image point in the ζ plane. Consider two line segments dz_1, dz_2 forming a wedge at z, and their images $d\zeta_1$ and $d\zeta_2$ in the ζ plane

$$d\zeta_1 = f'(z)dz_1, \qquad d\zeta_2 = f'(z)dz_2,$$

as sketched in Figure 11.1. Since $f(z)$ is analytic in the neighborhood of z, $d\zeta/dz = f'(z)$ exists and is unique. The ratio of the two segments and their images are related by

$$\frac{d\zeta_1}{d\zeta_2} = \frac{dz_1}{dz_2}. \tag{11.1.1}$$

In polar form let $dz_1 = |dz_1|e^{i\theta_1}$ and $d\zeta_1 = |d\zeta_1|e^{i\varphi_1}$, etc. Clearly,

$$\frac{|dz_1|}{|dz_2|} = \frac{|d\zeta_1|}{|d\zeta_2|} \quad \text{and} \quad \theta_1 - \theta_2 = \varphi_1 - \varphi_2 + 2n\pi.$$

It follows that the length ratio of two segments and the angle between them are preserved under mapping. In this sense the mapping by an analytic function is conformal.

Note first that (11.1.1) holds only when $f(z)$ is analytic. At any point where $f'(z)$ is either 0 or ∞, the mapping is not conformal. Also, in general the ratio $d\zeta/dz = f'(z)$ changes from one z to another, hence

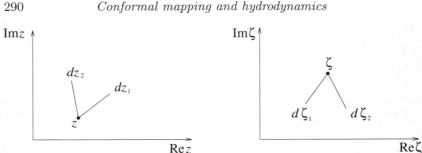

Fig. 11.1. Mapping of a small wedge by an analytic function.

the similarity is strictly a local feature. For any finite-sized configuration in one plane, the image need not look similar at all.

What is the use of conformal mapping in engineering problems? In this chapter examples will be selected from one area of applied mechanics that has been enriched by the mathematical technique: irrotational flows of incompressible fluids in a plane.

11.2 Relevance to plane potential flows

In steady irrotational flows of inviscid and incompressible fluids, one often wishes to find the complex potential $w(z)$, which is an analytic function of z in part of the complex z plane with the boundary condition that $\operatorname{Im} w = $ constant on a curve C. In general, this is not easy to achieve directly. However, if we can find an analytic relation $\zeta = f(z)$ that maps C to a much simpler curve C' in ζ plane, it may be easy to find an analytic function $w(\zeta)$ such that $\operatorname{Im} w = $ constant on C'. Since w is analytic in ζ, which is analytic in z, $w(\zeta(z))$ must be analytic in z, and is the desired solution.

As an elementary example let us find a potential flow in a right-angle corner, i.e., find a $w(z)$ analytic in the corner so that $\operatorname{Im} w(z) = 0$ on AOA', as shown in Figure 11.2. We use the simple analytic function $\zeta = z^2$ to map the corner in z to the upper half plane of ζ. Clearly, in the ζ plane $w = U\zeta$ with U real is analytic in ζ and satisfies the boundary condition $\operatorname{Im} w = 0$ on $\eta = 0$; $w = Uz^2$ is a solution. Thus the potential of a uniform flow in ζ corresponds to a corner flow in z.

The key to solving this type of potential flow problem is to find the analytic function that maps a region of complicated shape in z to the upper half plane of ζ. For reasonably simple geometries this task can be accomplished analytically by various *ad hoc* and ingeneous means,

11.3 Schwarz–Christoffel transformation

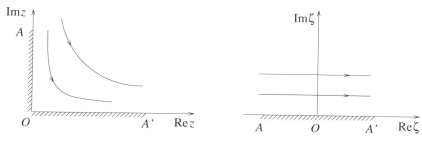

Fig. 11.2. Mapping for a potential flow in a corner.

but numerical methods must be turned to in general. For a collection of known transformations the reader may consult the dictionary compiled by Kober (1957). In the next section we shall discuss the celebrated Schwarz–Christoffel transformation, which is particularly suited for polygonal boundaries. Before going into specifics let us first explain how singularities representing sources and vortices are affected under conformal mapping.

Consider a source at z_o in the z plane. In the neighborhood of the source the complex potential is

$$w(z) = \frac{m}{2\pi} \ln(z - z_o) + \text{regular terms}.$$

Let $\zeta = f(z)$ map the domain in z onto the upper ζ plane. By definition the image of the source point is at $\zeta_o = f(z_o)$ near which one can expand

$$\zeta - \zeta_o = (z - z_o) f'(z_o) + \cdots.$$

As long as $f'(z_o) \neq 0$ or ∞, we have

$$w = \frac{m}{2\pi} \ln \left(\frac{\zeta - \zeta_o}{f'(z_o)} \right) + \text{regular terms}$$

$$= \frac{m}{2\pi} \ln(\zeta - \zeta_o) + \text{regular terms}.$$

Thus the singular part is unchanged. It may be concluded that sources, sinks and vortices for which m is replaced by $i\Gamma$ do not change their singular nature or strength under conformal mapping.

11.3 Schwarz–Christoffel transformation

Schwarz–Christoffel transformation is a systematic way of mapping the interior of a polygon in the z plane to an upper half plane of ζ; see Figure 11.3.

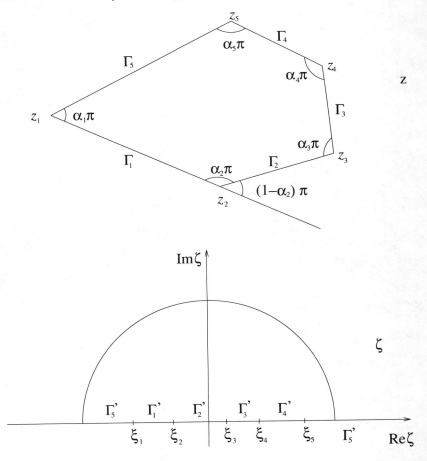

Fig. 11.3. Mapping the interior of a polygon in z plane to the upper half plane of ζ.

Let the vertices of the polygon in z be designated by $z_1 \ldots z_n$ and their images by $\xi_1 \ldots \xi_n$, whose positions are yet unknown. Note that for an n-sided polygon, the sum of all internal angles is

$$\sum_{k=1}^{n} \alpha_n \pi = (n-2)\pi,$$

hence

$$\sum_{k=1}^{n} \alpha_n = n-2. \tag{11.3.1}$$

11.3 Schwarz–Christoffel transformation

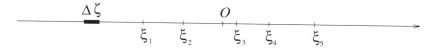

Fig. 11.4. $\Delta \zeta$ on Γ'_5.

Let us try a mapping function that is the solution of the differential equation

$$\frac{dz}{d\zeta} = \lim_{\Delta \to 0} \frac{\Delta z}{\Delta \zeta} = A \prod_{k=1}^{n} (\zeta - \xi_k)^{(\alpha_k - 1)}. \qquad (11.3.2)$$

In general the right-hand side is either 0 or ∞ at ξ_k so that $\zeta = \xi_k$ is a singular point about which mapping is not conformal. This singular behavior is a positive attribute that will enable us to flatten the corners of the polygon.

To establish the intended mapping it is enough to show that the real ξ axis can be mapped onto the sides of the polygon in the z plane. We proceed in the order of (i) the interior angles, (ii) the length ratio of the sides, (iii) the orientation of the polygon, and (iv) the absolute position of the polygon.

For any line segment Δz whose image is $\Delta \zeta$, we get from (11.3.2) that

$$\arg \Delta z - \arg \Delta \zeta = \arg A + \sum_{k=1}^{n} (\alpha_k - 1) \arg (\zeta - \xi_k).$$

Consider a segment $\Delta \zeta$ on the segment Γ'_5, as shown in Figure 11.4. Since the segment is to the left of all ξ_i, $i = 1, 2, \ldots, 5$,

$$\arg (\zeta - \xi_k) = \pi \quad \text{for} \quad k = 1, 2, \ldots, 5.$$

From (11.3.2)

$$(\arg \Delta z)_{\Gamma_5} = \arg A + \sum_{k=1}^{5} (\alpha_k - 1) \pi,$$

which is a constant. Thus the image Δz on Γ_5 of any $\Delta \zeta$ on Γ'_5 has the same argument. All parts of the line Γ_5 must have the same argument, and Γ_5 is a straight-line segment.

Next we consider any $\Delta \zeta$ on Γ'_1, as shown in Figure 11.5. Now ζ is to the right of ξ_1 but to the left of $\xi_2, \ldots, \xi_5$, hence

$$\arg (\zeta - \xi_1) = 0, \qquad \arg (\zeta - \xi_k) = \pi, \quad k = 2, 3, 4, 5$$

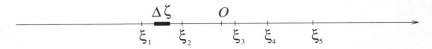

Fig. 11.5. $\Delta\zeta$ on Γ'_1.

and

$$(\arg \Delta z)_{\Gamma_1} = \arg A + \sum_{k=2}^{5} (\alpha_k - 1)\pi.$$

Again, the image of Γ'_1, i.e., Γ_1, is a straight line but its direction is different from that of Γ_5 by the counterclockwise angle

$$(\arg \Delta z)_{\Gamma_1} - (\arg \Delta z)_{\Gamma_5} = -\pi(\alpha_1 - 1) = (1 - \alpha_1)\pi.$$

Thus we have shown that $\zeta(z)$, which is the solution of the differential equation (11.3.2), maps a typical corner of the polygon in z onto a part of the real axis in the ζ plane.

Similar reasonings show that (11.3.2) maps the segments $\Gamma'_2, \Gamma'_3 \ldots$ of the real ζ axis onto sides of a certain polygon in the z plane with the same internal angles as the intended polygon.

We now take measure that the image polygon is similar to the intended one in the z plane. Let $L_n(L'_n)$ be the length of the nth side of the intended (image) polygon. To ensure similarity we require

$$\frac{L_1}{L'_1} = \frac{L_2}{L'_2} = \cdots \frac{L_{n-2}}{L'_{n-2}}, \qquad (11.3.3)$$

i.e., that the length ratio for $n - 2$ sides be the same. When these conditions are met, the proportionality of the other two sides is then guaranteed. Thus $n - 3$ conditions can be used to fix $n - 3$ of the points ξ_k on the real axis; three among them can be arbitrary.

By proper choice of the complex constant A the absolute size and orientation of the image polygon can be made identical to that of the given polygon.

To complete the mapping it is necessary to make the image polygon coincide with the given polygon. This goal can be accomplished by integrating (11.3.2)

$$z = z_o + \int A \prod_{k=1}^{n} (\zeta - \xi_k)^{\alpha_k - 1} d\zeta \qquad (11.3.4)$$

and fixing the integration constant z_o so that the mapped polygon, i.e., the image of ξ axis, coincides exactly with the intended polygon.

Suppose that the original polygon is open with one of its corners mapped to $\xi \to \pm\infty$. How does this affect the Schwarz–Christoffel transformation? Recall (11.3.2)

$$\frac{dz}{d\zeta} = A\left(\zeta - \xi_1\right)^{\alpha_1 - 1} \left(\zeta - \xi_2\right)^{\alpha_2 - 1} \ldots$$

If ξ_1 is the image point that will be moved to infinity, we redefine first

$$A = B\left(-\xi_1\right)^{\alpha_1 + 1}$$

so that

$$\lim_{\xi_1 \to \infty} \frac{dz}{d\zeta} = B \left(\frac{\zeta - \xi_1}{-\xi_1}\right)^{\alpha_1 - 1} \left(\zeta - \xi_2\right)^{\alpha_2 - 1} \ldots$$
$$= B\left(\zeta - \xi_2\right)^{\alpha_2 - 1} \left(\zeta - \xi_2\right)^{\alpha_3 - 1} \ldots$$

Thus if $\xi_1 \to \infty$, we need only omit the factor $\zeta - \xi_1$ in the mapping function; ξ_1 disappears altogether from the formula.

Now a few examples.

11.4 An infinite channel

11.4.1 Mapping onto a half plane

Referring to Figure 11.6, let us consider $ABCC'B'A'$ as a degenerate polygon with interior angles $0, \pi, 0, \pi$. First we choose to put three image points $(A, A'), (C, C')$ and B' arbitrarily on the ξ axis: A' at $\xi = \infty$, B' at $\xi = 1, \eta = 0$ and C' at the origin. By symmetry, A is mapped to $-\infty$, B to $\xi = -1, \eta = 0$ and C to the origin also. By Schwarz–Christoffel formula

$$\frac{dz}{d\zeta} = K\left(\zeta - \xi_B\right)^{1-1} \left(\zeta - \xi_C\right)^{0-1} \left(\zeta - \xi'_B\right)^{1-1} = \frac{K}{\zeta},$$

therefore,

$$z = K \ln \zeta + z_o. \qquad (11.4.1)$$

Let us define $\ln \zeta$ by cutting the ζ plane along $B'A'$ and $\ln \zeta = \ln \rho e^{i\theta} = \ln \rho + i\theta$ with $0 \leq \theta < 2\pi$. Substituting $\zeta_{B'} = 1, z_{B'} = 0$ in (11.5.6)

$$0 = K \ln 1 + z_o,$$

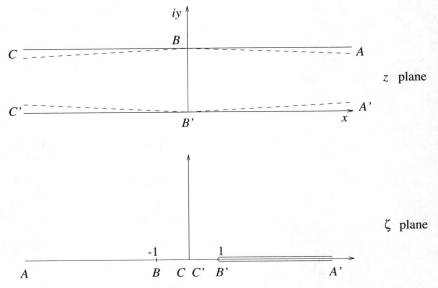

Fig. 11.6. An infinite channel and its image.

we get $z_o = 0$. For any point along the upper edge of $C'B'A'$, $y = 0$ in the z plane, while $\xi > 0, \eta = 0$ in the ζ plane. Hence,

$$x + i0 = K \ln |\xi|,$$

implying that K is real. On the other hand, for any point along ABC, $y = ia$ in the z plane, while $\xi < 0, \eta = 0$ in the ζ plane, or $\zeta = |\xi|e^{i\pi}$. Hence,

$$x + ia = K(\ln |\xi| + i\pi).$$

Equating the imaginary parts of the two sides, we get $Ki\pi = ia$, therefore, $K = a/\pi$. The final result is

$$z = \frac{a}{\pi} \ln \zeta \quad \text{or} \quad \zeta = e^{\pi z/a}. \tag{11.4.2}$$

11.4.2 Source in an infinite channel

Let us use the result just found to solve the potential flow problem of a source in the middle of a long channel. Let the source of strength m be located at $x = 0, y = a/2$. Mass conservation requires that half of the flux $m/2$ goes to $x \sim \infty$ and half to $x \sim -\infty$. Thus, in the z plane there is a sink of strength $-m/2$ at point CC' and an equal sink $-m/2$ at point AA'. Using the fact that the source strength is invariant under conformal mapping, we put a source of strength m at the image of the source and a sink of strength $-m/2$ at CC'. The location of the source is at $z_S = ia/2$ in the z plane. The image source is located at $\zeta_S = e^{\pi i/2} = i$ according to (11.4.2). Let us regard the upper half plane of ζ plane as one-half of another symmetric flow problem in the entire ζ plane, with a source m at $(0,i)$, a source m at $(0,-i)$ and a sink $(-m)$ at $(0,0)$, as shown in Figure 11.7. The complex potential for the entire ζ plane is then easily written

$$\begin{aligned} w(\zeta) &= \frac{m}{2\pi} \left[\ln(\zeta - i) + \ln(\zeta + i) - \ln \zeta \right] \\ &= \frac{m}{2\pi} \ln \frac{\zeta^2 + 1}{\zeta} = \frac{m}{2\pi} \ln \left(\zeta + \frac{1}{\zeta} \right) \\ &= \frac{m}{2\pi} \ln \left(e^{\pi z/a} + e^{-\pi z/a} \right) \\ &= \frac{m}{2\pi} \ln \left(2 \cosh \frac{\pi z}{a} \right). \end{aligned} \qquad (11.4.3)$$

This result is the desired solution for the potential of a source in an infinite channel.

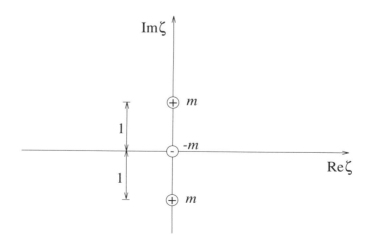

Fig. 11.7. Distribution of sources in the entire ζ plane.

Let us verify that the velocity field satisfies the no-flux boundary condition on the walls. The complex velocity is

$$\frac{dw}{dz} = u - iv = \frac{m}{2\pi} \frac{\pi}{a} \frac{\sinh \frac{\pi z}{a}}{\cosh \frac{\pi z}{a}} = \frac{m}{2a} \tanh \frac{\pi z}{a}.$$

Since $\tanh \pi z/a \to \pm 1$ as $z \to \pm \infty$, we have

$$\frac{dw}{dz} = u - iv = \pm \frac{m}{2a},$$

which means that the velocity at infinity is uniform and parallel to the walls

$$u \to \pm \frac{m}{2a}, \quad v \to 0 \quad \text{as} \quad x \to \pm \infty.$$

11.5 A semi-infinite channel

Referring to Figure 11.8, we consider the strip as the limit of a triangle, ABC, which is mapped onto the upper ζ plane, with the images of

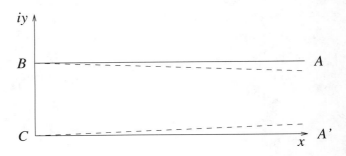
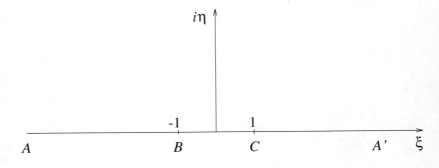

Fig. 11.8. A semi-infinite channel and its image.

11.5 A semi-infinite channel

A, B, C at $\xi = -\infty, -1, 1$, respectively, along the real axis. The point A' then falls on $\xi = \infty$ by symmetry.

Since the interior angles of the two corners in the finite part of the z plane are $\pi/2$, the Schwarz–Christoffel formula gives

$$\frac{dz}{d\zeta} = K(\zeta+1)^{\frac{1}{2}-1}(\zeta-1)^{\frac{1}{2}-1} = \frac{K}{\sqrt{\zeta^2-1}},$$

which can be integrated to give

$$z = K\int \frac{d\zeta}{\sqrt{\zeta^2-1}} + z_o = K\ln\left(\zeta + \sqrt{\zeta^2-1}\right) + z_o. \qquad (11.5.1)$$

The function $\ln\left(\zeta + \sqrt{\zeta^2-1}\right)$ is multi-valued unless cuts are introduced and a branch chosen. Let us first define $\sqrt{\zeta^2-1}$ by a cut along $-1 < \xi < 1$, as shown in Figure 11.9, and take the branch so that $\sqrt{\zeta^2-1} \to \zeta$ as $\zeta \to \infty$. Along the real ξ axis $\sqrt{\zeta^2-1}$ takes the following values:

$$\sqrt{\zeta^2-1} = \begin{cases} -\sqrt{\xi^2-1}, & -\infty < \xi < -1; \\ \pm i\sqrt{1-\xi^2}, & -1 < \xi < 1, \quad \eta = \pm 0, \\ \sqrt{\xi^2-1}, & 1 < \xi < \infty. \end{cases}$$

It can be shown that $\zeta + \sqrt{\zeta^2-1} \neq 0$ anywhere, hence the argument of $\ln\left(\zeta + \sqrt{\zeta^2-1}\right)$ does not vanish in the finite ζ plane, and there is no

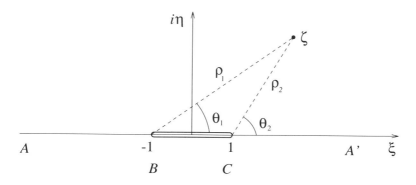

Fig. 11.9. Branch cut of the square root.

logarithmic singularity. Along the real ξ axis the values of the logarithmic function are

$$\ln\left(\zeta+\sqrt{\zeta^2-1}\right) = \begin{cases} \ln\left(\xi-\sqrt{\xi^2-1}\right), & -\infty < \xi < -1 \\ \ln\left(\xi+i\sqrt{1-\xi^2}\right), & -1 < \xi < 1 \\ \ln\left(\xi+\sqrt{\xi^2-1}\right), & 1 < \xi < \infty, \end{cases}$$

where the value in $-1 < \xi < 1$ is for the upper edge of the cut. As $\xi \to 1$ either from the left or from the right

$$\ln(\zeta+\sqrt{\zeta^2-1}) \to \ln 1 = 0,$$

and as $\xi \to -1$

$$\ln(\zeta+\sqrt{\zeta^2-1}) \to \ln(-1) = i\pi.$$

Since $\zeta_B = -1$ must correspond to $z_B = ia$, we require from (11.5.1)

$$ia = i\pi K + z_o.$$

Similarly, $\zeta_C = 1$ must be the image of $z_C = 0$

$$0 = 0 + z_o.$$

It follows that

$$z_o = 0, \qquad K = a/\pi$$

so that

$$z = \frac{a}{\pi}\ln\left(\zeta+\sqrt{\zeta^2-1}\right) \tag{11.5.2}$$

is the desired analytic function that maps the semi-infinite strip in z onto the upper half plane of ζ.

To give an alternative form for (11.5.2), we let

$$\zeta = \frac{1}{2}\left(e^s + e^{-s}\right) = \cosh s.$$

The first equality is a quadratic equation for e^s, which can be solved to yield

$$e^s = \zeta \pm \sqrt{\zeta^2 - 1}.$$

In order that $\zeta \to \infty$ corresponds to $s \to \infty$, the positive sign must be chosen so that

$$\cosh^{-1}\zeta = s = \ln\left(\zeta+\sqrt{\zeta^2-1}\right).$$

Therefore, (11.5.2) can also be written as

$$z = \frac{a}{\pi} \cosh^{-1} \zeta. \tag{11.5.3}$$

11.6 An estuary

Here we wish to transform a slotted half plane to a full half plane. As shown in Figure 11.10, the images of points A, B, C in the z plane are chosen for convenience to be at $-\infty, -1$ and 0 in the ζ plane. By symmetry, C', B', A' are also fixed, as shown in Figure 11.10. Corresponding to the three finite image points, the internal angles are $3\pi/2, 0$ and $3\pi/2$. By Schwarz–Christoffel transformation

$$\frac{dz}{d\zeta} = A(\zeta+1)^{3/2-1}\zeta^{0-1}(\zeta-1)^{3/2-1} = \frac{A}{\zeta}\sqrt{\zeta^2-1}, \tag{11.6.1}$$

which can be integrated to give

$$z = z_o + A\left[\sqrt{\zeta^2-1} + i\ln\frac{\zeta}{\sqrt{\zeta^2-1}+i}\right]. \tag{11.6.2}$$

For $\sqrt{\zeta^2-1}$ we choose the cut along $-1 < \xi < 1$ and the branch such that $\sqrt{\zeta^2-1} \to \zeta$ as $\zeta \to \infty$. This choice amounts to defining $\zeta-1 = \rho_1 e^{i\theta_1}$, $\zeta+1 = \rho_2 e^{i\theta_2}$ and restricting ζ on the top sheet of the Riemann surface with $0 \le \theta_1, \theta_2 < 2\pi$. As for the logarithmic function we

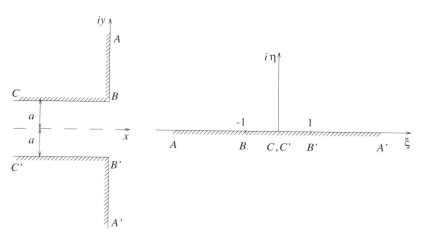

Fig. 11.10. An estuary in the z plane and its image in the ζ plane.

make the term $\ln \zeta$ single-valued by cutting the ζ plane along the positive real axis $\xi > 0$ and choosing $0 \le \arg \zeta < 2\pi$. The term $\ln\left(\sqrt{\zeta^2 - 1} + i\right)$ is singular at $0, \infty$ in addition to the square root branch points at $-1, 1$. Hence, we introduce a cut from 0 to ∞ and define $0 \le \arg \zeta < 2\pi$.

To fix A and z_o we use the fact that for the corner point B, $z_B = ia$, and $\zeta_B = e^{i\pi} = -1$, hence,

$$ia = z_o + A\left[i \ln \frac{e^{i\pi}}{e^{i\pi/2}}\right]$$

$$= z_o - \frac{A\pi}{2},$$

implying that

$$z_o = ia + \frac{A\pi}{2}.$$

For the corner point B' we substitute $z_{B'} = -ia$ and $\zeta_{B'} = e^{i0}$ in (11.6.2) to get

$$-ia = z_o + A\left[0 + i \ln \frac{e^{i0}}{e^{i\pi/2}}\right]$$

or

$$z_o = -ia - \frac{A\pi}{2},$$

hence,

$$z_o = 0, \qquad A = -\frac{2ia}{\pi}.$$

Finally,

$$z = -\frac{2ia}{\pi}\left(\sqrt{\zeta^2 - 1} + i \ln \frac{\zeta}{\sqrt{\zeta^2 - 1} + i}\right). \qquad (11.6.3)$$

We now return to physical applications.

11.7 Seepage flow under an impervious dam

An impervious concrete dam of thickness $2a$ rests on a porous river bed. The unequal water depths on the left and right sides of the dam are H and h, respectively. What is the flow through the river bed under the dam? Referring to Figure 11.11, we wish to find a complex potential $w(z) = \phi + i\psi$ that is analytic in the lower half plane of z and satisfies

$$\phi = -kH \quad \text{along} \quad AB \qquad (11.7.1)$$

11.7 Seepage flow under an impervious dam

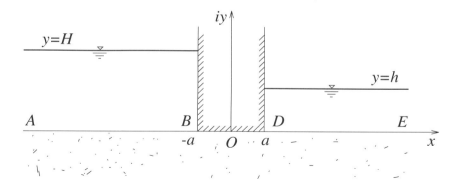

Fig. 11.11. Dam on a porous foundation in the physical plane of z.

$$\phi = -kh \quad \text{along} \quad DE \tag{11.7.2}$$

and

$$\psi = 0 \quad \text{along} \quad BOD. \tag{11.7.3}$$

The boundary-value problem for the velocity potential is of the mixed type since (11.7.3) amounts to prescribing $\partial\phi/\partial y$. In the complex plane of $w = \phi + i\psi$, the boundary curves correspond to the sides of a semi-infinite strip, as shown in Figure 11.12. Therefore, we must find the mapping function that transforms the interior of the semi-infinite strip in w to the lower half plane of z. Let us first rotate the w plane to the t plane by

$$t = i(w + kH), \tag{11.7.4}$$

as shown in Figure 11.13, and then rotate the z plane by π so that the lower half plane of z becomes the upper half plane of ζ, as shown in Figure 11.14.

The mapping function between w and ζ

$$\begin{aligned} t &= \frac{k(H-h)}{\pi} \ln\left(\zeta + \sqrt{\zeta^2 - 1}\right) \\ &= \frac{k(H-h)}{\pi} \cosh^{-1}\zeta \end{aligned} \tag{11.7.5}$$

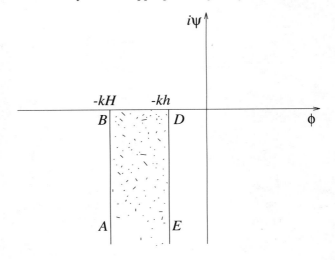

Fig. 11.12. The complex potential plane.

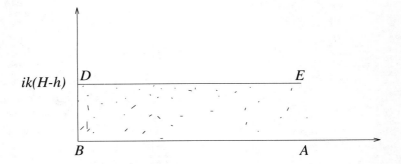

Fig. 11.13. The t plane.

is known from §11.5. Since

$$\cosh^{-1}\zeta = \ln\left(\zeta + \sqrt{\zeta^2 - 1}\right)$$
$$= \frac{\pi t}{k(H-h)} = \frac{i\pi(w+kH)}{k(H-h)}, \quad (11.7.6)$$

we obtain

$$\zeta = \cosh\frac{i\pi(w+kH)}{k(H-h)} = \cos\frac{\pi(w+kH)}{k(H-h)}. \quad (11.7.7)$$

11.7 Seepage flow under an impervious dam

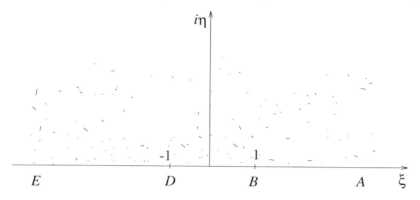

Fig. 11.14. The ζ plane.

Now the z plane is easily mapped onto the upper half of the ζ plane by

$$z = -a\zeta. \tag{11.7.8}$$

Eliminating ζ from (11.7.6) and (11.7.7), we get

$$-\frac{z}{a} = \zeta = \cos\frac{\pi(w+kH)}{k(H-h)}, \tag{11.7.9}$$

implying

$$\frac{\pi(w+kH)}{k(H-h)} = \cos^{-1}\left(-\frac{z}{a}\right)$$

$$= -i\ln\left(-\frac{z}{a} + \sqrt{\left(\frac{z}{a}\right)^2 - 1}\right), \tag{11.7.10}$$

which is the desired solution.

To see the physical picture of the flow we examine the far field $|z| \gg a$, which corresponds to $\zeta \gg 1$. By approximating (11.7.10), we get

$$\frac{\pi(w+kH)}{k(H-h)} = \frac{1}{i}\ln\left(\zeta + \sqrt{\zeta^2 - 1}\right)$$

$$\sim \frac{1}{i}\ln 2\zeta \sim -i\ln\left(-\frac{2z}{a}\right)$$

$$\sim -i\ln z.$$

Therefore, in the far field the potential is approximately that of a point vortex

$$w + kH \sim -\frac{i}{\pi}k(H-h)\ln z.$$

To find the streamlines and equipotential lines everywhere we first rewrite (11.7.9) as

$$-\frac{z}{a} = \cos\alpha(\phi' + i\psi) \qquad (11.7.11)$$

or

$$-\frac{x + iy}{a} = \cos\alpha\phi' \cosh\alpha\psi - i\sin\alpha\phi' \sinh\alpha\psi,$$

where

$$\phi' = \phi + kH, \qquad \alpha = \frac{\pi}{k(H - h)}.$$

By separating the real and imaginary parts, we obtain

$$-\frac{x}{a} = \cos\alpha\phi' \cosh\alpha\psi, \qquad -\frac{y}{a} = -\sin\alpha\phi' \sinh\alpha\psi. \qquad (11.7.12)$$

Clearly, ϕ' can be eliminated by using the identity $\cos^2\alpha\phi' + \sin^2\alpha\phi' = 1$, yielding

$$\left(\frac{x}{a\cosh\alpha\psi}\right)^2 + \left(\frac{y}{a\sinh\alpha\psi}\right)^2 = 1. \qquad (11.7.13)$$

Thus the streamlines are ellipses. Similarly, the equipotential lines are found by eliminating ψ

$$\left(\frac{x}{a\cos\alpha\phi'}\right)^2 - \left(\frac{y}{a\sin\alpha\phi'}\right)^2 = 1 \qquad (11.7.14)$$

and are hyperbolas.

The complex velocity field at any z is

$$\frac{dw}{dz} = \frac{-i}{\alpha\sqrt{z^2 - a^2}}.$$

Choosing the branch of the square root so that $\sqrt{z^2 - a^2} \to z$ as $|z| \sim \infty$, we get along DE

$$\frac{dw}{dz} = u - iv = -\frac{ik(H - h)}{\pi a} \frac{1}{\sqrt{\frac{x^2}{a^2} - 1}}.$$

Hence,

$$u = 0, \qquad v = \frac{k(H - h)}{\pi a} \frac{1}{\sqrt{\frac{x^2}{a^2} - 1}}.$$

The vertical velocity is upward from the bed into the river. Along AB

$$\frac{dw}{dz} = u - iv = -i\frac{k(H - h)}{a} \frac{1}{-\sqrt{\frac{x^2}{a^2} - 1}},$$

hence,
$$u = 0, \quad v = -\frac{k(H-h)}{\pi a} \frac{1}{\sqrt{\frac{x^2}{a^2} - 1}}.$$

Directly beneath the dam along BD, $\sqrt{z^2 - a^2} = -i\sqrt{a^2 - x^2}$ so that
$$\frac{dw}{dz} = \frac{k(H-h)}{\pi a} \frac{1}{\sqrt{1 - \frac{x^2}{a^2}}},$$
which is real, implying that $v = 0$.

The pressure distribution along the base of the dam is found by setting $\psi = 0$ in (11.7.12)
$$-\frac{x}{a} = \cos\alpha(\phi + kH),$$
then,
$$\phi + kH = \frac{k(H-h)}{\pi} \cos^{-1}\left(-\frac{x}{a}\right).$$

Hence the pressure is
$$p = -\rho g \frac{\phi}{k} = \rho g \left[H - \frac{H-h}{\pi} \cos^{-1}\left(-\frac{x}{a}\right)\right], \tag{11.7.15}$$
which varies from $\rho g H$ to $\rho g h$ from $x = \alpha$ to $x = a$. This pressure distribution can be used to calculate the overturning torque on the dam.

11.8 Water table above an underground line source

From a line source in porous ground, water is emitted steadily to wet the neighboring soil, which is otherwise dry. Let us calculate the profile of the water table (phreatic surface). With reference to Figure 11.15, let us choose the line source A to be the origin of the z plane. By symmetry only one-half of the picture $x \geq 0$ needs to be considered. The boundary conditions are
$$\psi = 0 \quad \text{on} \quad A'B' \tag{11.8.1}$$
$$\phi + ky = 0 \quad \text{on} \quad DB \tag{11.8.2}$$
$$\psi = \frac{Q}{2} \quad \text{on} \quad AD, DB. \tag{11.8.3}$$

Note that $A'B'$ is one streamline while ADB is another. In the plane of the complex potential $w = \phi + i\psi$ the flow region corresponds to the

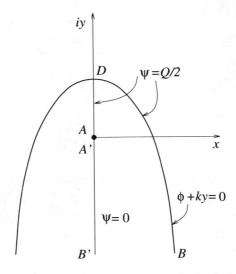

Fig. 11.15. Water table above a line source.

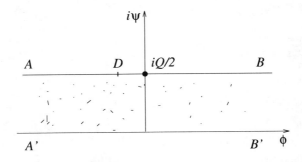

Fig. 11.16. Plane of the complex potential w.

interior of the infinite strip, as shown in Figure 11.16, in which the value of ϕ_D is unknown.

Let us introduce the Zhukovsky function

$$\theta = w - ikz \qquad (11.8.4)$$

so that

$$\theta_1 + i\theta_2 = \phi + i\psi - ik(x + iy)$$

11.8 Water table above an underground line source

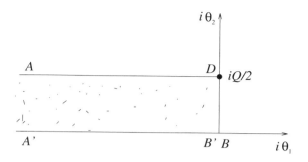

Fig. 11.17. The Zhukovsky plane.

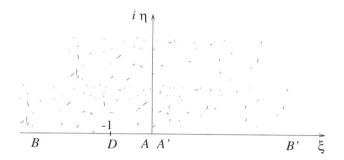

Fig. 11.18. The ζ plane, $\zeta = \xi + i\eta$.

and
$$\theta_1 = \phi + ky, \qquad \theta_2 = \psi - kx.$$

The boundary conditions (11.8.1–11.8.3) become, respectively,

$$\theta_2 = 0 \quad \text{on} \quad A'B' \tag{11.8.5}$$

$$\theta_2 = \frac{Q}{2} \quad \text{on} \quad AD \tag{11.8.6}$$

$$\theta_1 = 0 \quad \text{on} \quad DB. \tag{11.8.7}$$

At AA', $w \sim (Q/2\pi) \ln z$, hence $\phi \sim -\infty$, and the flow field corresponds to the interior of the semi-infinite strip, as shown in Figure 11.17. To find $w(z)$ we shall map both w and θ onto the upper half plane of ζ; see Figure 11.18. In this way the relation $\theta = \theta(w)$ will be found through ζ. Since $\theta = w(z) - ikz = \theta(w)$, $w = w(z)$ will then follow.

In the upper half plane of ζ, one needs an analytic function $w(\zeta)$ subject to the boundary conditions (11.8.1) and (11.8.3), which can be met by a source of strength Q emitting fluid from AA' to the entire ζ plane

$$w = \phi_D + \frac{Q}{2\pi} \ln \zeta, \qquad (11.8.8)$$

where ϕ_D is yet unknown. Similar to the example in §11.5, it may be shown that

$$\theta = \frac{Q}{2\pi} \left[2 \ln \left(\sqrt{\zeta+1} - 1 \right) - \ln \zeta \right]. \qquad (11.8.9)$$

For the square root $\sqrt{\zeta+1}$, the ζ plane is cut along DB; for the two logarithmic functions, the cut is along the positive ξ axis $A'B'$. Hence,

$$w - ikz = \frac{Q}{2\pi} \left[2 \ln \left(\sqrt{\zeta+1} - 1 \right) - \ln \zeta \right]. \qquad (11.8.10)$$

The difference of (11.8.8) and (11.8.10) is

$$\phi_D - ikz = \frac{Q}{\pi} \ln \left(\sqrt{\zeta+1} - \ln \zeta \right). \qquad (11.8.11)$$

To determine the value of ϕ_D, we consider the source at AA': $z_A = 0$, $\zeta_A = 0$. Letting $\zeta \to 0$ in (11.8.11), we get

$$\phi_D = \lim_{\tau \to 0} \frac{Q}{\pi} \left[\ln \left(1 + \frac{\zeta}{2} - 1 \right) - \ln \zeta \right] = -\frac{Q}{\pi} \ln 2.$$

It follows that

$$-ikz = \frac{Q}{\pi} \ln 2 + \frac{Q}{\pi} \left[\ln \left(\sqrt{\zeta+1} - 1 \right) - \ln \zeta \right] \qquad (11.8.12)$$

and, from (11.8.8), that

$$w = -\frac{Q}{\pi} \ln 2 + \frac{Q}{2\pi} \ln \zeta. \qquad (11.8.13)$$

Equations (11.8.12) and (11.8.13) together give $w = w(z)$ with ζ as a parameter. The complex potential is complete.

To find the water table, we know from the ζ plane that DB corresponds to $\zeta = -|\xi|, |\xi| > 1, \eta = 0$. From (11.8.12)

$$-ikx + ky = \frac{Q}{\pi} \ln 2 + \frac{Q}{\pi} \left[\ln \left(i\sqrt{|\xi|-1} - 1 \right) \right] - (\ln |\xi| + i\pi)$$

$$= \frac{Q}{\pi} \ln 2 + \frac{Q}{\pi} \left(\frac{1}{2} \ln |\xi| + i\pi - i \tan^{-1} \sqrt{|\xi|-1} - \ln |\xi| - i\pi \right).$$

The real and imaginary parts give, respectively,

$$ky = \frac{Q}{\pi} \ln 2 - \frac{Q}{2\pi} \ln |\xi|, \quad |\xi| > 1 \tag{11.8.14}$$

$$kx = \frac{Q}{\pi} \tan^{-1} \sqrt{|\xi| - 1}, \quad |\xi| > 1, \tag{11.8.15}$$

which are the parametric equations for the water table. The highest point D of the water table corresponds to $\xi = -1$ and

$$ky_{max} = \frac{Q}{\pi} \ln 2. \tag{11.8.16}$$

For the maximum width of the wetted zone we take the limit $|\xi| \to \infty$ so that $\tan^{-1} \sqrt{|\xi| - 1} \to \pi/2$, yielding

$$kx_{max} = \frac{Q}{2}. \tag{11.8.17}$$

This result can be confirmed heuristically. The points BB' are very far beneath the source, i.e., $y \sim -\infty$, where the velocity potential $\phi \sim -ky$ is dominated only by gravity. The velocity is vertically downward and uniform with the magnitude

$$v = \frac{\partial \phi}{\partial y} = -k.$$

Equation (11.8.17) then follows by mass conservation.

Many interesting applications of conformal mapping can be found in Milne-Thomson (1955) on classical hydrodynamics and in Polubarinova-Kochina (1962), Strack (1989) and Yih (1965) on flow through porous media.

Exercises

11.1   The Joukowski transformation

$$z = \zeta + \frac{c^2}{\zeta} \tag{E11.1}$$

is the best known in aerodynamics. Derive the following properties of this mapping function:

(i) The exterior of the circle $|\zeta| = c$ in the ζ plane is mapped onto the entire z plane cut along the real axis

$$-2c \leq x \leq 2c, \quad y = 0.$$

(ii) The point ζ outside the c circle and the image point c^2/ζ inside are mapped onto the same point in the z plane. This double-valuedness corresponds to the two numbers of the inverse transform

$$\zeta = \frac{1}{2}\left(z \pm \sqrt{z^2 - 4c^2}\right).$$

(iii) A circle $|\zeta| = c'$ with $c' > c$ is mapped onto an ellipse in the z plane.

(iv) A circle passing through $\zeta = c$ and centered at ζ_o away from the origin is mapped onto an airfoil with a cusped trailing edge at $z = 2C$. In particular: (a) If $\zeta_o = -m$ with m being real and positive $m/c = \epsilon \ll 1$, the image in the z plane is a symmetric airfoil. (b) If $\zeta_o = in$ with n being real and positive and $n/C = \epsilon \ll 1$, the image is a circular arc concaving downward and with leading and trailing edges at $z = -2a$ and $2a$, respectively. (c) If $\zeta_o = -m + in$ with $m > 0, n > 0$ but $\sqrt{m^2 + n^2} = \epsilon C$, $\epsilon \ll 1$, the image is a cambered airfoil with a thick leading edge.

Hint: The circle has the equation

$$|\zeta - \zeta_o| = |c - \zeta_o|. \tag{E11.2}$$

Near the leading edge use Taylor expansions to approximate (E11.1) and (E11.2) for small ϵ.

11.2 Show that the function $w = 1/z$ always maps a circle in one plane to a circle in the other, with straight lines as special cases.

11.3 The 600-mile-long pipeline transporting oil across Alaska is a monumental engineering accomplishment. Along much of the way the pipe carrying warm oil is buried in the frozen ground. Heat from the pipe causes thawing of the permafrost and weakens the soil. Consider the simplest heat transfer problem of a buried pipe (Lunardini, 1981). Let the y axis coincide with the ground surface and the x axis point into the ground. Show that the following bilinear (or Möbius) transformation

$$\zeta(z) = \frac{A + Bz}{C + Dz}$$

can be used to map the right half z plane outside a circle of radius a to a ring in the w plane $1 < |w| < R$. Let a = pipe radius and h = depth of pipe center below the ground surface.

Show that
$$\frac{A}{D} = -R\frac{C}{D} = -R(h-a)\frac{(R+1)}{R-1}, \quad \frac{B}{D} = R,$$
and
$$R = \frac{h}{a} + \sqrt{\left(\frac{h}{a}\right)^2 - 1}.$$

If the ground temperature is T_0 and the pipe temperature is T_1, what is the temperature distribution in the ground?

Hint: Rewrite the mapping function as
$$\zeta(z) = \frac{B}{D} + \frac{B}{D}\left(\frac{A}{B} - \frac{C}{D}\right)\left(z + \frac{C}{D}\right)^{-1}$$
and use the last exercise.

11.4 Consider a plane potential flow past an idealized offshore island of circular form. The coastline coincides with the y axis. The island is centered at $x = h, y = 0$ and has the radius a with $a < h$. A uniform flow of complex potential $w_o = -iVz$ arrives from $y \sim -\infty$. Use the mapping function in the last problem to analyze this Neuman problem. For uniqueness insist that the velocity potential is single-valued along a closed contour encircling the island. Physically, this condition implies that there is no *circulation* around the island.

Hint: First identify the image of the point $z \sim \infty$ in the ζ plane and show that the uniform flow corresponding to ϕ_o becomes a doublet. Define
$$w = w_o + w'$$
and solve the boundary-value problem for w' in the ring of the ζ plane.

11.5 Find the complex potential for the irrotational flow out of a rectangular estuary if the total discharge Q is given.

Hint: The upstream end of the flow is a source of strength Q.

11.6 Show that (11.8.9) maps the semi-infinite strip in the θ plane to the upper half plane of ζ, as shown in Figure 11.18.

11.7 Due to a large pressure difference $p_\infty - p_o$ between a reservoir and the atmosphere, a water jet is emitted out of a two-dimensional funnel. The funnel walls are inclined at the angle β from the axis; see Figure 11.19. Find the flow field in the jet and

314   *Conformal mapping and hydrodynamics*

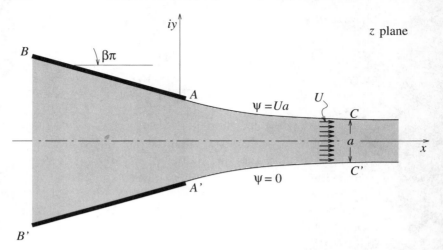

Fig. 11.19. The z plane for a water jet from a funnel.

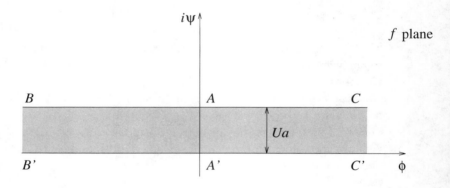

Fig. 11.20. The complex potential plane of $w = \phi + i\psi$.

the jet width a far downstream by using a sequence of conformal mappings suggested below.

Use the Bernoulli equation to express the ultimate jet velocity

$$U = \sqrt{2\left(p_\infty - p_o\right)/\rho}.$$

Show first that along the free streamlines, $p = p_o$ so that $(u^2 + v^2)/2 = \text{constant} = U^2/2$ and $\psi = Ua$.

In the complex potential plane of $w = \phi + i\psi$ the flow is inside an infinite strip; see Figure 11.20.

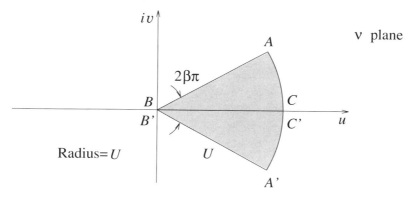

Fig. 11.21. The hodograph plane $\nu = u - iv$.

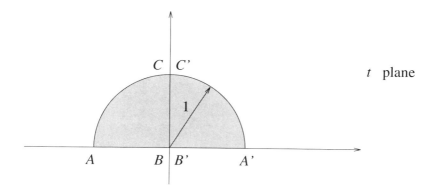

Fig. 11.22. The t plane.

(i) Show that in the complex velocity (hodograph) plane of $\nu = u - iv$ (see Figure 11.21) the flow is the interior of a fan of apex angle $2\beta\pi$ and radius U.

(ii) Show that

$$t = i \left(\frac{\nu}{U}\right)^{\frac{1}{2\beta}}$$

maps the fan in the ν plane to the interior of a semicircle of unit radius in the t plane; see Figure 11.22.

(iii) Show that $\tau = -1/t$ maps the interior of the semicircle to the exterior of the semicircle in the τ plane; see Figure 11.23.

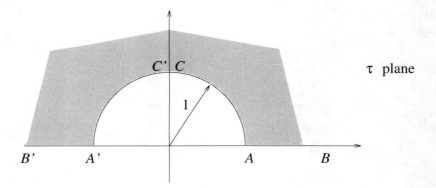

Fig. 11.23. The τ plane.

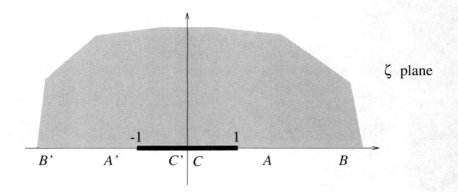

Fig. 11.24. The ζ plane.

(iv) Show that

$$\zeta = \frac{1}{2}\left(\tau + \frac{1}{\tau}\right)$$

maps the τ plane to the upper half plane of ζ; see Figure 11.24.

(v) Show that

$$w = \frac{Ua}{\pi}(i\pi - \ln\zeta)$$

maps the strip in w plane to the upper half plane of ζ.

(vi) Integrate $\nu = dw/dz$ to get $z = z(w)$, and determine the integration constant by requiring that the width is a, i.e.,

$$z_C - z_{C'} = ia.$$

(vii) For $\beta = 1$, show that $z_A - z_{A'} = 2ia$. This is the case of Borda's mouth piece.

(viii) For $\beta = 1/2$ show that

$$z = z_o - \frac{ia}{\pi}\left(\frac{1}{t} + 2\tan^{-1}\frac{1}{t}\right).$$

Show that $z_A - z_{A'} = (ia/\pi)(2+\pi)$. This case corresponds to a jet from a slot in the wall.

12
Riemann–Hilbert problems in hydrodynamics and elasticity

In this chapter we shall apply the basic techniques of complex function theory to some slightly more advanced problems involving mixed boundary conditions and biharmonic functions. The common mathematical thread is the so-called Riemann–Hilbert problem whose mathematical solution involves the clever use of Cauchy's integral formula. More interestingly, in these applications a highly ingenious use of ordinary complex analysis is required to reduce the physical problem to Riemann–Hilbert form. We shall illustrate this ingenuity through two physically very different examples: cavity flows and elastic contact in a plane. The mathematical idea of the Schwarz principle of analytical continuation for treating half-plane problems is also discussed.

12.1 Riemann–Hilbert problem and Plemelj's formulas

The Riemann–Hilbert problem is defined as follows. Find a function $f(z)$ that is analytic in the entire z plane except along a curve L across which

$$f_+(t) - g(t)f_-(t) = f_o(t), \quad t \in L, \qquad (12.1.1)$$

where t is the value of z on L, and $f_o(t)$ is given. For a chosen direction of traversing L, $f_+(t)$ is the limit of f when $z \to t$ from the left, while $f_-(t)$ is the limit of f when $z \to t$ from the right.

The basic mathematical tool for the solution of a Riemann–Hilbert problem is a pair of formulas due to Plemelj. Let $f(z)$ denote the line integral

$$f(z) = \frac{1}{2\pi i} \int_L \frac{f_o(t)\,dt}{t - z}, \qquad (12.1.2)$$

12.1 Riemann–Hilbert problem and Plemelj's formulas

then as z approaches the point t_o from the left $(+)$ or right $(-)$,

$$\begin{matrix} f_+(t_o) \\ f_-(t_o) \end{matrix} = \pm\frac{1}{2}f(t_o) + \frac{1}{2\pi i}\int_L \frac{f_o(t_o)\,dt}{t-t_o}. \qquad (12.1.3)$$

The proof is a straightforward exercise of the residue theorem. Referring to Figure 12.1a, let z approach t_o on L from the right $(-)$. In order not to change the value of the integral we indent the path by a semicircle C_ϵ of radius ϵ to the left of t_o. The integral is first broken into three parts

$$\lim_{z\to t_o}\frac{1}{2\pi i}\int_L \frac{f_o(t)\,dt}{t-z} = \lim_{\epsilon\to 0}\frac{1}{2\pi i}\left\{\int_A^{t_o-\epsilon_-} + \int_{C_\epsilon} + \int_{t_o+\epsilon_+}^B\right\}\frac{f_o(t)\,dt}{t-z},$$

where ϵ_+ and $\epsilon_- = \epsilon_+ e^{i\pi}$ are, respectively, the complex vectors from t_o to the end and beginning points of the semicircle, and the second integral traverses the semicircle C_ϵ clockwise. The sum of the first and third integrals on the right is just the principal-valued integral in the sense of Cauchy. The integral along C_ϵ is

$$\frac{1}{2\pi i}\int_\pi^0 \epsilon i e^{i\theta}\,d\theta\,\frac{f_o(t_o)}{\epsilon e^{i\theta}} = -\frac{1}{2}f_o(t_o).$$

It follows that

$$f_-(t_o) = -\frac{1}{2}f_o(t_o) + \frac{1}{2\pi i}\int_L \frac{f_o(t)\,dt}{t-z}. \qquad (12.1.4)$$

Similarly, if z approaches t_o from the left, Figure 12.1b, we must indent the path by a semicircle on the right of t_o. The direction of integration

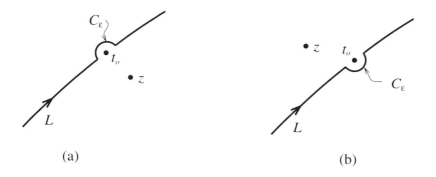

Fig. 12.1. Path L and indentations.

along the semicircle C_ϵ is then counterclockwise, thus

$$\frac{1}{2\pi i}\int_{C_\epsilon}\frac{f_o(t)\,dt}{t-z}=\frac{1}{2\pi i}\int_\pi^{2\pi}\frac{i\epsilon e^{i\theta}f_o(t_o)\,d\theta}{\epsilon e^{i\theta}}=\frac{1}{2}f_o(t_o).$$

Hence

$$f_+(t_o)=\frac{1}{2}f_o(t_o)+\frac{1}{2\pi i}\int_L\frac{f_o(t)\,dt}{t-z}. \qquad (12.1.5)$$

By subtracting and adding (12.1.4) and (12.1.5), we get

$$f_+(t_o)-f_-(t_o)=f_o(t_o) \qquad (12.1.6)$$

and

$$f_+(t_o)+f_-(t_o)=\frac{1}{\pi i}\int_L\frac{f_o(t)\,dt}{t-t_o}. \qquad (12.1.7)$$

Equation pairs (12.1.3) and (12.1.6–12.1.7) are called Plemelj's formulas.

An immediate application of these formulas is that if in the entire plane of z, $f(z)$ satisfies the jump condition (12.1.6) along L and behaves as z^m for large z, with m being a positive integer, then

$$f(z)=\frac{1}{2\pi i}\int_L\frac{f_o(t)\,dt}{t-x}+P_m(z), \qquad (12.1.8)$$

where P_m is a polynomial of order m. Use has been made of the extended Liouville theorem of §9.15 and the analyticity of $P_m(z)$.

12.2 Solution to the Riemann–Hilbert problem

To solve the Riemann–Hilbert problem defined by (12.1.1), the trick is to find a multi-valued analytic function $X(z)$, which takes on different limits on two sides of L such that

$$\frac{X_+(t)}{X_-(t)}=g(t).$$

If this function is found, (12.1.1) may be written as

$$\frac{f_+(t)}{X_+(t)}-\frac{f_-(t)}{X_-(t)}=\frac{f_o(t)}{X_+(t)},$$

which may be solved by (12.1.8)

$$\frac{f(z)}{X(z)}=\frac{1}{2\pi i}\int_L\frac{f_o(t)\,dt}{X_+(t)(t-z)}+P_m(x). \qquad (12.2.1)$$

In this chapter, it will be assumed that L is a straight-line segment

12.2 Solution to the Riemann–Hilbert problem

and $g = $ constant. The search for $X(z)$ is then easy. Let L be the segment $-a < x < a$ on the x axis and

$$X(z) = \frac{(z+a)^{-\gamma}}{(z-a)^{1-\gamma}}, \quad (12.2.2)$$

where $\gamma = \alpha + i\beta$ is a complex number yet to be determined. Let us define X in the complex plane cut along L and choose the branch such that $X \sim 1/z$ as $|z| \sim \infty$. Referring to Figure 12.2, we introduce the polar coordinates

$$z + a = r_1 e^{i\theta_1}, \qquad z - a = r_2 e^{i\theta_2},$$

then as z approaches t from above (left) and below (right)

$$X_\pm(t) = \frac{r_1^{-\gamma}}{r_2^{1-\gamma}} e^{-i\pi(1\mp\gamma)}.$$

Thus

$$\frac{X_+(t)}{X_-(t)} = e^{2\pi i \gamma}.$$

Clearly, γ must be chosen such that

$$e^{2\pi i \gamma} = g \quad \text{or} \quad \gamma = \frac{1}{2\pi i} \ln g. \quad (12.2.3)$$

The final solution is then given by (12.2.1) and (12.2.2).

In the special case where $g = 1$, there is no need for the function X, i.e., $X = 1$. If $g = -1$, then $\gamma = 1/2$ and

$$X(z) = \frac{1}{\sqrt{z^2 - a^2}}. \quad (12.2.4)$$

We first demonstrate the application of the mathematical theory to a problem in hydrodynamics.

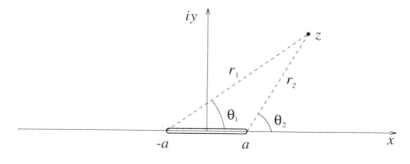

Fig. 12.2. Branch cut for $X(z)$.

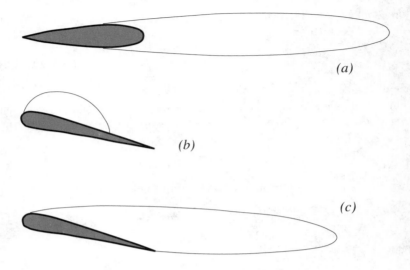

Fig. 12.3. Types of cavitied hydrofoils.

12.3 Linearized theory of cavity flow

When a hydrofoil advances at a high speed in water the pressure at the leading edge can be so low that the passing streamline is forced to separate from the hydrofoil and forms a cavity. This phenomenon is called *cavitation*† and is important to flows around hydraulic turbine blades, ship propellers, and stabilizing fins of torpedoes. Cavitation affects the lift and drag on the body and can be a source of underwater noise. To predict cavitated flows one needs a theory for the cavity shape and size as well as for the pressure and velocity fields in water. In Figure 12.3 three types of two-dimensional cavities are sketched. For a symmetrical hydrofoil with zero angle of attack, the cavity can start from somewhere along the surface and extends beyond the trailing edge (Figure 12.3a). If the hydrofoil is inclined, the flow around the body can be either partially cavitated (Figure 12.3b) or fully cavitated (Figure 12.3c). We shall give a brief account of the first case when the thickness of the hydrofoil is very small compared to the chord so that a linearized treatment suffices. The linearized cavitation theory was initiated by Tulin (1953). Authoritative surveys of the general topic of cavities can be found in Birkhoff and Zarantonello (1957), Gilbarg (1960) and Wu (1968).

---

† The present section is based on lectures given at California Institute of Technology by Professor T. Y. Wu.

12.3 Linearized theory of cavity flow

We shall treat water as an inviscid and incompressible fluid and assume the flow to be irrotational. Let P and U denote the upstream pressure and flow velocity, and p_c and q_c denote the pressure and tangential velocity on the cavity boundary. By Bernoulli's equation (9.6.9), we have

$$p_c + \frac{1}{2}\rho q_c^2 = P + \frac{1}{2}\rho U^2,$$

which can be rewritten as

$$\frac{q_c}{U} = \sqrt{1+\sigma}, \qquad (12.3.1)$$

where

$$\sigma = \frac{P - p_c}{\frac{1}{2}\rho U^2} \qquad (12.3.2)$$

is called the cavitation number.

Let us normalize the velocity components at any point by q_c so that

$$(U + u', v') = q_c(1 + u, v). \qquad (12.3.3)$$

In particular, at far upstream we have

$$q_c(1 + u_\infty, v_\infty) = (U, 0)$$

so that

$$u_\infty = \frac{U}{q_c} - 1 = \frac{1}{\sqrt{1+\sigma}} - 1 \equiv -\beta, \qquad v_\infty = 0. \qquad (12.3.4)$$

Consider a symmetrical hydrofoil at zero angle of attack. Along the wetted surface $y = \pm h(x)$, the flow must be tangential so that

$$\frac{v}{1+u} = \pm h'(x) \equiv \pm s(x),$$

where $\pm s(x)$ is the local slope of the body. For zero angle of attack and a thin body, $u, v \ll 1$ so that we have, to the first order of approximation,

$$v = \pm s(x), \qquad 0 < x < c, \qquad y = \pm 0, \qquad (12.3.5)$$

where c is the start of the cavity. Let the length of the cavity be denoted by L. On the cavity boundary, we have by definition

$$q_c[(1+u)^2 + v^2]^{1/2} = q_c, \qquad c < x < L,$$

hence to the leading order,

$$u = 0, \qquad c < x < L, \qquad y = \pm 0. \qquad (12.3.6)$$

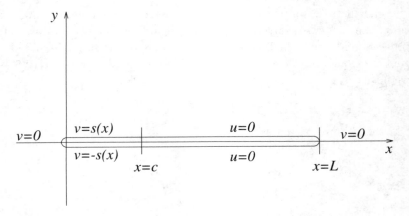

Fig. 12.4. The linearized boundary conditions for a symmetrical cavity.

Clearly, by symmetry

$$v = 0, \quad -\infty < x < 0, \quad y = 0. \qquad (12.3.7)$$

These conditions are summarized in Figure 12.4. The starting point c as well as the total length L of the cavity are parts of the solution; for their determination two additional conditions are needed. One is that the presence of the cavity does not create net mass. In the linearized approximation, this implies

$$\oint_C v \, ds = 0, \qquad (12.3.8)$$

where C is a closed contour wrapping around the cavity. The second condition is that at the starting point of the cavity the flow should be finite.

Let us introduce the complex velocity

$$w(z) = u(x,y) - iv(x,y) \qquad (12.3.9)$$

in terms of which the linearized boundary-value problem becomes the search of the analytic function $w(z)$ such that

$$\operatorname{Im} w = 0, \quad -\infty < x < 0, \quad L < x < \infty, \quad y = 0; \qquad (12.3.10)$$

$$\operatorname{Im} w = \pm s(x), \quad 0 < x < c, \quad y = \pm 0 \qquad (12.3.11)$$

$$\operatorname{Re} w = 0, \quad c < x < L, \quad y = \pm 0 \qquad (12.3.12)$$

$$\operatorname{Im} \oint_C w \, dz = 0. \qquad (12.3.13)$$

12.3 Linearized theory of cavity flow

Note that w is continuous across the x axis along $-\infty < x < 0$ and $L < x < \infty$, but discontinuous across the x axis along $0 < x < L$. Specifically, we have from the boundary conditions

$$w_+(x) - w_-(x) = -2is(x), \quad 0 < x < c \quad (12.3.14)$$

and

$$w_+(x) + w_-(x) = 0, \quad c < x < L. \quad (12.3.15)$$

We now begin the application of the Riemann–Hilbert technique. To change (12.3.15) to the form of (12.3.14) we use the factor

$$H(z) = \sqrt{\frac{z-L}{z-c}} \quad (12.3.16)$$

defined in the z plane cut along $c < x < L, y = 0$, so that

$$H(z) = \begin{cases} \sqrt{(L-x)/(c-x)} & \text{if } -\infty < x < c \\ \pm\sqrt{(L-x)/(x-c)} & \text{if } c < x < L, \ y = \pm 0 \\ \sqrt{(x-L)/(x-c)} & \text{if } L < x < \infty. \end{cases}$$

Now define

$$G(z) = H(z)w(z) = \sqrt{\frac{z-L}{z-c}}\, w(z), \quad (12.3.17)$$

then

$$G_+(x) - G_-(x) = -2i\sqrt{\frac{L-x}{c-x}}\, s(x), \quad -\infty < x < c, \quad (12.3.18)$$

with the understanding that $s(x) = 0$ for $-\infty < x < 0$. Along the cavity, however,

$$H_+(x) = -H_-(x) = \sqrt{\frac{L-x}{x-c}},$$

hence

$$G_+ = H_+ w_+, \quad G_- = H_- w_- = -H_+ w_-$$

and

$$G_+ - G_- = H_+(w_+ + w_-) = 0, \quad c < x < L. \quad (12.3.19)$$

The problem is now to find $G(z)$, which must satisfy (12.3.18) and (12.3.19). The solution is

$$G(z) = -\frac{1}{\pi}\int_0^c \sqrt{\frac{L-x'}{c-x'}}\frac{s(x')}{x'-z}\, dx' + P_m(z).$$

The degree of the polynomial has to be determined by considering the asymptotic behavior of G or w. Since as $z \sim \infty$, w is analytic and bounded, G must approach a constant, say A. The integral dies out as $1/z$, hence

$$G(z) = A - \frac{1}{\pi} \int_0^c \sqrt{\frac{L-x'}{c-x'}} \frac{s(x')}{x'-z} dx' \qquad (12.3.20)$$

and

$$w = \sqrt{\frac{z-c}{z-L}} \left\{ A - \frac{1}{\pi} \int_0^c \sqrt{\frac{L-x'}{c-x'}} \frac{s(x')}{x'-z} dx' \right\}. \qquad (12.3.21)$$

We must now determine the constants A, c and L.

Because of (12.3.4), $w \to -\beta$ as $|z| \to \infty$. It is necessary to take $A = -\beta$, thus

$$w = -\sqrt{\frac{z-c}{z-L}} \left\{ \beta + \frac{1}{\pi} \int_0^c \sqrt{\frac{L-x'}{c-x'}} \frac{s(x')}{x'-z} dx' \right\}. \qquad (12.3.22)$$

To apply the condition that there cannot be a net flux from the cavity, we let the contour C be replaced by a large contour encircling the body and the cavity; w can then be expanded for large z. Since

$$\sqrt{\frac{z-c}{z-L}} \cong 1 - \frac{L-c}{2z} + O\left(\frac{1}{z^2}\right),$$

we find

$$w = -\beta + \frac{1}{z}\left[-\frac{L-c}{2}\beta + \frac{1}{\pi}\int_0^c \sqrt{\frac{L-x'}{c-x'}} s(x')\,dx'\right] + O\left(\frac{1}{z^2}\right).$$

By the residue theorem

$$\oint_C w\,dz = 2\pi i \left[-\frac{L-c}{2}\beta + \frac{1}{\pi}\int_0^c \sqrt{\frac{L-x'}{c-x'}} s(x')\,dx'\right],$$

hence

$$\frac{L-c}{2}\beta = \frac{1}{\pi}\int_0^c \sqrt{\frac{L-x'}{c-x'}} s(x')\,dx', \qquad (12.3.23)$$

which is a condition between c and L.

To make sure that the flow is not singular at $x = c$ we observe from (12.3.22) that $w \sim \sqrt{z-c}$, and that

$$\frac{dw}{dz} \sim \frac{1}{(z-c)^{1/2}}$$

is singular near $z = c$. To remove this singularity we expand the curley brackets in (12.3.22) as a Laurent series of $\sqrt{z-c}$ and insist that the coefficient of the term $\sqrt{z-c}$ vanish. This condition leads to the second relation

$$\frac{s(c)}{\pi}\left[2\sqrt{\frac{L}{c}}+\ln\frac{\sqrt{L}-\sqrt{c}}{\sqrt{L}+\sqrt{c}}\right] = -\beta + \frac{1}{\pi}\int_0^c \sqrt{\frac{L-x'}{c-x'}}\frac{s(x')-s(c)}{x'-c}dx'.$$
(12.3.24)

The problem is now completely solved. The integrals in (12.3.24) can be numerically evaluated once the slope $s(x)$ is specified.

According to this linearized theory, the velocity field is singular at the end of the cavity $z = L$, $w \propto 1/\sqrt{z-L}$. In fact, it is precisely this singularity that accounts for the drag force (Wu, 1968).

Many aspects of linearized cavity flow theory for hydrofoils at an angle of attack have been beautifully analyzed by conformal mapping and the Riemann–Hilbert techniques. Indeed, the general topic of two-dimensional cavities could not have been developed to the present state without the powerful techniques of complex functions.

12.4 Schwarz's principle of reflection

Many physical problems are initially defined in a half plane with conditions specified on the straight boundary. However, it is sometimes simpler to solve a fictitious problem in the whole plane first. For example, consider the steady conduction of heat in the lower half plane $y < 0$ due to a point heat source of strength q at (x_o, y_o) with $y_o < 0$. The temperature $T(x, y)$ satisfies Laplace's equation and the condition

$$T(x, 0, t) = 0 \qquad (12.4.1)$$

on the upper boundary $y = 0$. A convenient trick is the so-called *method of images*. The steady temperature in the entire x, y plane due to a point source is well known: $(q/2\pi) \ln |\mathbf{r}-\mathbf{r}_o|$, where $\mathbf{r} = (x, y)$ and $\mathbf{r}_o = (x_o, y_o)$. If we add this to the solution for a sink at the image point $\bar{\mathbf{r}}_o = (x_o, -y_o)$ in the upper half plane: $-(q/2\pi) \ln |\mathbf{r} - \bar{\mathbf{r}}_o|$, the sum satisfies Laplace's equation in the entire plane and the boundary condition on the x axis, and has the right singularity, hence gives the desired steady temperature in the lower half plane.

Now this trick can be worded more formally as follows. In view of (12.4.1) let us continue $T(x, y)$ from the lower half plane to the upper half plane by the relation

$$T'(x, y) = -T(x, -y).$$

By differentiation $T'(x,y)$ can be shown to satisfy Laplace's equation in the upper half plane. If we define $\hat{T}$ to be equal to $T(x,y)$ for $y < 0$ and to $T'(x,y)$ for $y > 0$, then $\hat{T}$ is harmonic in the whole x,y plane, odd with respect to, and zero along the x axis. If there is a source at $\mathbf{r}_o$, there must be a sink at $\bar{\mathbf{r}}_o$. The value of $\hat{T}$ due to the source/sink pair gives T in the lower half plane.

Extended to complex functions, the idea of continuation can be more powerful for treating less trivial problems. First let us define *analytical continuation* in general. Let two regions D_1 and D_2 overlap in a subregion D. If $f_1(z)$ is analytic in D_1, while $f_2(z)$ is analytic in D_2 and $f_1 = f_2$ in the overlapping region D, then $f_2(z)$ is said to be the analytical continuation of $f_1(z)$ in D_2, and $f_1(z)$ is the analytical continuation of $f_2(z)$ in D_1.

In theory there are many ways to construct the continuation of $f(z)$ from one region to a larger region. We discuss here only the Schwarz principle of reflection, which is useful for continuation across a segment of a straight line.

Let us first introduce the notation $\overline{\phi}(z)$ by the definition

$$\overline{\phi}(z) \equiv \overline{\phi(\bar{z})}. \tag{12.4.2}$$

More specifically, if

$$\phi(z) = u(x,y) + iv(x,y),$$

then

$$\overline{\phi}(z) = u(x,-y) - iv(x,-y).$$

Thus, if $\phi(z)$ is defined in the lower half plane, then $\overline{\phi}(z)$ is defined in the upper half plane. With this convention we state the following:

Schwarz's principle of reflection *In the lower half plane, let D be a closed region with part of its boundary L' on the real axis. Let $\phi(z)$ be analytic in D, continuous on its boundary and have a vanishing imaginary part on L'. Then $\overline{\phi}(z)$, defined by (12.4.2), is the analytical continuation of $\phi(z)$ across L' from D to its mirror reflection $\overline{D}$.*

First let us check that $\phi(z)$ and $\overline{\phi}(z)$ are continuous across L'. Since $\phi(x - i0) = u(x,0) + iv(x,0)$, we have $\overline{\phi}(x + i0) = u(x,0) - iv(x,0)$. It follows from $v(x,0) = 0$ along L' that

$$\phi(x) = \overline{\phi}(x), \quad x \in L'$$

and continuity is verified.

12.4 Schwarz's principle of reflection

Next we show that $\overline{\phi}(z)$ is an analytical function of z for $z \in \overline{D}$. Let us denote

$$u(x, -y) = \hat{u}(x, y), \qquad -v(x, -y) = \hat{v}(x, y). \qquad (12.4.3)$$

Since ϕ is analytic in z for $z \in D$, $u(x, y)$ and $v(x, y)$ satisfy the Cauchy–Riemann conditions

$$\frac{\partial u(x, y)}{\partial x} + \frac{\partial v(x, y)}{\partial y} = 0, \qquad \frac{\partial u(x, y)}{\partial y} = \frac{\partial v(x, y)}{\partial x}.$$

By changing y to $-y$ and using (12.4.3), it follows that

$$\frac{\partial \hat{u}(x, y)}{\partial x} + \frac{\partial \hat{v}(x, y)}{\partial y} = 0, \qquad \frac{\partial \hat{u}(x, y)}{\partial y} = \frac{\partial \hat{v}(x, y)}{\partial x},$$

and $\overline{\phi}(z) = u(x, -y) - iv(x, -y)$ is analytic in z for $z \in \overline{D}$.

Let us denote $\Phi(z)$ by

$$\Phi(z) = \begin{cases} \phi(z), & z \in D + L' \\ \overline{\phi}(z), & z \in \overline{D}. \end{cases}$$

$\Phi(z)$ has just been shown to be analytic in D and $\overline{D}$ and continuous across L'. As the last step we shall show that $\Phi(z)$ is also analytic on L'.

Referring to Figure 12.5, we pick any closed contour $C + L'_-$ in D, where L'_- is slightly beneath L'. By Cauchy's theorem, the following

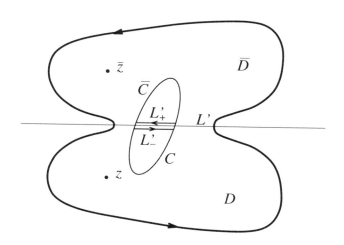

Fig. 12.5. Contours for reflection.

integral vanishes

$$\frac{1}{2\pi i}\oint_{C+L'_-}\frac{\Phi(t)dt}{t-z}=0$$

for z in and on $C+L'_-$. Similarly, we pick any other contour $\overline{C}+L'_+$ in $\overline{D}$, where L'_+ is slightly above L', and get

$$\frac{1}{2\pi i}\oint_{\overline{C}+L'_+}\frac{\Phi(t)dt}{t-z}=0$$

for z in and on $\overline{C}+L'_+$. Let L'_- and L'_+ approach L' from both sides and add these two integrals. Their sum gives

$$\frac{1}{2\pi i}\oint_{C+\overline{C}}\frac{\Phi(t)dt}{t-z}=0$$

because the contributions from L'_- and L'_+ cancel in the limit. Now since $C+\overline{C}$ is arbitrary, by Morea's theorem (§9.8), $\Phi(z)$ is analytic in $D+\overline{D}$ including the real axis. Thus $\overline{\phi}(z)$ is the analytical continuation of $\phi(z)$ in $\overline{D}$ and vise versa.

This idea of analytical continuation will be applied in the next section.

12.5 *Complex formulation of plane elasticity

We now turn to the second example of applications, the forced contact between a rigid strip and the surface of an elastic half space. Because Airy's stress function is biharmonic, the use of complex functions is far less straightforward, and the complex formulation of plane elasticity is itself a jewel of mathematical acrobatics.†

12.5.1 Airy's stress function

Recall from (7.5.11) that the Airy stress function F satisfies the biharmonic equation

$$\nabla^2\nabla^2 F=0. \qquad (12.5.1)$$

Let us change from the real variables x,y to the complex variables via $z=x+iy$ and $\overline{z}=x-iy$. Since

$$\frac{\partial}{\partial x}=\frac{\partial}{\partial z}+\frac{\partial}{\partial \overline{z}}, \qquad \frac{\partial}{\partial y}=i\frac{\partial}{\partial z}-i\frac{\partial}{\partial \overline{z}},$$

† Materials presented here are extracted from the treatise by Muskhelishvili (1962).

12.5 Complex formulation of plane elasticity

the Laplacian operator may be rewritten as

$$\nabla^2 = 4\frac{\partial^2}{\partial z \partial \bar{z}}.$$

Equation (12.5.1) becomes

$$\frac{\partial^4 F}{\partial z^2 \partial \bar{z}^2} = 0,$$

which may be integrated to give

$$2F = \bar{z}\phi(z) + z\phi_1(\bar{z}) + \chi(z) + \chi_1(\bar{z}),$$

where ϕ, ϕ_1, χ and χ_1 are yet unknown. In order that F is real we must choose

$$\phi_1(\bar{z}) = \overline{\phi(z)}, \qquad \chi_1(\bar{z}) = \overline{\chi(z)},$$

hence

$$2F = \bar{z}\phi(z) + z\overline{\phi(z)} + \chi(z) + \overline{\chi(z)}, \qquad (12.5.2)$$

where ϕ and χ are called the Goursat functions.

12.5.2 Stress components

Since

$$2\frac{\partial F}{\partial x} = \phi(z) + \bar{z}\phi'(z) + \overline{\phi(z)} + z\overline{\phi'(z)} + \psi(z) + \overline{\psi(z)}$$

and

$$2\frac{\partial F}{\partial y} = -i\phi(z) + i\bar{z}\phi'(z) + i\overline{\phi(z)} - iz\overline{\phi'(z)} + i\psi(z) - i\overline{\psi(z)},$$

where we have written

$$\psi(z) \equiv \chi'(z)$$

for later convenience, we obtain by addition

$$\frac{\partial F}{\partial x} + i\frac{\partial F}{\partial y} = \phi(z) + z\overline{\phi'(z)} + \overline{\psi(z)}. \qquad (12.5.3)$$

In view of (7.5.10), further differentiation with respect to x yields the stress components

$$\sigma_y - i\tau_{xy} = \phi'(z) + \overline{\phi'(z)} + z\overline{\phi''(z)} + \overline{\psi'(z)}. \qquad (12.5.4)$$

Similarly, the y derivative of (12.5.3) gives

$$-\tau_{xy} + i\sigma_x = i\phi'(z) + i\overline{\phi'(z)} - iz\overline{\phi''(z)} - i\overline{\psi'(z)}$$

or
$$\sigma_x + i\tau_{xy} = \phi'(z) + \overline{\phi'(z)} - z\overline{\phi''(z)} - \overline{\psi'(z)}. \tag{12.5.5}$$

Equation (12.5.4) is particularly convenient for specifying the boundary values of the traction forces σ_y and τ_{xy} along a horizontal boundary, while (12.5.5) is useful for specifying σ_x and τ_{xy} along a vertical boundary. The sum of (12.5.4) and (12.5.5) gives

$$\sigma_x + \sigma_y = 2\left[\phi'(z) + \overline{\phi'(z)}\right] = 4\text{Re}\left[\phi(z)\right]. \tag{12.5.6}$$

The difference of (12.5.4) and (12.5.5) gives

$$\sigma_y - \sigma_x - 2i\tau_{xy} = 2\left[z\overline{\phi''(z)} + \overline{\psi'(z)}\right]. \tag{12.5.7}$$

Equation (12.5.4) may also be written as

$$\sigma_y - i\tau_{xy} = \frac{\partial}{\partial x}\left[\phi(z) + z\overline{\phi'(z)} + \overline{\psi(z)}\right] \tag{12.5.8}$$

and can be integrated to give the complex traction force acting on a segment L of the horizontal boundary

$$\mathcal{F}_y - i\mathcal{F}_x = \int_L (\sigma_y - i\tau_{xy})\, dx = \left[\phi(z) + z\overline{\phi'(z)} + \overline{\psi(z)}\right]_A^B, \tag{12.5.9}$$

where $(\mathcal{F}_x, \mathcal{F}_y)$ are the horizontal and vertical forces on L, and A, B are the two end points of L.

12.5.3 Displacement components

For plane strain problems ($\varepsilon_z = 0$), Hooke's law (7.5.6) between stress and strain components can be written in terms of the Airy function

$$2\mu\frac{\partial U}{\partial x} = -\frac{\partial^2 F}{\partial x^2} + \frac{\lambda + 2\mu}{2(\lambda + \mu)}\nabla^2 F \tag{12.5.10}$$

and

$$2\mu\frac{\partial V}{\partial y} = -\frac{\partial^2 F}{\partial y^2} + \frac{\lambda + 2\mu}{2(\lambda + \mu)}\nabla^2 F, \tag{12.5.11}$$

where

$$\lambda = \frac{E\nu}{(1+\nu)(1-2\nu)} \quad \text{and} \quad \mu = \frac{E}{2(1+\nu)}$$

are called the Lamé constants (see, e.g., Fung, 1965). Unlike in §7.5 the solid displacements (U, V) are expressed in capital symbols. Since

12.5 Complex formulation of plane elasticity

(12.5.6) may be expressed in different ways

$$\sigma_x + \sigma_y = \nabla^2 F = 2\left[\phi'(z) + \overline{\phi'(z)}\right]$$

$$= 2\frac{\partial}{\partial x}\left[\phi(z) + \overline{\phi(z)}\right] \quad (12.5.12)$$

$$= -2i\frac{\partial}{\partial y}\left[\phi(z) - \overline{\phi(z)}\right], \quad (12.5.13)$$

we may integrate (12.5.10) and (12.5.11) to get

$$2\mu U = -\frac{\partial F}{\partial x} + \frac{(\lambda + 2\mu)}{\lambda + \mu}\left[\phi(z) + \overline{\phi(z)}\right]$$

and

$$2\mu V = -\frac{\partial F}{\partial y} + \frac{(\lambda + 2\mu)}{i(\lambda + \mu)}\left[\phi(z) - \overline{\phi(z)}\right].$$

Further use of (12.5.3) gives the complex displacement

$$2\mu D \equiv 2\mu(U + iV) = \kappa\phi(z) - z\overline{\phi'(z)} - \overline{\psi(z)}, \quad (12.5.14)$$

where

$$\kappa = \frac{\lambda + 3\mu}{\lambda + \mu} = 3 - 4\nu. \quad (12.5.15)$$

12.5.4 Half-plane problems

Let us consider the lower half plane of z, i.e., $y \leq 0$, and denote by L' the part of the ground surface that is traction-free and by L the part under direct loading either by applied stresses or by contact with another body. Along L' the boundary conditions are

$$\tau_{xy} = \sigma_y = 0, \quad x \in L', \quad y = 0. \quad (12.5.16)$$

Recall

$$\sigma_y - i\tau_{xy} = \phi'(z) + \overline{\phi'(z)} + z\overline{\phi''(z)} + \overline{\psi'(z)}.$$

It is convenient to introduce the following abbreviations in the rest of this section:

$$\Phi(z) \equiv \phi'(z), \quad \Psi(z) \equiv \psi'(z) \quad (12.5.17)$$

so that

$$\sigma_y - i\tau_{xy} = \Phi(z) + \overline{\Phi(z)} + z\overline{\Phi'(z)} + \overline{\Psi(z)} \quad (12.5.18)$$

from now on. On the segment L', $\tau_{xy} = \sigma_y = 0$, hence

$$\Phi_-(x) = -\overline{\Phi_-(x)} - x\overline{\Phi'_-(x)} - \overline{\Psi_-(x)}, \quad x \in L', \qquad (12.5.19)$$

where

$$f_-(x) \equiv \lim_{z \to x - i0} f(z).$$

Let us define for the upper half plane an analytic function $\Phi(z)$ by

$$\Phi(z) = -\overline{\Phi(z)} - z\overline{\Phi'(z)} - \overline{\Psi(z)}, \qquad (12.5.20)$$

which may be written

$$\Phi(z) = -\overline{\Phi(\bar z)} - z\overline{\Phi'(\bar z)} - \overline{\Psi(\bar z)}$$

in view of (12.4.2). As z approaches a point $x + i0 \in L'$ from above, $\bar z = x - i0$ approaches L' from below, and

$$\Phi_+(x) = -\overline{\Phi_-(x)} - x\overline{\Phi'_-(x)} - \overline{\Psi_-(x)}, \quad x \in L'. \qquad (12.5.21)$$

It follows by comparing (12.5.19) and (12.5.21) that

$$\Phi_+(x) = \Phi_-(x), \quad x \in L'. \qquad (12.5.22)$$

This crucial observation implies that $\Phi(z)$ is analytically continued to the upper half plane by (12.5.20). Let us interchange z and $\bar z$ in (12.5.20)

$$\Phi(\bar z) = -\overline{\Phi(z)} - \bar z\overline{\Phi'(z)} - \overline{\Psi(z)}$$

and take the complex conjugate

$$\overline{\Phi}(z) = -\Phi(z) - z\Phi'(z) - \Psi(z),$$

which may be solved for $\Psi(z)$

$$\Psi(z) = -\Phi(z) - \overline{\Phi}(z) - z\Phi'(z). \qquad (12.5.23)$$

Thus for a half plane with zero traction on part of the boundary, the mathematical problem is reduced to the search for one analytic function $\Phi(z)$, now defined in the entire z plane. In terms of Φ alone, (12.5.18) and (12.5.14) become

$$\sigma_y - i\tau_{xy} = \Phi(z) - \Phi(\bar z) + (z - \bar z)\overline{\Phi'(z)} \qquad (12.5.24)$$

and

$$2\mu\frac{\partial D}{\partial x} = \kappa\Phi(z) + \Phi(\bar z) - (z - \bar z)\overline{\Phi'(z)}. \qquad (12.5.25)$$

The seeds are now sown; we shall reap the fruits by considering a specific example.

12.6 *A strip footing on the ground surface

12.6.1 General solution to the boundary-value problem

We assume that on the top surface of the lower half plane ($y \leq 0$) the displacement components are prescribed along the segment L: $-a < x < a, y = 0$; the rest of the surface is free of traction. The boundary conditions of this problem are therefore mixed. Putting $z = x - i0$ in (12.5.25) and (12.5.24), we obtain

$$\Phi_+(x) + \kappa \Phi_-(x) = 2\mu D'(x), \quad x \in L, \quad y = 0 \qquad (12.6.1)$$

and

$$\Phi_+(x) - \Phi_-(x) = 0, \quad x \notin L, \quad y = 0. \qquad (12.6.2)$$

The search for $\Phi(z)$, defined in the entire z plane subject to (12.6.1) and (12.6.2) along L, is a Riemann–Hilbert problem. The solution, which vanishes at infinity, is

$$\Phi(z) = \frac{X(z)}{2\pi i} \int_L \frac{2\mu D'(t) dt}{X^+(t)(t-z)} + C_o X(z), \qquad (12.6.3)$$

where

$$X(z) = (z+a)^{-1/2+i\alpha}(z-a)^{-1/2-i\alpha} \qquad (12.6.4)$$

with

$$\alpha = \frac{\ln \kappa}{2\pi} \qquad (12.6.5)$$

in accordance with §12.2. The coefficient C_o has an important physical meaning. Recall from (12.5.9) that the complex traction force on any segment of a horizontal line from A to B is

$$\mathcal{F}_x + i\mathcal{F}_y = i \left[\phi(z) + \overline{z\phi'(z)} + \overline{\psi(z)} \right]_A^B. \qquad (12.6.6)$$

Because of (12.5.23), $\Psi(z)$ can be found from $\Phi(z)$, which is known from (12.6.3). For large z the reader may verify that $X \sim O(1/z)$ so that $\Phi \sim C_o/z$ and $\Psi \sim C_1/z$, where C_1 is some constant. It follows by integration that

$$\phi \cong C_o \ln z, \qquad \psi \cong C_1 \ln z.$$

Now we let A be any point to the left and B any point to the right of the footing so that (12.6.6) represents the total force from the footing to the ground. Substituting these results into (12.6.6), we get

$$\mathcal{F}_x + i\mathcal{F}_y \cong i\left[C_o \ln z + z\overline{\left(\frac{C_o}{z}\right)} + \overline{C_1 \ln z}\right]_A^B.$$

Since $x_A = |x_A|\exp(-i\pi)$ and $x_B = |x_B|$,

$$\mathcal{F}_x + i\mathcal{F}_y \cong i\left[(C_o + \overline{C}_1)\ln\frac{|x_B|}{|x_A|} + i\pi(C_o - \overline{C}_1)\right].$$

In order that the force be independent of the specific values of $|x_A|$ and $|x_B|$, it is necessary that $C_o + \overline{C}_1 = 0$.† This implies immediately that

$$\mathcal{F}_x + i\mathcal{F}_y = -2\pi C_o \qquad (12.6.7)$$

so that the constant C_o is proportional to the total complex force acted by the footing on the half space.

We now further specify the strip displacement D in order to evaluate the integrals in (12.6.3).

12.6.2 Vertically pressed flat footing

Directly beneath a rigid footing of width $2a$ pressed down by a vertical force P_o, the ground displacement is purely vertical and uniform in x. If no slipping is allowed, then the horizontal displacement vanishes so that

$$D = iV_o = \text{constant}, \quad -a < x < a. \qquad (12.6.8)$$

The integral in (12.6.3) vanishes, and the solution for Φ is simply

$$\Phi = \frac{iP_o}{2\pi}\frac{1}{\sqrt{z^2 - a^2}}\left(\frac{z+a}{z-a}\right)^{i\alpha}. \qquad (12.6.9)$$

The branch is chosen in the z plane cut along $-a < x < a, y = 0$ such that $X \sim 1/z$ for large z.

The stresses beneath the footing are, from (12.5.24),

$$\sigma_y - i\tau_{xy} = \Phi_+(x) - \Phi_-(x), \quad x \in L.$$

† The same result can be obtained from (12.5.23) by using the asymptotic expressions of $\Phi(z)$ and $\Psi(z)$ for large z.

12.6 A strip footing on the ground surface

The displacement condition implies $D' = 0$, hence from (12.5.25) we get $\kappa \Phi_- + \Phi_+ = 0$ on L, and

$$\sigma_y - i\tau_{xy} = -\frac{1+\kappa}{\kappa}\Phi_+(x), \quad -a < x < a.$$

Along the upper edge of the branch cut

$$z - a = (a-x)e^{i\pi}, \quad z + a = a + x,$$

hence

$$\sigma_y - i\tau_{xy} = -\frac{iP_o}{2\pi}\frac{1+\kappa}{\kappa}\frac{1}{\sqrt{a^2-x^2}}\left(\frac{a+x}{a-x}\right)^{i\alpha} e^{(1/2+i\alpha)\pi i}$$

$$= -\frac{P_o}{2\pi}\frac{1+\kappa}{\sqrt{\kappa}}\frac{1}{\sqrt{a^2-x^2}}\left(\frac{a+x}{a-x}\right)^{i\alpha},$$

where use has been made of (12.6.5). Note that

$$\left(\frac{a+x}{a-x}\right)^{i\alpha} = \exp\left(i\alpha \ln \frac{a+x}{a-x}\right),$$

hence

$$\sigma_y - i\tau_{xy} = -\frac{P_o}{2\pi}\frac{1+\kappa}{\sqrt{\kappa}}\left\{\cos\left(\alpha \ln \frac{a+x}{a-x}\right) + i\sin\left(\alpha \ln \frac{a+x}{a-x}\right)\right\}.$$

(12.6.10)

The stresses are highly oscillatory near the edges $x = \pm a$, owing to the assumption of no slipping. These results are due to Abramov (1937); see also Muskhelishvili (1962).

To a foundation engineer, the ground settlement due to a footing supporting a heavy load is also of interest. Let us calculate the deformation of the ground surface outside the footing: $z = x - i0$ with $|x| > a$. From (12.5.25) and (12.5.22) we get

$$2\mu \frac{\partial D}{\partial x} = \kappa \Phi_-(x) + \Phi_+(x) = (1+\kappa)\Phi^+(x), \quad |x| > a,$$

or

$$2\mu \frac{\partial D}{\partial x} = \frac{iP_o}{2\pi}\frac{1+\kappa}{\sqrt{x^2-a^2}}\left(\frac{x+a}{x-a}\right)^{i\alpha},$$

which can be integrated from a to x to give

$$2\mu\{U(x) + i[V(x) - V_0]\}$$
$$= \frac{iP_o}{2\pi}(1+\kappa) \int_a^x \frac{dx'}{\sqrt{x'^2 - a^2}} \left(\frac{x'+a}{x'-a}\right)^{i\alpha}$$
$$= \frac{iP_o}{2\pi}(1+\kappa) \int_1^{x/a} \frac{dx'}{\sqrt{x'^2 - 1}} \left(\frac{x'+1}{x'-1}\right)^{i\alpha}$$
$$= \frac{iP_o}{2\pi}(1+\kappa) \int_1^{x/a} \frac{dx'}{\sqrt{x'^2 - 1}} \left[\cos\left(\alpha \ln \frac{x'+1}{x'-1}\right) + i\sin\left(\alpha \ln \frac{x'+1}{x'-1}\right)\right].$$

Note first that the vertical displacement of the footing V_o is indeterminate. For $x/a \gg 1$, the integrand behaves as $1/x'$ so that the integral behaves as $\ln x/a$. Let us rewrite the preceding result, for any finite x/a,

$$2\mu\{U(x) + i[V(x) - V(a)]\} = \frac{iP_o}{2\pi}(1+\kappa)\left\{\ln\frac{x}{a} + \int_1^{x/a} dx' \left\{\frac{1}{\sqrt{x'^2 - 1}} \left[\cos\left(\alpha \ln \frac{x'+1}{x'-1}\right) + i\sin\left(\alpha \ln \frac{x'+1}{x'-1}\right)\right] - \frac{1}{x'}\right\}\right\}.$$

Separating the real and imaginary parts, we get for $a < x < \infty, y = 0$,

$$2\mu U(x) = -\frac{P_o}{2\pi}(1+\kappa) \int_1^{x/a} \frac{dx'}{\sqrt{x'^2 - 1}} \sin\left(\alpha \ln \frac{x'+1}{x'-1}\right) \quad (12.6.11)$$

and

$$2\mu[V(x) - V_o] = \frac{P_o}{2\pi}(1+\kappa)\left\{\ln\frac{x}{a} + \int_1^{x/a} \left[\frac{1}{\sqrt{x'^2 - 1}} \cos\left(\alpha \ln \frac{x'+1}{x'-1}\right) - \frac{1}{x'}\right] dx'\right\}. \quad (12.6.12)$$

Far from the footing $x/a \gg 1$, the horizontal displacement is

$$2\mu U(x) \sim -\frac{P_o}{2\pi}(1+\kappa) \int_1^\infty \frac{dx'}{\sqrt{x'^2 - 1}} \sin\left(\alpha \ln \frac{x'+1}{x'-1}\right), \quad (12.6.13)$$

which is a finite number and does not diminish to zero! Worse still,

$$2\mu[V(x) - V_o] \sim \frac{P_o}{2\pi}(1+\kappa)\left\{\ln\frac{x}{a} + \int_1^\infty \left[\frac{1}{\sqrt{x'^2-1}}\cos\left(\alpha\ln\frac{x'+1}{x'-1}\right) - \frac{1}{x'}\right]dx'\right\}, \quad (12.6.14)$$

which is unbounded at infinity! These results (Tuck and Mei, 1983) are clearly unacceptable on physical grounds. As usual, paradoxical deductions are caused by shortcomings in the original assumptions. A possible remedy is to consider the more realistic three-dimensional model of a slender footing of finite length, as discussed in Kalker (1977) and Tuck and Mei (1983).

In this chapter we have only discussed one of the simplest applications of the Riemann–Hilbert problems to plane elasticity. A few similar examples are included in the exercises, which should better acquaint the reader with the basic ideas. For further information and inspiration, the reader will find the treatise by Muskhelishvili unmatchable. For applications to the modern topic of fracture mechanics, where one examines the stress concentration near the tips of thin cracks in an elastic solid, the article by Rice (1968), and the books by Sih (1973) and Kanninen and Popelar (1985) are highly recommended.

Exercises

12.1 Solve for the complex function $F(z) = u(x,y) + iv(x,y)$ analytic in the lower half plane ($y \leq 0$) and subject to the mixed boundary conditions

$$u(x,0) = a(x), \quad |x| < 1$$
$$v(x,0) = b(x), \quad |x| > 1$$

with $b \sim x^{-2}$ as $|x| \sim \infty$. Introduce first an analytic function for the entire z plane by

$$F(z) = \begin{cases} f(z), & y \leq 0 \\ \overline{f(\bar{z})}, & y \geq 0, \end{cases}$$

where $\bar{f}(z) = \overline{f(\bar{z})}$ is the analytic continuation of $f(z)$ in the upper half plane. Then show that

$$F_+(x) + F_-(x) = 2a(x), \quad |x| < 1$$
$$F_+(x) - F_-(x) = -2ib(x), \quad |x| > 1$$

and work out the result.

12.2 Apply the general result of the last problem to find the pressure in a porous soil beneath a concrete dam with a flat base ($-L < x < L, y = 0$). The water depth is h in the reservoir ($x < -L$) and zero outside the reservoir ($x > L$).

12.3 The dynamic pressure on a cylinder oscillating near the sea surface is important in ship hydrodynamics. If the amplitude is small and the frequency of oscillation high, the velocity potential ϕ and stream function ψ satisfy Laplace's equation and the following boundary conditions

$$\phi(x, 0) = 0, \quad x \text{ on } S_f$$

$$\psi(x, y) = \psi_o(x, y), \quad (x, y) \text{ on } S_B,$$

where S_f denotes the sea surface and S_B the mean surface of the cylinder. Consider specifically a semicircular cylinder of unit radius heaving vertically at unit amplitude. Show first that the stream function on the cylinder surface is $\psi = -x$. Then map the lower half of the z plane outside the semicircle onto the lower half plane of $\zeta = \xi + i\eta$ by Joukowski transformation

$$\zeta = \frac{1}{2}\left(z + \frac{1}{z}\right)$$

or

$$z = \zeta + \sqrt{\zeta^2 - 1}.$$

Deduce the following mixed boundary conditions on the real axis $\operatorname{Im} \zeta = 0$

$$\phi = 0, \quad |\xi| > 1; \qquad \psi = -\xi, \quad |\xi| < 1.$$

Finally, show that $f(z) = \phi + i\psi = 1/z$. (This simple solution can be obtained in much quicker ways, but the Riemann–Hilbert technique can be effective for cylinders of more general form if the mapping function from z to ζ is first found.)

12.4 Let the traction (normal and shear stresses) be specified along the top of the elastic half plane $y \leq 0$

$$\sigma_y = -P(x) \quad \text{and} \quad \tau_{xy} = T(x), \quad -\infty < x < \infty,$$

where

$$P(x) = T(x) = 0, \quad x \in L'.$$

Show that

$$\Phi_+(x) - \Phi_-(x) = P(x) + iT(x)$$

on $y = 0$. Use Plemelj's formulas to obtain the general solution for the stresses and strains.

12.5 Verify that if the applied stresses on the top of an elastic half space are

$$P = \frac{P_o}{\sqrt{a^2 - x^2}}, \quad -a < x < a; \quad T = 0, \quad -\infty < x < \infty,$$

then the Goursat stress function is

$$\phi(z) = \frac{iP_o}{2\pi} \ln\left(z + \sqrt{z^2 - a^2}\right)$$

for $y \leq 0$. Show that the displacement components on $|x| < a, y = 0$ are

$$U = \frac{P_o}{4\pi\mu}(1 - \kappa) \sin^{-1}\frac{|x|}{a}, \quad V = V_o = \text{constant}.$$

Thus the solution applies to a rigid and frictionless base pressed on the half space by the normal force P_o. Show that the displacement of the unloaded part of the ground surface is

$$U = \pm\frac{P_o}{4}(1 - \kappa), \quad V - V_o = \frac{P_o}{4\pi\mu}(1 + \mu)\ln\left(\frac{|x|}{a} + \sqrt{\frac{x^2}{z^2} - 1}\right)$$

and comment on the physical implications.

12.6 Consider the linearized two-dimensional problem of radiation of surface water waves by an oscillating thin ship (Levine and Rodemich, 1958). The spatial factor of the velocity potential is governed by

$$\nabla^2 \phi = 0, \quad |x| < \infty, \quad y < 0$$

$$\frac{\partial \phi}{\partial y} - \nu\phi = 0, \quad |x| < \infty, \quad y = 0$$

$$\frac{\partial \phi}{\partial x} = U(y), \quad x = \pm 0, \quad -a < y < 0,$$

where $\nu = \omega^2/g$. In addition ϕ behaves as outgoing waves at infinity. Follow the steps suggested below and reduce the above problem to a Riemann–Hilbert problem.

Let $f(z) = \phi + i\psi$ and show that the boundary-value problem becomes the search for an analytic function in the lower half z plane subject to

$$\text{Im}\,(f' - i\nu f) = 0, \quad |x| < \infty, \quad y = 0$$
$$\text{Re}\,(f' - i\nu f) = W(y), \quad x = \pm 0, \quad -a < y < 0,$$

where

$$W(y) = U(y) + \int_{-a}^{y} U(y')dy'.$$

Introduce $F(z) = f'(z) - i\nu f(z)$ so that the boundary conditions become

$$\text{Im}\,F = 0, \quad |x| < \infty, \quad y = 0$$
$$\text{Re}\,F = W(y), \quad x = \pm 0, \quad -a < y < 0.$$

Continue F to the upper half plane by $\overline{F(\bar z)}$, i.e., define

$$\Omega(z) = \begin{cases} F(z), & y \le 0 \\ \overline{F(\bar z)}, & y \ge 0. \end{cases}$$

Show that $\Omega(z)$ must be analytic in the entire z plane and satisfies

$$\text{Re}\,\Omega(z) = W(-|y|), \quad -a < y < a.$$

Finally, show that

$$\Omega_+(z_o) + \Omega_-(z_o) = 2W(-|y|), \quad -a < y < a, \quad x = 0$$

and

$$\Omega_+(z_o) - \Omega_-(z_o) = 0, \quad |y| > a, x = 0,$$

where $z_o = 0 + iy$ is a point on the vertical plane, and $\Omega_\pm(z_o)$ is the limit of $\Omega(z)$ as $x \to \pm\infty$ and $-a < y < a$. The complete analysis is lengthy and involves the use of the radiation condition (see Mei, 1966; and Wehausen and Laitone, 1960).

13
Perturbation methods – the art of approximation

13.1 Introduction

In previous chapters we have only discussed techniques of getting exact solutions. Clearly, the problems must be sufficiently idealized for these techniques to be effective. For more practical problems either the boundary geometry or the governing equations are less simple, and one must often be content with approximate solutions. Among methods of approximation two are the most important. If the problem is close to one that is solvable exactly, perturbation methods are powerful tools for getting analytical answers. If, however, the problem is far from anything that can be solved exactly, strictly numerical methods via discretization must be employed. In general, analytical perturbation methods are much more effective in gaining qualitative insight, while numerical methods are good in producing quantitative information. Sometimes the two can be mixed for studying small departures from a basic state that must itself be solved numerically.

In this chapter we shall give an introductory account of the analytical approach of perturbation methods. To have a bird's-eye view of the subject, let us first outline the typical ideas and procedure of a perturbation analysis.

(i) Identify a small parameter. This is a very important first step which must be taken by recognizing the physical scales relevant to the problem. One then normalizes all variables with respect to these characteristic scales. In the normalized form, the governing equations will display certain dimensionless parameters, each of which represents the relative importance of certain physical mechanisms. If one of the parameters, say ϵ, is much less than unity,† then ϵ can be chosen as the perturbation parameter.

(ii) Expand the solution as an ascending series of the small parameter ϵ, for example, a power series

$$u = u_0 + \epsilon u_1 + \epsilon^2 u_2 + \cdots, \qquad (13.1.1)$$

where u_n is called the nth-order term. The form of the series may vary according to the manner that ϵ appears in the equations. If, for example, $\epsilon^{1/2}, \epsilon, \ldots$ are present, try a series in integral powers of $\epsilon^{1/2}$. If only $\epsilon^2, \epsilon^4, \ldots$ appear, try a series involving integral powers of ϵ^2, etc.

(iii) Collect terms of the same order in all governing equations and auxiliary conditions, and get perturbation equations at each order.

(iv) Starting from the lowest order, solve the problems at each order successively, up to certain order, say $O(\epsilon^m)$.

(v) Substitute the results for $u_n, n = 0, 1, 2, \ldots$ back into (13.1.1) to get the final solution, which is accurate up to some desired order, say $O(\epsilon^m)$.

This straightforward procedure is called *regular perturbation analysis*.

There are, however, many situations where the regular perturbation series fails in some range of the independent variable. Then one must turn to the *singular perturbation analysis*, which involves the following additional steps:

(i) Diagnose the failure of the regular expansion. Check which of the original assumptions are violated when failure occurs. Are the original scales still valid? Which terms that were supposed to be small are now large? Examine the qualitative nature of breakdown.

(ii) Choose new scales and new normalization to resurrect the terms that should be important near breakdown and start a new perturbation analysis. Sometimes the new solution may reveal the need to include ordering terms such as $\epsilon \ln \epsilon$, $\epsilon^2 \ln \epsilon, \ldots$, etc.

The above scheme can be applied to almost any problem. Indeed, the governing equations can be algebraic or transcendental equations, ordinary or partial differential equations or integral equations.

We stress again the importance of identifying the right small parameter by first finding the relevant scales, which is a matter of physics. Perturbation analysis itself is a mathematical exercise. Without some physical foresight, it is difficult to make effective use of the mathematics to be discussed. We shall explain these ideas and tactics through

† If the parameter happens to be large, its reciprocal is small.

examples. In §13.2 algebraic equations are discussed. Regular perturbations are then demonstrated in §13.3 to §13.4 by two problems in heat transfer governed by ordinary or partial differential equations. The first singular perturbation method of multiple scales is explained in §13.5 to §13.7. Breakdown of regular perturbations is first diagnosed to motivate the remedy. Physical applications include the recent method of homogenization, which is designed for deducing effective equations on the macroscale if variations on the microscale are periodic. §13.8 to §13.11 then deal with the boundary-layer technique as the second singular perturbation method.

In general the execution of perturbation analysis can be tedious. We explain in Chapter 14 the basics and the use of a popular symbolic computation software called MASCYMA, which can be as helpful in manipulating formulas as FORTRAN is in crunching numbers.

13.2 Algebraic equations

13.2.1 Regular perturbations

Let us examine the quadratic equation

$$u^2 + \epsilon u - 1 = 0, \tag{13.2.1}$$

where ϵ is much smaller than unity. The exact solution is well known and will be used for comparison later. We shall use this simple example to illustrate the procedure that can be extended to equations that cannot be solved exactly.

Let us propose to find the solution as a perturbation series

$$u = u_0 + \epsilon u_1 + \epsilon^2 u_2 + \epsilon^3 u_3 + \cdots \tag{13.2.2}$$

and substitute this series into (13.2.1)

$$(u_0 + \epsilon u_1 + \epsilon^2 u_2 + \epsilon^3 u_3 + \cdots)^2 +$$
$$\epsilon(u_0 + \epsilon u_1 + \epsilon^2 u_2 + \epsilon^3 u_3 + \cdots) - 1 = 0.$$

Expanding the square and collecting terms of equal powers, we get

$$(u_0^2 - 1) + \epsilon(2u_0 u_1 + u_0) + \epsilon^2(2u_0 u_2 + u_1^2 + u_1) + \cdots = 0.$$

With the coefficient of each power of ϵ set to zero, a sequence of perturbation equations is obtained at various orders

$$O(\epsilon^0): \quad u_0^2 - 1 = 0 \tag{13.2.3}$$

$$O(\epsilon): \quad 2u_0 u_1 + u_0 = 0 \tag{13.2.4}$$

$$O(\epsilon^2): \quad 2u_0 u_2 + u_1^2 + u_1 = 0 \tag{13.2.5}$$

$$\vdots$$

From (13.2.3) the lowest-order solution is

$$u_0 = \pm 1.$$

With this result higher-order problems are solved successively

$$u_1 = -\frac{1}{2}$$

and

$$u_2 = -\frac{u_1^2 + u_1}{2u_0} = \pm\frac{1}{8}.$$

Summarizing, the approximation up to $O(\epsilon^2)$ is

$$u = \pm 1 - \frac{\epsilon}{2} \pm \frac{\epsilon^2}{8} + \cdots. \tag{13.2.6}$$

Higher-order corrections can be continued if desired.

In this case the efficacy of the approximate result can be judged by comparing with the exact solution

$$u = \frac{1}{2}(-\epsilon \pm \sqrt{\epsilon^2 + 4}),$$

which can be expanded for small ϵ in Taylor series

$$u = \frac{1}{2}\left(-\epsilon \pm 2\sqrt{1 + \frac{\epsilon^2}{4}}\right) = \frac{1}{2}\left[-\epsilon \pm 2\left(1 + \frac{\epsilon^2}{8} + O(\epsilon^4)\right)\right]$$

$$= \pm 1 - \frac{\epsilon}{2} \pm \frac{\epsilon^2}{8} + O(\epsilon^4).$$

Clearly, this result confirms the perturbation series to the accuracy calculated.

Note that since the small parameter does not multiply the term with the highest power, the perturbation equation at the leading order for $u^{(0)}$ is still quadratic and has two solutions. Higher-order solutions simply improve the accuracy of the two. This feature is typical of regular perturbations.

13.2.2 Singular perturbations

The following cubic equation

$$u = 1 + \epsilon u^3 \qquad (13.2.7)$$

can also be solved exactly. For small ϵ let us try the straightforward expansion

$$u = u_0 + \epsilon u_1 + \epsilon^2 u_2 + \cdots. \qquad (13.2.8)$$

Substituting this series into (13.2.7) and expanding the cubic term, we get

$$u_0 + \epsilon u_1 + \epsilon^2 u_2 + \epsilon^3 u_3 + O(\epsilon^4)$$
$$= 1 + \epsilon \left[u_0^3 + \epsilon 3 u_0^2 u_1 + \epsilon^2 (3 u_0^2 u_2 + 3 u_0 u_1^2) + O(\epsilon^3) \right].$$

Equating like powers of ϵ yields the perturbation equations

$$O(\epsilon^0): \qquad u_0 = 1$$
$$O(\epsilon): \qquad u_1 = u_0^3$$
$$O(\epsilon^2): \qquad u_2 = 3 u_0^2 u_1$$
$$\vdots$$

The solutions are obviously

$$u_0 = 1, \quad u_1 = 1, \quad u_2 = 3, \ldots,$$

hence the final solution is

$$u = 1 + \epsilon + 3\epsilon^2 + O(\epsilon^3). \qquad (13.2.9)$$

Why did the two other solutions of the original cubic equation disappear? The reason is that in (13.2.7) the term u^3 of highest power is multiplied by the small parameter. The straightforward perturbation series annihilates the highest power at the leading order, hence only one solution is left; higher-order analysis merely improves the accuracy of this solution. In similar situations the problem is called *singular*, and the straightforward expansion is sometimes called the *naive* expansion.

After diagnosing the source of error we seek a cure by rescaling the unknown so as to shift the small parameter to a lower-order term in the new equation. Let $u = x\epsilon^m$, where m is yet unknown. Equation (13.2.7) then becomes

$$x\epsilon^m = 1 + x^3 \epsilon^{3m+1}. \qquad (13.2.10)$$

We now face several choices: (i) All three terms are equally important.

This choice would require $m = 0$ and $3m + 1 = 0$, which cannot be satisfied at the same time. (ii) Only one of the three terms dominates. Clearly, the results are full of inconsistencies. (iii) Two out of three terms dominate over the remaining one. One must now identify the pair by trial and error.

Let us first assume that the second and third terms in (13.2.10) are more important than the first. Equating the powers of ϵ, we get $3m+1 = 0$, implying that $m = -1/3$. But (13.2.10) becomes $x\epsilon^{-1/3} = 1 + x^3$, where the first term appears to be the greatest, contradicting the original assumption. Hence the choice is not acceptable. The second choice that the first and second terms dominate must also be ruled out, since this corresponds to the naive expansion. The remaining choice is to balance the first and third. Equating their powers of ϵ, we get $m = 3m + 1$ or $m = -1/2$. Equation (13.2.10) becomes

$$x\epsilon^{-1/2} = 1 + x^3 \epsilon^{-1/2}.$$

Indeed, the second term is much smaller. Substituting the new expansion

$$x = x_0 + \epsilon^{1/2} x_1 + \epsilon x_2 + \epsilon^{3/2} x_3 + \cdots$$

into (13.2.10) and collecting like powers of ϵ, we get the perturbation equations

$$O(1): \quad x_0^3 - x_0 = 0$$
$$O(\epsilon^{1/2}): \quad 3x_0^2 x_1 - x_1 + 1 = 0$$
$$O(\epsilon): \quad 3x_0^2 x_2 + 3x_0 x_1^2 - x_2 = 0$$
$$\vdots$$

The solutions at successive orders are

$$x_0 = (0, 1, -1)$$

$$x_1 = -\frac{1}{3x_0^2 - 1}$$

$$x_2 = -\frac{3x_0 x_1^2}{3x_0^2 - 1}$$

$$\vdots$$

To find $x_1, x_2, \ldots$ explicitly the three solutions for x_0 must be taken one at a time. For $x_0 = 0$, we have $x_1 = 1$ and $x_2 = 0$, hence

$$x = \epsilon^{1/2} + O(\epsilon^{3/2})$$

and
$$u = \epsilon^{-1/2} x = 1 + O(\epsilon). \qquad (13.2.11)$$

This result is just the solution found by naive expansion. For the second root $x_0 = 1$, $x_1 = -1/2$ and $x_2 = -3/8$, hence
$$x = 1 - \frac{\epsilon^{1/2}}{2} - \frac{3}{8}\epsilon + \cdots$$
and
$$u = \epsilon^{-1/2} x = \epsilon^{-1/2} - \frac{1}{2} + \frac{3}{8}\epsilon^{1/2} + \cdots. \qquad (13.2.12)$$

For the third root $x_0 = -1$, $x_1 = -1/2$ and $x_2 = 3/8$, hence
$$x = -1 - \frac{\epsilon^{1/2}}{2} + \frac{3\epsilon}{8} + \cdots,$$
$$u = -\epsilon^{-1/2} - \frac{1}{2} + \frac{3}{8}\epsilon^{1/2} + \cdots. \qquad (13.2.13)$$

Improvement for all the roots can be pursued in an obvious manner.

In the following two sections we discuss examples of regular perturbations for differential equations. These examples are but a few of many interesting applications of perturbation methods to heat transfer collected in the book by Aziz and Na (1984).

13.3 Parallel flow with heat dissipation

Consider a steady laminar flow between two parallel walls. The lower wall is fixed while the upper wall moves at the speed U_o in its own plane. The temperature of the two walls is kept constant at T_o. If the speed U_o is sufficiently high, dissipation caused by viscous shear can affect the flow and temperature distribution. We now wish to calculate this mutual influence.

In the region $0 < y < a$, the flow is one-dimensional in the x direction. The viscous stress $\mu du/dy$ cannot vary along y if there is no pressure gradient. Conservation of momemtum in the x direction demands that
$$\frac{d}{dy}\left(\mu \frac{du}{dy}\right) = 0. \qquad (13.3.1)$$

On the other hand, energy conservation requires that the rate of heat diffusion be balanced by the rate of viscous dissipation
$$k\frac{d^2 T}{dy^2} + \mu \left(\frac{du}{dy}\right)^2 = 0. \qquad (13.3.2)$$

We shall consider a fluid whose viscosity varies with the local temperature according to

$$\mu = \mu_o e^{-\alpha(T-T_o)}, \tag{13.3.3}$$

where T_o is the wall temperature.

As boundary conditions along $y = 0$ and a, the fluid must adhere to the walls, and the temperature is fixed

$$u = 0, \quad T = T_o \quad \text{on} \quad y = 0 \tag{13.3.4}$$

$$u = U_o, \quad T = T_o \quad \text{on} \quad y = a. \tag{13.3.5}$$

Thus the unknowns u and T are nonlinearly coupled by ordinary differential equations and (13.3.3).

In order to see whether there are small parameters we introduce the scaled variables

$$\theta = \frac{T - T_o}{T_o}, \quad Y = \frac{y}{a}, \quad U = \frac{u}{U_o}. \tag{13.3.6}$$

The governing equations become

$$\frac{d}{dY}\left(e^{-\beta\theta}\frac{dU}{dY}\right) = 0 \tag{13.3.7}$$

and

$$\frac{d^2\theta}{dY^2} + \epsilon e^{-\beta\theta}\left(\frac{dU}{dY}\right)^2 = 0, \tag{13.3.8}$$

where

$$\beta = \alpha T_o, \quad \epsilon = \frac{\mu_o U_o^2}{kT_o}. \tag{13.3.9}$$

The dimensionless boundary conditions are

$$U = 0, \quad \theta = 0 \quad \text{on} \quad Y = 0 \tag{13.3.10}$$

and

$$U = 1, \quad \theta = 0 \quad \text{on} \quad Y = 1. \tag{13.3.11}$$

Clearly, if the wall velocity U_o is sufficiently low, the parameter ϵ is very small, $\epsilon \ll 1$. We may introduce the expansions

$$U = U_0 + \epsilon U_1 + \epsilon^2 U_2 + \epsilon^3 U_3 + \cdots \tag{13.3.12}$$

$$\theta = \theta_0 + \epsilon\theta_1 + \epsilon^2\theta_2 + \epsilon^3\theta_3 + \cdots. \tag{13.3.13}$$

13.3 Parallel flow with heat dissipation

Note that

$$\exp\left[-\beta\left(\theta_0 + \epsilon\theta_1 + \epsilon^2\theta_2 + \cdots\right)\right]$$
$$= e^{(-\beta\theta_0)} e^{(-\epsilon\beta\theta_1)} e^{(-\epsilon^2\beta\theta_2)} \cdots$$
$$= e^{-\beta\theta_0}\left[1 - \epsilon\beta\theta_1 + \frac{\epsilon^2\beta^2\theta_1^2}{2} + \cdots\right]\left[1 - \epsilon^2\beta\theta_2 + \cdots\right]\left[1 - O(\epsilon^3)\right]$$
$$= e^{-\beta\theta_0}\left[1 - \epsilon\beta\theta_1 + \epsilon^2\left(\frac{(\beta\theta_1)^2}{2} - \beta\theta_2\right)\right] + O(\epsilon^3).$$

Note also that

$$\left(\frac{dU}{dY}\right)^2 = \left(\frac{dU_0}{dY} + \epsilon\frac{dU_1}{dY} + \epsilon^2\frac{dU_2}{dY} + \cdots\right)^2$$
$$= \left(\frac{dU_0}{dY}\right)^2 + \epsilon\left(2\frac{dU_0}{dY}\frac{dU_1}{dY}\right) + \epsilon^2\left[2\frac{dU_0}{dY}\frac{dU_1}{dY} + \left(\frac{dU_1}{dY}\right)^2\right] + O(\epsilon^3).$$

The perturbation problem at $O(1)$ is governed by

$$\frac{d}{dY}\left(e^{-\beta\theta_0}\frac{dU_0}{dY}\right) = 0$$

$$\frac{d^2\theta_0}{dY^2} = 0$$

with the inhomogeneous boundary conditions

$$U_0 = 0, \quad \theta_0 = 0 \quad \text{on} \quad Y = 0$$
$$U_0 = 1, \quad \theta_0 = 0 \quad \text{on} \quad Y = 1.$$

The solution is

$$\theta_0 = 0, \quad U_0 = Y. \tag{13.3.14}$$

Because of this result, $e^{-\beta\theta_0} = 1$ and the $O(\epsilon)$ problem is governed by

$$\frac{d}{dY}\left[\left(\frac{dU_1}{dY} - \beta\theta_1\frac{dU_0}{dY}\right)\right] = 0$$

$$\frac{d^2\theta_1}{dY^2} + \left(\frac{dU_0}{dY}\right)^2 = 0$$

and the homogeneous boundary conditions

$$U_1 = 0, \quad \theta_1 = 0 \quad \text{on} \quad Y = 0, 1.$$

The solution is

$$\theta_1 = \frac{1}{2}Y(1-Y) \qquad (13.3.15)$$

$$U_1 = -\frac{\beta}{2}(Y - 4Y^2 + 3Y^2). \qquad (13.3.16)$$

The $O(\epsilon^2)$ problem is defined by

$$\frac{d}{dY}\left\{\frac{dU_2}{dY} - \beta\theta_1\frac{dU_1}{dY} + \left[\frac{(\beta\theta_1)^2}{2} - \beta\theta_2\right]\frac{dU_0}{dY}\right\} = 0$$

$$\frac{d^2\theta_2}{dY^2} + 2\frac{dU_0}{dY}\frac{dU_1}{dY} - \beta\theta_1\left(\frac{dU_0}{dY}\right)^2 = 0$$

and the homogeneous boundary conditions

$$U_2 = 0, \qquad \theta_2 = 0 \quad \text{on} \quad Y = 0, 1.$$

The solution is

$$\theta_2 = -\frac{\beta}{24}(Y - 2Y^2 + 2Y^3 - Y^4) \qquad (13.3.17)$$

$$U_2 = \frac{\beta^2}{120}(Y - 5Y^2 + 10Y^3 - 10Y^4 + 4Y^5). \qquad (13.3.18)$$

The final solution is obtained by substituting (13.3.14–13.3.18) into the expansions (13.3.12–13.3.13).

The original differential equations for U and T can be solved exactly (Turian and Bird, 1963). Comparison of the exact and approximate solutions for fairly large $\epsilon = 0.4, 0.6, 0.8$ have been made by Aziz and Na (1984). The agreement is excellent, as shown in Figure 13.1. This kind of unexpected accuracy is common in perturbation theories.

13.4 Freezing of water surface

As a simple model of the freezing of water surface beneath a cold atmosphere, we consider the temperature variation in the half space $x > 0$, as sketched in Figure 13.2. At the begining, $t = 0$, there is no ice and the boundary temperature is suddenly lowered to below-freezing temperature T_0. At any later time t, the temperature in ice and water is governed, respectively, by

$$k\frac{\partial^2 T}{\partial x^2} = \frac{\partial T}{\partial t}, \quad 0 < x < y(t) \qquad (13.4.1)$$

13.4 Freezing of water surface

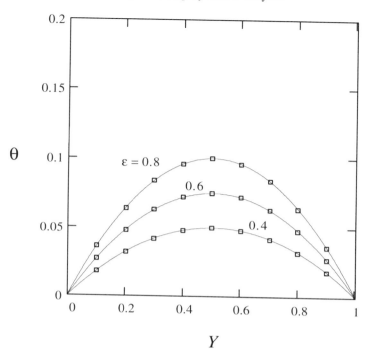

Fig. 13.1. Exact (solid line) and approximate (squares) in parallel flow, $\beta = 1$.

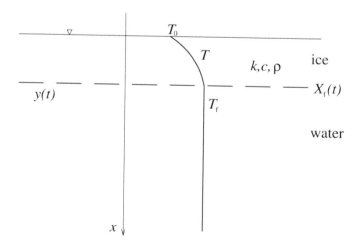

Fig. 13.2. Freezing near the water surface.

and
$$k'\frac{\partial^2 T'}{\partial x^2} = \frac{\partial T'}{\partial t}, \quad y(t) < x < \infty. \qquad (13.4.2)$$

The boundary conditions are
$$T(0,t) = T_0 \qquad (13.4.3)$$

and
$$T'(\infty, t) = T_1. \qquad (13.4.4)$$

On the interface, where the phase changes from liquid to solid, the temperature must be continuous
$$T(y(t), t) = T'(y(t), t) = T_f, \qquad (13.4.5)$$

where T_f is the temperature of freezing. In addition the net loss of heat must be expended at the rate proportional to dy/dt
$$K\frac{\partial T}{\partial x} - K'\frac{\partial T'}{\partial x} = \rho L\frac{dy}{dt}, \quad x = y(t), \qquad (13.4.6)$$

where L denotes the latent heat of phase transition and
$$K = kc\rho, \qquad K' = k'c'\rho'.$$

The initial conditions are
$$T = T_1, \quad 0 < x < \infty, \qquad (13.4.7)$$

implying that
$$y(0) = 0. \qquad (13.4.8)$$

While the diffusion equations in both ice and water are linear, the interface conditions are not. The position of the interface is unknown and must be solved as part of the solution. This type of problem is called the free boundary problem or Stefan problem and is notoriously difficult. We shall examine a simpler version of it by assuming that water is kept at a constant temperature T_f. Therefore the reduced problem is governed by (13.4.1), subject to the condition (13.4.3),
$$T = T_f, \qquad x = y(t) \qquad (13.4.9)$$

on the free surface and
$$K\frac{\partial T}{\partial x} = \rho L\frac{dy}{dt}, \qquad x = y(t) \qquad (13.4.10)$$

on the interface. The initial condition is still given by (13.4.8).

13.4 Freezing of water surface

We follow Aziz and Na (1984) and introduce the following normalization:

$$\theta = \frac{T_f - T}{T_f - T_0}, \quad X = \frac{x}{a}, \quad Y = \frac{y}{a}, \quad \tau = \frac{kt}{a^2}, \qquad (13.4.11)$$

where a is some unspecified length. The heat equation becomes

$$\frac{\partial^2 \theta}{\partial X^2} = \frac{\partial \theta}{\partial \tau}. \qquad (13.4.12)$$

The boundary conditions are now

$$\theta = 1, \qquad X = 0 \qquad (13.4.13)$$

$$\theta = 0, \qquad X = Y(\tau) \qquad (13.4.14)$$

$$\frac{dY}{d\tau} = -\epsilon \frac{\partial \theta}{\partial X}, \quad X = Y(\tau), \qquad (13.4.15)$$

while the initial condition becomes

$$Y(0) = 0. \qquad (13.4.16)$$

The parameter ϵ denotes the ratio of sensible heat to latent heat

$$\epsilon = \frac{C(T_f - T_0)}{L}. \qquad (13.4.17)$$

Since $Y(\tau)$ is expected to be a monotonic function of τ, we replace τ by Y as the second independent variable, i.e., $\theta(X, \tau)$ is changed to $\theta(X, Y)$. Since

$$\frac{\partial \theta}{\partial \tau} = \frac{\partial \theta}{\partial Y} \frac{dY}{d\tau} = -\epsilon \frac{\partial \theta}{\partial Y} \left(\frac{\partial \theta}{\partial X} \right)_{X=Y},$$

the heat equation now reads

$$\frac{\partial^2 \theta}{\partial X^2} = -\epsilon \frac{\partial \theta}{\partial Y} \left(\frac{\partial \theta}{\partial X} \right)_{X=Y}. \qquad (13.4.18)$$

Only one boundary condition is needed on the interface

$$\theta(X, Y) = 0, \qquad X = Y. \qquad (13.4.19)$$

We now seek a perturbation approximation for small ϵ and let

$$\theta = \theta_0(X, Y) + \epsilon \theta_1(X, Y) + \epsilon^2 \theta_2(X, Y) + \cdots.$$

At the order $O(1)$, we have
$$\frac{\partial^2 \theta_0}{\partial X^2} = 0, \quad 0 < X < Y;$$
$$\theta_0(0, Y) = 0, \quad \theta_0(Y, Y) = 0.$$
With Y being just a parameter, the solution is
$$\theta_0 = 1 - \frac{X}{Y}.$$
At the next order $O(\epsilon)$, the heat equation gives
$$\frac{\partial^2 \theta_1}{\partial X^2} = -\frac{\partial \theta_0}{\partial Y}\left(\frac{\partial \theta_0}{\partial X}\right)_{X=Y} = \frac{X}{Y^3}.$$
On the boundaries $X = 0$ and Y, $\theta_1 = 0$. The solution is
$$\theta_1 = -\frac{1}{6}\frac{X}{Y}\left[1 - \left(\frac{X}{Y}\right)^2\right].$$
At the order $O(\epsilon^2)$, the heat equation is
$$\frac{\partial^2 \theta_2}{\partial X^2} = -\frac{\partial \theta_0}{\partial Y}\left(\frac{\partial \theta_1}{\partial X}\right)_{X=Y} - \frac{\partial \theta_1}{\partial Y}\left(\frac{\partial \theta_0}{\partial X}\right)_{X=Y}$$
$$= -\frac{1}{6}\left(\frac{X}{Y^3} + \frac{1}{2}\frac{X^3}{Y^5}\right)$$
after using the known results of θ_0 and θ_1. On $X = 0$ and Y, $\theta_2 = 0$ also. The solution is easily found to be
$$\theta_2 = \frac{1}{360}\frac{X}{Y}\left[19 - 10\left(\frac{X}{Y}\right)^2 - 9\left(\frac{X}{Y}\right)^4\right].$$
In summary, the approximate solution is
$$\theta = 1 - \frac{X}{Y} - \frac{\epsilon}{6}\frac{X}{Y}\left[1 - \left(\frac{X}{Y}\right)^2\right] +$$
$$\frac{\epsilon^2}{360}\frac{X}{Y}\left[19 - 10\left(\frac{X}{Y}\right)^2 - 9\left(\frac{X}{Y}\right)^4\right] + O(\epsilon^3). \quad (13.4.20)$$

The flux condition on the interface is then used to get the position of the interface $Y(\tau)$
$$\frac{dY}{d\tau} = -\epsilon\left(\frac{\partial \theta}{\partial X}\right)_{X=Y} = \frac{1}{Y}\left[\epsilon - \frac{\epsilon^2}{3} + \frac{7}{45}\epsilon^3 + \cdots\right].$$

Therefore,
$$Y\frac{dY}{d\tau} = \frac{d}{d\tau}\left(\frac{Y^2}{2}\right) = \epsilon - \frac{\epsilon^2}{3} + \frac{7}{45}\epsilon^3 + \cdots$$

so that3
$$Y^2 = 2\tau\left(\epsilon - \frac{\epsilon^2}{3} + \frac{7}{45}\epsilon^3\right) + \cdots$$

or
$$Y = \left[2\tau\left(\epsilon - \frac{\epsilon^2}{3} + \frac{7}{45}\epsilon^3 + \cdots\right)\right]^{1/2}. \qquad (13.4.21)$$

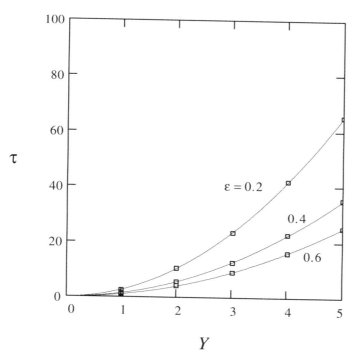

Fig. 13.3. Exact (solid line) and approximate (squares) freezing level.

Note that the normalizing length scale a does not appear in the dimensional solutions; this is because the present problem does not have any characteristic length. Aziz and Na have compared the approximate solution above with the exact solution available in Carslaw and Jaeger (1959)
$$\theta = 1 - \frac{\mathrm{erf}(\lambda X/Y)}{\mathrm{erf}(\lambda)},$$

where λ is the root of
$$\sqrt{\pi}\lambda e^{\lambda^2}\operatorname{erf}\lambda = \epsilon$$
and
$$Y = 2\lambda\tau^{1/2}, \qquad \operatorname{erf}(z) = \frac{2}{\sqrt{\pi}}\int_0^z e^{-z^2}dz.$$

The agreement is truly remarkable even for $\epsilon = 0.6$, as shown in Figure 13.3.

In both examples studied in §§13.3–13.4, the perturbation series are valid everywhere and are regular. We now turn to the first of two singular perturbation methods for differential equations, the *method of multiple scales*.

13.5 Method of multiple scales for an oscillator

When applied to differential equations, a regular perturbation series often breaks down when one (or more) of the independent coordinates becomes large. The series solution is not uniformly valid for the entire range of that coordinate, which implies that the original scale chosen for that coordinate is no longer adequate. The method of multiple scales is a scheme to extend the range of validity by seeking to describe slow variations over much larger scales, where different physical elements become prominent. This method is of the singular perturbation variety.

13.5.1 Weakly damped harmonic oscillator

Our first example is taken from freshman physics: a spring with elastic constant k, a mass m and a damper with coefficient c, are connected, as sketched in Figure 13.4. The mass rolls on frictionless wheels. The displacement $X(t)$ of the mass from the equilibrium position is governed by
$$m\frac{d^2X}{dt^2} + c\frac{dX}{dt} + kX = 0, \qquad (13.5.1)$$
subject to the initial conditions on the displacement and velocity
$$X(0) = a, \qquad \frac{dX(0)}{dt} = 0. \qquad (13.5.2)$$
Let
$$\omega_o = \sqrt{\frac{k}{m}}$$

13.5 Method of multiple scales for an oscillator

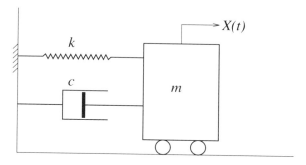

Fig. 13.4. A spring-mass-damper system.

denote the natural frequency of the undamped system. The obvious scale for time is $1/\omega_o$ so we introduce the dimensionless time $\tau = \omega_o t$. If the displacement is normalized by a, the governing equation for $x = X/a$ becomes

$$\ddot{x} + 2\epsilon \dot{x} + x = 0, \quad \tau = 0, \tag{13.5.3}$$

where

$$\dot{x} \stackrel{\text{def}}{=} \frac{dx}{d\tau},$$

and $2\epsilon = c/m\omega_o$ is the dimensionless measure of damping. The initial conditions are

$$x(0) = 1, \quad \frac{dx(0)}{d\tau} = 0. \tag{13.5.4}$$

The exact solution is well known

$$x = e^{-\epsilon \tau} \left[\cos \sqrt{1 - \epsilon^2}\tau + \frac{\epsilon}{\sqrt{1 - \epsilon^2}} \sin \sqrt{1 - \epsilon^2}\tau \right]. \tag{13.5.5}$$

Consider weak damping, $\epsilon \ll 1$. Let us start with a regular expansion

$$x = x_0 + \epsilon x_1 + \epsilon^2 x_2 + \cdots.$$

The perturbation problems at orders $O(1), O(\epsilon)$ and $O(\epsilon^2)$ are governed by

$$\ddot{x}_0 + x_0 = 0, \quad x_0(0) = 1, \quad \dot{x}_0(0) = 0;$$

$$\ddot{x}_1 + x_1 = -2\dot{x}_0, \quad x_1(0) = 0, \quad \dot{x}_1(0) = 0;$$

and

$$\ddot{x}_2 + x_2 = -2\dot{x}_1, \quad x_2(0) = 0, \quad \dot{x}_2(0) = 0.$$

The leading-order solution is
$$x_0 = \frac{1}{2}\left(e^{-i\tau} + e^{i\tau}\right).$$
The forcing term for x_1 is easily calculated so that x_1 satisfies
$$\ddot{x}_1 + x_1 = -2\dot{x}_0 = ie^{-i\tau} - ie^{i\tau}$$
with zero initial displacement and velocity. The solution is
$$x_1 = -\frac{\tau}{2}\left(e^{-i\tau} + e^{i\tau}\right),$$
hence
$$x = x_0 + \epsilon x_1 + \cdots = \frac{1}{2}(1 - \epsilon\tau + \cdots)\left(e^{-i\tau} + e^{i\tau}\right)$$
$$= (1 - \epsilon\tau + \cdots)\cos\tau.$$

This solution is indeed the two-term expansion of (13.5.5) as long as $\epsilon\tau \ll 1$. However, the $O(\epsilon)$ term ϵx_1 becomes comparable to the leading-order term x_0 if $\epsilon\tau = O(1)$ because x_1 is resonated at its own natural frequency by the forcing terms $-2\dot{x}_0$. Thus the naive expansion is not uniformly valid for all time.

The symptom that the solution fails when $O(\epsilon\tau) = 1$ suggests that there is a new time scale $\tau = O(1/\epsilon)$ over which something is not properly accounted for in the naive expansion. Taking a cue from this observation, let us introduce two time coordinates
$$t_0 = \tau \quad \text{and} \quad t_1 = \epsilon\tau$$
and pretend that $x(t_0, t_1)$ is a function of two independent variables. The original time derivative must be reinterpreted according to the chain rule
$$\frac{dx}{d\tau} = \frac{\partial x}{\partial t_0} + \epsilon\frac{\partial x}{\partial t_1}.$$
Now we assume the perturbation series
$$x = x_0 + \epsilon x_1 + \epsilon^2 x_2 + \cdots$$
with $x_n = x_n(t_0, t_1)$. Then (13.5.3) becomes
$$\left(\frac{\partial}{\partial t_0} + \epsilon\frac{\partial}{\partial t_1}\right)^2 \left(x_0 + \epsilon x_1 + \epsilon^2 x_2 + \cdots\right)$$
$$+ 2\epsilon\left(\frac{\partial}{\partial t_0} + \epsilon\frac{\partial}{\partial t_1}\right)\left(x_0 + \epsilon x_1 + \epsilon^2 x_2 + \cdots\right)$$
$$+ x_0 + \epsilon x_1 + \epsilon^2 x_2 + \cdots = 0.$$

13.5 Method of multiple scales for an oscillator

The initial conditions must also be expanded

$$\left(x_0 + \epsilon x_1 + \epsilon^2 x_2 + \cdots\right) = 0, \qquad t_0 = 0, \quad t_1 = 0$$

$$\left(\frac{\partial}{\partial t_0} + \epsilon \frac{\partial}{\partial t_1}\right)\left(x_0 + \epsilon x_1 + \epsilon^2 x_2\right) = 0, \qquad t_0 = 0, \quad t_1 = 0.$$

The $O(1)$ problem is governed by

$$\frac{\partial^2 x_0}{\partial t_0^2} + x_0 = 0, \qquad t_0, t_1 > 0 \tag{13.5.6}$$

with the initial conditions

$$x_0(0,0) = 1, \qquad \frac{\partial x_0}{\partial t_0}(0,0) = 0. \tag{13.5.7}$$

At the order $O(\epsilon)$, the governing conditions are

$$\frac{\partial^2 x_1}{\partial t_0^2} + x_1 = -2\frac{\partial^2 x_0}{\partial t_0 \partial t_1} - 2\frac{\partial x_0}{\partial t_0}, \qquad t_0, t_1 > 0 \tag{13.5.8}$$

with

$$x_1(0,0) = 0, \qquad \frac{\partial x_1}{\partial t_0}(0,0) + \frac{\partial x_0}{\partial t_1}(0,0) = 0. \tag{13.5.9}$$

The $O(\epsilon^2)$ problem is governed by

$$\frac{\partial^2 x_2}{\partial t_0^2} + x_2 = -2\frac{\partial^2 x_1}{\partial t_0 \partial t_1} - \frac{\partial^2 x_0}{\partial t_1^2} - 2\frac{\partial x_1}{\partial t_0} - 2\frac{\partial x_0}{\partial t_1}, \qquad t_0, t_1 > 0 \tag{13.5.10}$$

with

$$x_2(0,0) = 0, \qquad \frac{\partial x_2}{\partial t_0}(0,0) + \frac{\partial x_1}{\partial t_1}(0,0) = 0. \tag{13.5.11}$$

The solution at $O(1)$ is

$$x_0 = A(t_1)e^{-it_0} + A^*(t_1)e^{it_0}, \tag{13.5.12}$$

where $A(t_1)$ is an undetermined function of t_1 and A^* denotes the complex conjugate of A. From the initial conditions for x_0 we get

$$A(0) + A^*(0) = 1, \qquad -iA(0) + iA^*(0) = 0.$$

Therefore,

$$A(0) = A^*(0) = 1/2. \tag{13.5.13}$$

The dependence of A on t_1 is still unknown and must be sought at $O(\epsilon)$. Substituting (13.5.12) into (13.5.8), we get

$$\frac{\partial^2 x_1}{\partial t_0^2} + x_1 = 2\left[i\frac{\partial A}{\partial t_1}e^{-it_0} + iAe^{-it_0} + *\right], \tag{13.5.14}$$

where $*$ denotes the complex conjugate of the preceding terms, i.e.,

$$f + * \stackrel{\text{def}}{=} f + f^*,$$

as introduced in Chapter 8. The initial conditions are

$$x_1(0,0) = 0 \qquad (13.5.15)$$

and

$$\frac{\partial x_1}{\partial t_0}(0,0) = -\frac{\partial x_0}{\partial t_1}(0,0) = -\frac{\partial A}{\partial t_1}(0,0) + *. \qquad (13.5.16)$$

To avoid resonance we set the forcing terms in (13.5.15) to zero

$$\frac{\partial A}{\partial t_1} + A = 0,$$

which is a differential equation for $A(t_1)$. In view of the initial condition (13.5.13), the solution is

$$A(t_1) = A(0)e^{-t_1} = \frac{1}{2}e^{-t_1}.$$

Therefore, to the leading order the solution is

$$x = x_0 = \frac{1}{2}e^{-t_1}e^{-it_0} + * = e^{-t_1}\cos t_0.$$

If we stop here and return to the original time variable, the leading-order approximation is

$$x = e^{-\epsilon\tau}\cos\tau + O(\epsilon). \qquad (13.5.17)$$

This first-order solution not only remains bounded at $\epsilon\tau = O(1)$ but duplicates the exponential damping behavior of the exact solution. The power of the multiple scale is evident.

Let us continue the solution for x_1 and go on to the next order. Now (13.5.14) is homogeneous. The general solution is

$$x_1 = Be^{-it_0} + B^*e^{it_0},$$

where $B(t_1)$ is undetermined. The slow-time dependence of $B(t_1)$ must be sought from the $O(\epsilon^2)$ problem. After using what is known of x_0 and x_1, we have

$$\frac{\partial^2 x_2}{\partial t_0^2} + x_2 = \left[2i\frac{\partial B}{\partial t_1} + \frac{\partial^2 A}{\partial t_1^2} + 2iB + 2\frac{\partial A}{\partial t_1}\right]e^{-it_0} + *$$

$$= \left[2i\frac{\partial B}{\partial t_1} + 2iB - A\right]e^{-it_0} + *.$$

13.5 Method of multiple scales for an oscillator

To remove resonance we require that the forcing terms vanish

$$2i\left(\frac{\partial B}{\partial t_1} + B\right) = A = \frac{1}{2}e^{-t_1}.$$

This result is an inhomogeneous differential equation for $B(t_1)$. From the initial conditions (13.5.15) and (13.5.16) we have

$$B(0) + B^*(0) = 0, \qquad -iB(0) + iB^*(0) = 1,$$

hence

$$B(t_1) = \frac{i}{2}\left(1 - \frac{t_1}{2}\right)e^{-t_1}.$$

The first two orders can now be added to give the approximate solution

$$x = x_0 + \epsilon x_1 + O(\epsilon^2)$$

$$= \frac{1}{2}e^{-t_1}e^{-it_0}\left\{1 + i\epsilon\left(1 - \frac{t_1}{2}\right)\right\} + * + O(\epsilon^2). \qquad (13.5.18)$$

If $\epsilon t_1 \ll 1$, the above result can be approximated by

$$x = e^{-t_1}\left[\frac{1}{2}\left(e^{-it_0}e^{-i\epsilon t_1/2} + *\right) - \epsilon\left(\frac{e^{-it_0}}{2i} + *\right)\right] + O(\epsilon^2). \qquad (13.5.19)$$

In terms of the original dimensionless time τ, the approximate solution is

$$x = e^{-\epsilon\tau}\left[\cos\left(\tau\left(1 - \frac{\epsilon^2}{2}\right)\right) + \epsilon\sin\left(\tau\left(1 - \frac{\epsilon^2}{2}\right)\right)\right] + O(\epsilon^2), \qquad (13.5.20)$$

which brings improvement on the oscillation frequency. The approximate solution agrees with the exact result (13.5.5) up to $O(\epsilon)$ within the time range $\epsilon t_1 \ll 1$ or $\tau \ll O(\epsilon^{-2})$.

Although higher-order improvement can be pursued, the point of diminishing return has been reached since the exponential attenuation representing damping makes greater precision in the frequency no longer worthwhile.

13.5.2 Elastic spring with weak nonlinearity

Consider the undamped free vibration of a spring-mass system with displacement $X(t)$. For sufficiently large displacement the relation between the restoring force and the displacement may no longer be linear. A common nonlinear model is represented by $-kx - k'x^3$ so that the equation of motion is

$$m\ddot{X} + kX + k'X^3 = 0, \qquad (13.5.21)$$

where overhead dots represent time derivatives. We shall restrict our attention to periodic oscillations with a small amplitude characterized by a. Let $\omega_o = \sqrt{k/m}$ denote the natural frequency of the linear system. In terms of the normalized displacement $x = X/a$ and time $\tau = \omega_o t$, (13.5.21) may be replaced by

$$\ddot{x} + x + \epsilon x^3 = 0, \qquad (13.5.22)$$

where $\epsilon = k'a^2/k$. If $\epsilon \ll 1$ the nonlinear term is small, and it is natural to try first the straightforward expansion

$$x = x_0 + \epsilon x_1 + \epsilon^2 x_2 + \cdots. \qquad (13.5.23)$$

The following perturbation equations are obtained:

$$O(1): \qquad \ddot{x}_0 + x_0 = 0$$

and

$$O(\epsilon): \qquad \ddot{x}_1 + x_1 + x_0^3 = 0.$$

Let us seek a periodic solution corresponding to free oscillations; the solution for x_0 is

$$x_0 = Ae^{-i\tau} + A^* e^{i\tau}.$$

Substituting this result into the $O(\epsilon)$ equation, we get

$$\ddot{x}_1 + x_2 = -x_0^3 = -\left[A^3 e^{-3i\tau} + 3A^2 A^* e^{-i\tau} + *\right].$$

Note that at $O(\epsilon)$ the forcing terms on the right-hand side include the terms

$$3A^2 A^* e^{-i\tau} + *,$$

which have the same frequency as the natural frequency of x_1, hence will force resonance. The solution for x_1 will grow linearly in τ to become as large as x_0 when $\epsilon\tau = O(1)$, and violates the assumed orders of magnitude. Therefore, the perturbation series is not valid except for $\tau < O(\epsilon^{-1})$. Terms that cause unbounded resonance are said to be *secular* (literally, observed in a long time).

13.5 Method of multiple scales for an oscillator

Let us introduce a fast time $t_0 = \tau$ and a slow time corresponding to resonant growth $t_1 = \epsilon\tau$ and allow x to depend on t_0 and t_1. The original time derivative must be replaced by

$$\frac{dx}{d\tau} = \frac{\partial x}{\partial t_0} + \epsilon \frac{\partial x}{\partial t_1}.$$

Let

$$x = x_0 + \epsilon x_1 + \epsilon^2 x_2 + \cdots,$$

where $x_n = x_n(t_0, t_1)$, then

$$\left[\left(\frac{\partial}{\partial t_0} + \epsilon \frac{\partial}{\partial t_1}\right)^2 + 1\right](x_0 + \epsilon x_1 + \epsilon^2 x_2 + \cdots)$$

$$+ \epsilon (x_0 + \epsilon x_1 + \cdots)^3 = 0.$$

The perturbation equations are

$$O(1): \quad \frac{\partial^2 x_0}{\partial t^2} + x_0 = 0 \tag{13.5.24}$$

and

$$O(\epsilon): \quad \frac{\partial^2 x_1}{\partial t_0^2} + x_1 = -2\frac{\partial^2 x_0}{\partial t_0 \partial t_1} - x_0^3. \tag{13.5.25}$$

The solution for x_0 is

$$x_0 = A(t_1)e^{-it_0} + *. \tag{13.5.26}$$

Now the right-hand side of (13.5.25) is

$$-2\left(-i\frac{\partial A}{\partial t_1}e^{-it_0} + *\right) - \left(Ae^{-it_0} + A^*e^{it_0}\right)^3$$

$$= 2i\frac{\partial A}{\partial t_1}e^{-it_0} + * - \left[(A^3 e^{-3it_0} + *) + (3A^2 A^* e^{-it_0} + *)\right]$$

$$= \left[\left(2i\frac{\partial A}{\partial t_1} - 3A^2 A^*\right)e^{-it_0} + *\right] - \left(A^3 e^{-3it_0} + *\right).$$

To avoid secular growth of x_1, the coefficients of the first harmonics $e^{\pm it_0}$ must be made to vanish, i.e.,

$$2i\frac{\partial A}{\partial t_1} - 3A^2 A^* = 0 \tag{13.5.27}$$

and

$$-2i\frac{\partial A^*}{\partial t_1} - 3A^{*2} A = 0. \tag{13.5.28}$$

These two equations are complex conjugates of each other. Multiplying the first equation by A^* and the second by A and subtracting the results, we get

$$\frac{\partial AA^*}{\partial t_2} = 0 \quad \text{or} \quad |A| = \text{constant} = 1. \tag{13.5.29}$$

The constant amplitude has been taken to be unity by choosing the normalizing scale a to be the first-order amplitude. Thus only the phase δ of A defined by $A = e^{-i\delta(t_1)}$ may depend on t_1. It follows from (13.5.27) that

$$2\frac{\partial \delta}{\partial t_1}e^{-i\delta} - 3e^{-i\delta} = 0,$$

hence

$$\delta = \frac{3}{2}t_1$$

and

$$A = e^{-i\frac{3}{2}\epsilon t_0}. \tag{13.5.30}$$

Now x_0 may be written as

$$x_0 = e^{-i\omega t_0} + *$$

with

$$\omega = \left(1 + \frac{3}{2}\epsilon\right). \tag{13.5.31}$$

In dimensional form it reads

$$\omega = \omega_0 \left(1 + \frac{3k'a^2}{2k}\right). \tag{13.5.32}$$

Thus an effect of nonlinearity is that the frequency depends on the amplitude; this feature is sometimes called *amplitude dispersion*.

To complete the solution to $O(\epsilon)$, we calculate the third harmonic forcing term in (13.5.25)

$$-\exp\left\{-3it_0\left(1 + \frac{3}{2}\epsilon\right)\right\} + * = -\left(e^{-3i\omega t_0} + *\right).$$

The solution for x_1 is obtained by assuming

$$x_1 = A_1 e^{-3i\omega t_0} + *$$

so that

$$\left[(3i\omega)^2 + 1\right]A_3 = -1$$

or
$$A_3 = \frac{1}{8} + O(\epsilon).$$

In summary, the dimensionless solution is

$$x = \left(e^{-i\omega t_0} + *\right) + \frac{\epsilon}{8}\left(e^{-3i\omega t_0} + *\right) + \cdots \tag{13.5.33}$$

with ω given by (13.5.32). In dimensional form the solution is

$$X = a\left[\left(e^{-i\omega t} + *\right) + \frac{k'a^2}{8k}\left(e^{-3i\omega t} + *\right) + \cdots\right]. \tag{13.5.34}$$

The above result is uniformly valid up to $\epsilon t_0 = O(1)$. The range of uniform validity can be enlarged to $\epsilon^2 t_0 = O(1)$ if one introduces another slow time $t_2 = \epsilon^2 t_0$. Removal of secular terms at $O(\epsilon^2)$ will then give the dependence of A on t_2. This procedure can, in principle, be extended by introducing a cascade of time scales: $t_0, t_1, t_2, \ldots, t_n = \epsilon^n t_0, \ldots$, justifying the name of the method.

13.6 Theory of homogenization

To predict the behavior of a fiber-reinforced composite, seepage flow in a porous media or wave propagation over a slowly varying medium, one is often interested primarily in the gross behavior on the large scale. It is therefore desirable to have equations governing the large-scale variations by averaging over the small scale. This goal is not easy to achieve in general, except for a medium with a periodic microstructure. Then the *theory of homogenization* based on multiple scales applies. We consider two examples: (i) an equation with fast-varying periodic coefficients, and (ii) the flow through a bed of packed grains.

13.6.1 Differential equation with periodic coefficient

Let us consider an ordinary differential equation

$$\frac{d}{dx}\left[k(x)\frac{dp}{dx}\right] = 0, \tag{13.6.1}$$

where the coefficient $k(x)$ is a known periodic function of x. This equation may describe, for instance, one-dimensional seepage flow through a porous stratum saturated with fluid. The stratum is composed of periodically alternating sand and clay layers with different permeabilities $k(x)$. The pressure in the pore fluid then corresponds to p.

Let the thickness of one period be ℓ and the overall thickness of the entire stratum be L with $\ell/L = \epsilon \ll 1$. Can we find an effective equation for the large-scale variation of p?

Introducing fast and slow coordinates x and $X = \epsilon x$, we expand

$$p = p_0 + \epsilon p_1 + \epsilon^2 p_2 + \cdots, \qquad (13.6.2)$$

where $p_n, n = 0, 1, 2, \cdots$ are functions of x and X. From (13.6.1) we get

$$\left(\frac{\partial}{\partial x} + \epsilon \frac{\partial}{\partial X}\right)\left[k\left(\frac{\partial}{\partial x} + \epsilon \frac{\partial}{\partial X}\right)(p_0 + \epsilon p_1 + \epsilon^2 p_2 + \cdots)\right] = 0.$$

The leading-order equation is

$$\frac{\partial}{\partial x}\left(k(x)\frac{\partial p_0}{\partial x}\right) = 0, \qquad (13.6.3)$$

subject to the condition that p is periodic over the microscale length ℓ (or, in a more condensed language, p is ℓ-periodic). The equation can be integrated twice with respect to the fast x:

$$k(x)\frac{\partial p_0}{\partial x} = A_0(X), \qquad p_0 = A_0 \int_0^x \frac{dx}{k(x)} + B_0(X).$$

To ensure ℓ-periodicity, $p_0(x) = p_0(x + l)$, we must take $A_0 = 0$ so that $p_0 = p_0(X)$ is independent of the fast x. Clearly, p_0 represents the layer-averaged pressure. At the order $O(\epsilon)$ the perturbation equation is

$$\frac{\partial}{\partial x}\left(k\frac{\partial p_1}{\partial x}\right) + \frac{\partial}{\partial x}\left(k\frac{\partial p_0}{\partial X}\right) + \frac{\partial}{\partial X}\left(k\frac{\partial p_0}{\partial x}\right) = 0,$$

which can be simplified to an inhomogeneous equation for the microscale variation of p_1

$$\frac{\partial}{\partial x}\left[k\left(\frac{\partial p_1}{\partial x} + \frac{\partial p_0}{\partial X}\right)\right] = 0. \qquad (13.6.4)$$

Successive integration gives

$$\frac{\partial p_1}{\partial x} = -\frac{\partial p_0}{\partial X} + \frac{D_1(X)}{k(x)} \qquad (13.6.5)$$

and

$$p_1 = -x\frac{\partial p_0}{\partial X} + D_1(X)\int_0^x \frac{dx}{k(x)} + D_2, \qquad (13.6.6)$$

where D_1 and D_2 are constants of integration. Periodicity requires

$$p_1(0) = p_1(l)$$

13.6 Theory of homogenization

so that

$$-l\frac{\partial p_0}{\partial X} + D_1 \int_0^l \frac{dx}{k(x)} = 0$$

or

$$D_1 = \frac{\partial p_0}{\partial x}\left[\frac{1}{l}\int_0^l \frac{dx}{k}\right]^{-1}.$$

Mathematically, this result ensures the solvability of the inhomogeneous boundary-value problem governing p_1, given that p_0 is the nontrivial homogeneous solution. Define the harmonic mean of k by

$$\frac{1}{K} \stackrel{\text{def}}{=} \left\langle\frac{1}{k}\right\rangle = \frac{1}{l}\int_x^{x+l} \frac{dx}{k}, \qquad (13.6.7)$$

where a pair of angle brackets denotes averaging over a microscale period. Then

$$D_1 = K\frac{\partial p_0}{\partial X} \qquad (13.6.8)$$

and (13.6.5) may be written as

$$\frac{\partial p_1}{\partial x} = -\frac{\partial p_0}{\partial X} + \frac{K}{k}\frac{\partial p_0}{\partial X}.$$

In (13.6.6) the coefficient D_2 can be defined by insisting that the l-average of p_1 be zero so that p_0 alone represents, up to $O(\epsilon^2)$, the average of the total pressure p, i.e.,

$$D_2 = \frac{l}{2}\frac{\partial p_0}{\partial X} - D_1(X)\left\langle\int_0^x \frac{dx}{k(x)}\right\rangle. \qquad (13.6.9)$$

Now the $O(\epsilon^2)$ perturbation equation reads

$$\frac{\partial}{\partial x}\left(k\frac{\partial p_2}{\partial x}\right) + \frac{\partial}{\partial X}\left(k\frac{\partial p_0}{\partial X}\right) + \frac{\partial}{\partial x}\left(k\frac{\partial p_1}{\partial X}\right) + \frac{\partial}{\partial X}\left(k\frac{\partial p_1}{\partial x}\right) = 0, \qquad (13.6.10)$$

which is an inhomogeneous equation for p_2. Since

$$\frac{\partial}{\partial X}\left(k\frac{\partial p_1}{\partial x}\right) = -\frac{\partial}{\partial X}\left(k\frac{\partial p_0}{\partial x}\right) + \frac{\partial}{\partial X}\left(K\frac{\partial p_0}{\partial X}\right),$$

it follows from (13.6.10) that

$$\frac{\partial}{\partial x}\left(k\frac{\partial p_2}{\partial x}\right) + \frac{\partial}{\partial X}\left(K\frac{\partial p_0}{\partial X}\right) + \frac{\partial}{\partial x}\left(k\frac{\partial p_1}{\partial X}\right) = 0.$$

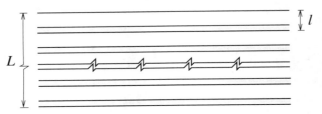

Fig. 13.5. A layered medium.

Taking the ℓ-average, we get the solvability condition for p_2

$$\frac{\partial}{\partial X}\left(K\frac{\partial p_0}{\partial X}\right) = 0, \qquad (13.6.11)$$

which is the governing equation for p_0 on the macroscale. In the context of seepage flow, K is the effective permeability. Once p_0 is solved from (13.6.11) under suitable global boundary conditions, p_1, which represents the local fluctuation within each layer from the mean, can be found from (13.6.6), with D_1 and D_2 given by (13.6.8) and (13.6.9).

As a simple example we let k be piece-wise constant with a period of thickness l, as shown in Figure 13.5. Then the effective coefficient (permeability) is

$$K = \left\langle\frac{1}{k}\right\rangle^{-1} = \left(\frac{1-n}{k_1} + \frac{n}{k_2}\right)^{-1} = \frac{k_1 k_2}{(1-n)k_2 + nk_1}.$$

If $n \ll 1$, $K \cong k_1$; if $n \approx 1$, $K \cong k_2$. The theory is derived on the premise that all derivatives are finite, thus the case of discontinuous coefficients may appear outside the range of validity of the homogenization theory. The special case treated here can be confirmed by direct solution of the layered problems, and can also be justified by the theory of generalized functions (see Bakhavlov and Pansenko, 1989).

This example typifies the homogenization analysis for a single equation with a fast varying coefficient. It can be generalized to two- and three-dimensional problems such as fiber-reinforced composite materials and other technically important topics, and to problems governed by a set of partial differential equations (Bensoussan et al., 1978; Bakhavlov and Pansenko, 1989).

13.6.2 *Darcy's law in seepage flow

The phenomenological law of Darcy describing seepage flow in a porous medium such as soils is based on experiments. A more theoretical

13.6 Theory of homogenization

approach is to derive this law and predict the hydraulic conductivity. This theory is possible if the granular structure is periodic. The method of homogenization can then be applied (Sanchez-Palencia, 1980; Bensoussan et al., 1978).

Let a rigid porous medium be saturated by an incompressible Newtonian fluid of constant density. Driven by an ambient pressure gradient, the steady flow velocity u_i and pressure p in the pores are governed by Navier-Stokes equations

$$\frac{\partial u_i}{\partial x_i} = 0 \tag{13.6.12}$$

and

$$\rho u_j \frac{\partial u_i}{\partial x_j} = -\frac{\partial p}{\partial x_i} + \mu \nabla^2 u_i. \tag{13.6.13}$$

On the wetted surface Γ of the solid matrix there is no slip

$$u_i = 0, \qquad x_i \in \Gamma. \tag{13.6.14}$$

For slow flows the two terms on the right-hand side of (13.6.13), representing the pressure gradient and the viscous force, are equally important. Both the change of pressure and the flow velocity vary according to two scales: the microscale ℓ-characteristic of the size of pores and grains, and the macroscale L imposed by global constraints. Again we assume that $\ell/L \ll 1$. Equating the order of magnitudes of the global pressure gradient P/L to the local viscous stress, we get

$$O\left(\frac{P}{L}\right) = O\left(\frac{\mu U}{\ell^2}\right),$$

which defines the velocity scale U. Let us normalize the space coordinates by the local scales and the unknowns according to the estimates just found

$$x_i = \ell x'_i, \qquad p = P p', \qquad u_i = \frac{\ell^2 P}{\mu L} u'_i, \tag{13.6.15}$$

where the primed quantities are dimensionless, and P here stands for the change in pressure since it only appears in differential form. Equation (13.6.13) becomes, formally,

$$Re \, u'_j \frac{\partial u'_i}{\partial x'_j} = -\frac{1}{\epsilon} \frac{\partial p'}{\partial x'_i} + \nabla'^2 u'_i, \tag{13.6.16}$$

where

$$Re \equiv \epsilon \frac{\rho \ell^2 P}{\mu^2} \tag{13.6.17}$$

is just the Reynolds number. The dimensionless continuity equation remains in the form of (13.6.12) and need not be repeated.

We assume the Reynolds number to be no greater than $O(\epsilon)$, i.e.,

$$Re \leq O(\epsilon) \quad \text{or} \quad \frac{\rho \ell^2 P}{\mu^2} \leq O(1). \tag{13.6.18}$$

Note that the pressure gradient term appears formally dominant in (13.6.16) because x_i is normalized by the microlength scale ℓ in every term.

From here on we shall return to dimensional variables but retain the ordering symbol ϵ to keep track of the bookkeeping

$$\epsilon^2 \rho u_j \frac{\partial u_i}{\partial x_j} = -\frac{\partial p}{\partial x_i} + \epsilon \mu \nabla^2 u_i. \tag{13.6.19}$$

The boundary condition on the wetted surface Γ of the pores is already given by (13.6.14).

Let us assume that the geometry of the porous matrix is periodic on the microscale, as depicted in Figure 13.6, although the structure may still change slowly over the macroscale L. Each periodic cell Ω is a rectangular box of dimension $O(\ell)$. We then expect u_i and p to

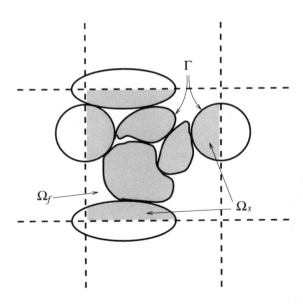

Fig. 13.6. Microstructure of a periodic porous medium.

13.6 Theory of homogenization

be spatially periodic from cell to cell, while changing slowly over the macroscale.

We now introduce the multiple-scale coordinates

$$x_i, \qquad X_i = \epsilon x_i \tag{13.6.20}$$

and the perturbation expansions

$$u_i = u_i^{(0)} + \epsilon u_i^{(1)} + \epsilon^2 u_i^{(2)} + \cdots \tag{13.6.21}$$

$$p = p^{(0)} + \epsilon p^{(1)} + \epsilon^2 p^{(2)} + \cdots, \tag{13.6.22}$$

where $u^{(j)}, p^{(j)}$ are functions of x_i and X_i.

From (13.6.12) we get at the first three orders $O(\epsilon^0)$, $O(\epsilon)$ and $O(\epsilon^2)$

$$\frac{\partial u_i^{(0)}}{\partial x_i} = 0 \tag{13.6.23}$$

$$\frac{\partial u_i^{(1)}}{\partial x_i} + \frac{\partial u_i^{(0)}}{\partial X_i} = 0 \tag{13.6.24}$$

$$\frac{\partial u_i^{(2)}}{\partial x_i} + \frac{\partial u_i^{(1)}}{\partial X_i} = 0 \tag{13.6.25}$$

$$\vdots$$

Similarly, we get from (13.6.19)

$$0 = -\frac{\partial p^{(0)}}{\partial x_i} \tag{13.6.26}$$

$$0 = -\frac{\partial p^{(0)}}{\partial X_i} - \frac{\partial p^{(1)}}{\partial x_i} + \mu \nabla^2 u_i^{(0)} \tag{13.6.27}$$

$$u_j^{(0)} \frac{\partial u_i^{(0)}}{\partial x_j} = -\frac{\partial p^{(1)}}{\partial X_i} - \frac{\partial p^{(2)}}{\partial x_i} + \mu \nabla^2 u_i^{(1)} \tag{13.6.28}$$

$$\vdots$$

On the grain/water interface Γ the velocity vanishes, hence

$$u_i^{(0)} = u_i^{(1)} = u_i^{(2)} = \cdots = 0, \quad x_i \in \Gamma. \tag{13.6.29}$$

In a typical Ω cell the flow must be periodic

$$u_i^{(0)}\, u_i^{(1)},\, u_i^{(2)},\ldots,\quad p^{(0)}\, p^{(1)}\, p^{(1)},\ldots \quad \text{are } \Omega \text{ periodic.} \tag{13.6.30}$$

It may be pointed out that it is the assumption (13.6.18) that renders the perturbation equations linear at each order. Had we assumed the Reynolds number to be finite, i.e., $Re = O(1)$, then the convective inertia would be of the order $O(\epsilon)$ and (13.6.27) would be replaced by

$$u_j^{(0)} \frac{\partial u_i^{(0)}}{\partial x_j} = -\frac{\partial p^{(0)}}{\partial X_j} + \frac{\partial p^{(1)}}{\partial x_i} + \mu \nabla^2 u_i^{(0)}. \tag{13.6.31}$$

Together with (13.6.23) the cell problem would be fully nonlinear (Ene and Sanchez-Palencia, 1975).

From (13.6.26) it is clear that

$$p^{(0)} = p^{(0)}(X_i). \tag{13.6.32}$$

Because of the linearity of (13.6.23) and (13.6.27), $u_i^{(0)}$ and $p^{(1)}$ can be formally represented by

$$u_i^{(0)} = -K_{ij}\frac{\partial p^{(0)}}{\partial X_j}, \quad p^{(1)} = -A_j\frac{\partial p^{(0)}}{\partial X_j} + \bar{p}^{(1)}, \tag{13.6.33}$$

where $\bar{p}^{(1)}(X_i)$ is independent of x_i. It then follows that the coefficients $K_{ij}(x_i, X_i)$ and $A_j(x_i, X_i)$ must satisfy

$$\frac{\partial K_{ij}}{\partial x_i} = 0 \tag{13.6.34}$$

$$-\frac{\partial A_j}{\partial x_i} + \mu \nabla^2 K_{ij} = -\delta_{ij}, \tag{13.6.35}$$

where

$$K_{ij} = 0 \quad \text{on } \Gamma \tag{13.6.36}$$

$$K_{ij},\ A_j \quad \text{are } \Omega \text{ periodic.} \tag{13.6.37}$$

These four equations define a linear boundary-value problem for K_{ij} and A_j in an Ω cell, which must be solved numerically for any prescribed microstructure. In the sequel we assume K_{ij} and A_j have been so obtained.

Defining the average over an Ω cell by

$$\langle f \rangle = \frac{1}{|\Omega|}\int_{\Omega_f} f\, d\Omega, \tag{13.6.38}$$

13.6 Theory of homogenization

where Ω_f is the fluid volume inside the Ω cell, we get

$$\langle u_i^{(0)} \rangle = -\langle K_{ij} \rangle \frac{\partial p^{(0)}}{\partial X_j} \tag{13.6.39}$$

$$\langle p^{(1)} \rangle = -\langle A_j \rangle \frac{\partial p^{(0)}}{\partial X_j} + n\bar{p}^{(1)}, \tag{13.6.40}$$

where n denotes the porosity, which is the ratio of fluid volume in the cell to the total cell volume

$$n = \frac{\Omega_f}{\Omega}. \tag{13.6.41}$$

To ensure that $p^{(0)}$ represents the mean pressure, we set $\langle p^{(1)} \rangle = 0$ so that

$$\bar{p}^{(1)} = \frac{\langle A_j \rangle}{n} \frac{\partial p^{(0)}}{\partial X_j}.$$

In view of (13.6.33), the pressure fluctuation from the mean is

$$p^{(1)} = \left(\frac{\langle A_j \rangle}{n} - A_j \right) \frac{\partial p^{(0)}}{\partial X_j}. \tag{13.6.42}$$

The Ω average of (13.6.25) gives

$$\frac{\partial \langle u_i^{(0)} \rangle}{\partial X_i} + \frac{1}{|\Omega|} \int_{\Omega_f} \frac{\partial u_i^{(1)}}{\partial x_i} d\Omega = 0.$$

By virtue of the Gauss theorem and the boundary condition (13.6.30), the volume integral vanishes, hence

$$\frac{\partial \langle u_i^{(0)} \rangle}{\partial X_i} = 0, \tag{13.6.43}$$

which implies, in turn, that

$$\frac{\partial}{\partial X_i} \left(\langle K_{ij} \rangle \frac{\partial p^{(0)}}{\partial X_j} \right) = 0. \tag{13.6.44}$$

Equations (13.6.39) and (13.6.43) or (13.6.44) govern the seepage flow in a rigid porous medium on the macroscale (Sanchez-Palencia, 1980; Bensoussan et al., 1978; Keller, 1980). In particular (13.6.39) is the classic Darcy's law, where $\langle K_{ij} \rangle$ is the tensor of hydraulic conductivity.

If the medium is isotropic and homogeneous on the L scale, we have

$$\langle K_{ij} \rangle = K \delta_{ij}, \qquad \langle A_j \rangle = 0, \tag{13.6.45}$$

where K is a constant. It follows from (13.6.44) that

$$\frac{\partial^2 p^{(0)}}{\partial X_k \partial X_k} = 0. \tag{13.6.46}$$

With proper boundary conditions on the macroscale, $p^{(0)}$ can then be found. If desired, the local velocity in the pores, and the local pressure fluctuation from the mean, can be found from (13.6.33).

Note that if the Reynolds number is of order unity, the local velocity $u^{(0)}$ and pressure $p^{(1)}$ then depend on the global pressure gradient $\partial p^{(0)}/\partial X_i$ in a nonlinear way through (13.6.31). Equations (13.6.33) and Darcy's law (13.6.39) no longer hold.

In the literature the term *homogenization* is sometimes used to mean any theoretical procedure for deriving macroscale equations by smoothing over the microscale. Homogenization by multiple scales is just one of the systematic procedures, limited to media with a periodic microstructure. In this procedure there is no need for additional closure assumptions. The derivation of effective coefficients is reduced to the solution of certain canonical boundary-value problems in a cell. Once the macroscale problem is solved, one also obtains the information on the microscale.

13.7 *Envelope of a propagating wave

The transverse vibration of a taut string supported laterally by nonlinear elastic springs may be modeled by

$$\rho \frac{\partial^2 y}{\partial t^2} - T \frac{\partial^2 y}{\partial x^2} + Ky = 0. \tag{13.7.1}$$

For a very long string the simplest solution is that of a propagating wave of single frequency

$$y = ae^{i\phi} + *, \tag{13.7.2}$$

where $a/2$ is the constant amplitude and $\phi = kx - \omega t$ is the phase. To ensure that (13.7.2) is a solution we substitute it into (13.7.1) to get

$$-\rho\omega^2 + Tk^2 + K = 0,$$

which is a quadratic equation for ω. Only the positive root corresponds to a right-going wave train

$$\omega = \sqrt{\frac{K + Tk^2}{\rho}}.$$

13.7 Envelope of a propagating wave

The phase velocity is

$$c(k) = \frac{\omega}{k},$$

which depends on the wave number k. Thus sinusoidal waves of different lengths propagate at different speeds. Waves with this property are called *dispersive*, and the relation between frequency and wave number is called the *dispersion relation*. If $K = 0$, then

$$\omega = \sqrt{\frac{T}{\rho}}k,$$

and the phase velocity is the same for all k. Thus without lateral support a taut string is a nondispersive medium.

How does a nearly sinusoidal wave train travel in a dispersive medium? In elementary physics it is well known that the superposition of two sinusoidal waves with nearly equal frequencies $\omega \pm \Delta\omega$ and wave numbers $k \pm \Delta k$ has a slowly and periodically modulated envelope, which moves at the so-called group velocity

$$c_g = \frac{d\omega}{dk}.$$

In general, for dispersive waves, c_g differs from the phase velocity. We now wish to generalize this classical result and consider slow modulations that need not be periodic. In principle, if the initial envelope diminishes sufficiently fast at infinities, Fourier transform can be used on (13.7.1) or similar linear equations with constant coefficients. The formally exact solution by inverse transform usually requires further asymptotic approximations in order to reveal the physics at large t. An effective alternative is to apply the method of multiple scales, whereby one approximates the equation asymptotically first, before working out the solution. The multiple-scales method is even more attractive when there are either weak nonlinearities or large-scale nonuniformities in the medium properties, or both. Then an exact solution is most likely impossible, but the asymptotic approximation of the governing equation can still simplify the analysis considerably.

Let $\lambda = \sqrt{T/K}, \omega_o = \sqrt{K/\rho}$, and a be the characteristic wavelength, frequency, and amplitude, respectively. The following normalization is natural:

$$y = aY, \quad t = T/\omega_o, \quad x = \lambda X.$$

Equation (13.7.1) becomes
$$\frac{\partial^2 Y}{\partial T^2} - \frac{\partial^2 Y}{\partial X^2} + Y = 0. \qquad (13.7.3)$$

Let us introduce fast and slow coordinates as follows:
$$t_0 = T, \quad t_1 = \epsilon T, \quad t_2 = \epsilon^2 T,$$
$$x_0 = X, \quad x_1 = \epsilon X, \quad x_2 = \epsilon^2 X$$

with ϵ characterizing the slowness of envelope modulation. The following expansion is then assumed
$$Y = y_0 + \epsilon y_1 + \epsilon^2 y_2 + \cdots$$

with $y_j = y_j(x_0, x_1, x_2; t_0, t_1, t_2), j = 0, 1, 2, \ldots$.

At order $O(\epsilon^0)$ we have a homogeneous equation
$$\frac{\partial^2 y_0}{\partial t_0^2} - \frac{\partial^2 y_0}{\partial x_0^2} + y_0 = 0. \qquad (13.7.4)$$

Let us restrict to a propagating wave
$$y_0 = A e^{i\phi} + *, \qquad (13.7.5)$$

where
$$\phi = k x_0 - \omega t_0$$

with k and ω being dimensionless. Equation (13.7.4) implies the dimensional dispersion relation
$$\omega = \sqrt{1 + k^2} \qquad (13.7.6)$$

for a right-going wave train. The dimensionless group velocity is
$$c_g = \frac{d\omega}{dk} = \frac{k}{\sqrt{1+k^2}} = \frac{k}{\omega}. \qquad (13.7.7)$$

The complex amplitude A is yet unknown and may depend on $x_1, x_2, \ldots$ and $t_1, t_2, \ldots$.

At $O(\epsilon)$ the perturbation equation is
$$\frac{\partial^2 y_1}{\partial t_0^2} - \frac{\partial^2 y_1}{\partial x_0^2} + y_1$$
$$= -2 \left(\frac{\partial^2 y_0}{\partial t_0 \partial t_1} - \frac{\partial^2 y_0}{\partial x_0 \partial x_1} \right)$$
$$= 2i \left(\omega \frac{\partial A}{\partial t_1} + k \frac{\partial A}{\partial x_1} \right) e^{i\phi} + *. \qquad (13.7.8)$$

13.7 Envelope of a propagating wave

If we let the inhomogeneous solution of y_1 be in the form $y_1 = y_1(\phi)$, then

$$\frac{dy_1}{d\phi} + y_1 = 2i\left(\omega\frac{\partial A}{\partial t_1} + k\frac{\partial A}{\partial x_1}\right)e^{i\phi} + *,$$

where use has been made of the dispersion relation (13.7.6). Clearly, the secular forcing terms $e^{\pm i\phi}$ would induce unbounded resonance unless their coefficients are set to zero, i.e.,

$$\frac{\partial A}{\partial t_1} + c_g\frac{\partial A}{\partial x_1} = 0 \qquad (13.7.9)$$

since $c_g = k/\omega$. Thus the solvability of the inhomogeneous problem for y_1 gives the evolution of the envelope A over the range $x_1, t_1 = O(1)$. It is evident that

$$A = A(x_1 - c_g t_1, x_2, \ldots; t_2 \ldots), \qquad (13.7.10)$$

i.e., the envelope advances at the group velocity c_g without an appreciable change of form.

Now that the secular forcing terms are removed, the equation for y_1 is homogeneous. Without loss of generality the homogeneous solution can be discarded since it can be absorbed by y_0. Let us now go to $O(\epsilon^2)$,

$$\frac{\partial^2 y_2}{\partial t_0^2} - \frac{\partial^2 y_2}{\partial x_0^2} + y_2$$

$$= -2\left(\frac{\partial^2 y_0}{\partial t_0 \partial t_2} - \frac{\partial^2 y_0}{\partial x_0 \partial x_2}\right) - \left(\frac{\partial^2 y_0}{\partial t_1^2} - \frac{\partial^2 y_0}{\partial x_1^2}\right). \qquad (13.7.11)$$

All terms on the right-hand side are secular, i.e., resonance-forcing, and their sum must be set to zero. Using (13.7.7) and (13.7.9), we may rewrite this solvability condition as

$$2i\left(\omega\frac{\partial A}{\partial t_2} + k\frac{\partial A}{\partial x_2}\right) + (1 - c_g^2)\frac{\partial^2 A}{\partial x_1^2} = 0$$

or

$$2i\left(\frac{\partial A}{\partial t_2} + c_g\frac{\partial A}{\partial x_2}\right) + \frac{1 - c_g^2}{\omega}\frac{\partial^2 A}{\partial x_1^2} = 0. \qquad (13.7.12)$$

This equation governs the evolution of A over the much larger scale $x_2, t_2 = O(1)$. Note that

$$\frac{1 - c_g^2}{\omega} = \frac{1}{\omega}\left(1 - \frac{k^2}{\omega^2}\right) = \frac{1}{\omega}\frac{1}{1 + k^2} = \frac{dc_g}{dk} \stackrel{\text{def}}{=} c_g'.$$

Equation (13.7.12) becomes

$$i\left(\frac{\partial A}{\partial t_2}+c_g\frac{\partial A}{\partial x_2}\right)+\frac{c'_g}{2}\frac{\partial^2 A}{\partial x_1^2}=0. \qquad (13.7.13)$$

Dependence on still larger scales $x_3, t_3 = O(1)$ may be pursued in the same manner at (ϵ^3).

Adding (13.7.9) and ϵ times (13.7.13), we get an equation valid for $x_2, t_2 \leq O(1)$

$$i\left(\frac{\partial A}{\partial t_1}+\epsilon\frac{\partial A}{\partial t_2}\right)+ic_g\left(\frac{\partial A}{\partial x_1}+\epsilon\frac{\partial A}{\partial x_2}\right)+\epsilon\frac{c'_g}{2}\frac{\partial^2 A}{\partial x_1^2}=0.$$

If we are content with the range of validity achieved so far, the artificial distinction among fast and slow coordinates is no longer needed. The following reversed transformations

$$\frac{\partial}{\partial t_1}+\epsilon\frac{\partial}{\partial t_2}\rightarrow\frac{\partial}{\partial t_1},\quad \frac{\partial}{\partial x_1}+\epsilon\frac{\partial}{\partial x_2}\rightarrow\frac{\partial}{\partial x_1}$$

imply that

$$i\left(\frac{\partial A}{\partial t_1}+c_g\frac{\partial A}{\partial x_1}\right)+\epsilon\frac{c'_g}{2}\frac{\partial^2 A}{\partial x_1^2}=0$$

is valid in the range $x_1, t_1 \leq O(1/\epsilon)$.

Further simplification may be made by employing a coordinate system moving with the group velocity

$$\xi = x_1 - c_g t_1, \quad \sigma = \epsilon t_1.$$

Since

$$\frac{\partial}{\partial t_1}=\epsilon\frac{\partial}{\partial \sigma}-c_g\frac{\partial}{\partial \xi},\quad \frac{\partial}{\partial x_1}=\frac{\partial}{\partial \xi},$$

we finally get

$$i\frac{\partial A}{\partial \sigma}+\frac{c'_g}{2}\frac{\partial^2 A}{\partial \xi^2}=0, \qquad (13.7.14)$$

which is the Schrödinger equation in quantum mechanics.

Based on this evolution equation, the long time evolution of A from a prescribed initial state can be conveniently examined. Extension to include weak nonlinearity is discussed in §14.8.

It may be remarked that the analysis in this section is also a process of homogenization, discussed in §13.5 and §13.6. Here the physical process is periodic in both space and time on the shortest scale. The resulting equation (13.7.14) describes the long-scale modulation of the wave envelope.

13.8 Boundary-layer technique

In some problems a straightforward expansion breaks down when the independent variable is in a special small neighborhood. This breakdown occurs usually when the small parameter multiplies a term with the highest derivative. Let us illustrate the symptom as well as the remedy by the following boundary-value problem† for $y(x)$, governed by the second-order equation

$$\epsilon y'' + y' + y = 0, \quad 0 < x < 1 \tag{13.8.1}$$

with the boundary conditions

$$y(0) = \alpha \tag{13.8.2}$$

and

$$y(1) = \beta. \tag{13.8.3}$$

Straightforward expansion

$$y = y_0 + \epsilon y_1 + \epsilon^2 y_2 + \cdots$$

yields to the leading order

$$y_0' + y_0 = 0,$$

which is only a differential equation of the first order. The solution

$$y_0 = c e^{-x} \tag{13.8.4}$$

contains only one unknown coefficient c and cannot satisfy the boundary conditions at both ends. If we choose to satisfy the boundary condition at the left end, then $c = \alpha$. The solution is

$$y_0(x) = \alpha e^{-x}, \tag{13.8.5}$$

whose boundary value at $x = 1$ is $y_0(1) = \alpha e^{-1}$ and is different from β in general. If instead we choose to satisfy (13.8.3) at the right end, then

$$c = \beta e^1$$

and

$$y_0 = \beta e^{1-x}, \tag{13.8.6}$$

whose value at $x = 0$ is $y_0(0) = \beta e$, which is in general not α and cannot match the boundary condition at the left end.

---

† This example was first used by L. Prandtl, who invented the boundary-layer theory for the flow of viscous fluids. The discussion here follows Nayfeh (1981).

Thus the solution (13.8.4) is a valid first approximation almost everywhere except near one of the two ends; it is called the *outer approximation*. We must now examine the small neighborhood of the end near which the outer approximation breaks down. Calling the neighborhood the inner region or the boundary layer, we now seek the correct inner approximation. Two questions arise: where is the boundary layer, and how thick is it?

Let us seek the answers to these questions by trial and errror. First, suppose that the boundary layer is near the right end $x = 1$. The outer approximation satisfying the boundary condition at the left end is given by (13.8.5). Within the boundary layer the approximation is best derived by finding the proper scale and renormalizing the coordinate

$$\zeta = \frac{1-x}{\epsilon^\lambda} \qquad \text{with} \quad \lambda > 0.$$

The thickness of the boundary layer is $O(\epsilon^\lambda)$, which is as yet unknown. The rescaled equation is

$$\frac{d^2 Y}{d\zeta^2} - \epsilon^{\lambda-1} \frac{dY}{d\zeta} + \epsilon^{2\lambda-1} Y = 0. \tag{13.8.7}$$

A choice of λ must first be made. Let us first assume an *inner expansion*

$$y = Y_0 + \epsilon Y_1 + \cdots.$$

The leading-order inner solution Y_0 must satisfy the boundary condition at the right end $\zeta = 0$ and match the outer solution in the intermediate region, i.e.,

$$\lim_{\zeta \gg 1} Y_0 = \lim_{1-x \ll 1} y_0. \tag{13.8.8}$$

There are several possibilities for λ. Let us first try $\lambda > 1$; the leading-order inner approximation is governed by

$$\frac{d^2 Y_0}{d\zeta^2} = 0,$$

hence

$$Y_0 = A + \beta \zeta.$$

The boundary condition at the right end (or $\zeta = 0$) requires that $A = \beta$ so that

$$Y_0 = \beta + B\zeta.$$

Now to determine the remaining coefficient we insist that y_0 given by

(13.8.5) should match with Y_0 according to (13.8.8). This requirement would imply that $\beta = \alpha e^{-1}$, which is impossible in general.

Next we try $0 < \lambda < 1$; the second term in (13.8.7) dominates so that

$$\frac{dY_0}{dy} = 0$$

and $Y_0 = $ constant in the boundary layer. To satisfy the boundary condition at the right end, we must have $Y_0 = \beta$. But this inner solution cannot match with the outer solution in general since $\beta \neq \alpha e^{-1}$. Finally, the choice $\lambda = 1$ implies from (13.8.7) that

$$\frac{d^2 Y_0}{d\zeta^2} - \frac{dY_0}{d\zeta} = 0$$

or

$$Y_0 = A + Be^\zeta.$$

To match the outer solution according to (13.8.8), we must have $B = 0$, $A = \alpha e^{-1}$, but then the boundary condition at the right end $\zeta = 0$ cannot be met.

Now we must give up the assumption that the boundary layer is at $x = 1$ and suppose that it is at $x = 0$. The outer solution satisfying the boundary condition at the right end is (13.8.6). In the boundary layer, let the inner coordinate be $\zeta = x/\epsilon^\lambda$ with $\lambda > 0$. The rescaled equation is

$$\frac{d^2 y}{d\zeta^2} + \epsilon^{\lambda-1}\frac{dy}{d\zeta} + \epsilon^{2\lambda-1}y = 0. \tag{13.8.9}$$

It can again be shown that when either $\lambda > 1$ or $\lambda < 1$ the leading-order inner approximation cannot satisfy the boundary condition at the left end and match with the outer solution. The only choice is $\lambda = 1$, implying that the boundary-layer thickness is $x = O(\epsilon)$ and

$$\frac{d^2 Y_0}{dy^2} + \frac{dY_0}{d\zeta} = 0.$$

The inner approximation is

$$Y_0 = A + Be^{-\zeta}.$$

The boundary condition at $x = 0$ requires that $B = \alpha - A$. Matching with the outer solution

$$\lim_{\zeta \gg 1}\left[A + (\alpha - A)e^{-\zeta}\right] = \lim_{1-x \ll 1} \beta e^{-x},$$

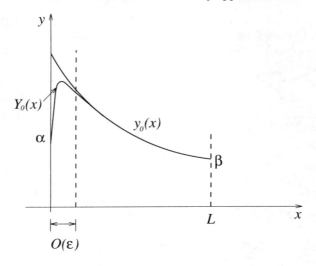

Fig. 13.7. The uniformly valid solution to leading order.

we find
$$A = \beta e$$
so that
$$Y_0 = \beta e + (\alpha - \beta e)e^{-\zeta} = \beta e + (\alpha - \beta e)e^{-x/\epsilon}. \qquad (13.8.10)$$

In summary, the solution is given by the outer approximation (13.8.6) for $0 < x \leq 1$ and by the inner approximation (13.8.10) for $0 \leq x/\epsilon = O(1)$.

A uniformly valid approximation can be constructed by adding the inner and outer approximations and then subtracting the common part, which is βe

$$y = y_0 + Y_0 - \beta e = \beta e^{1-x} + (\alpha - \beta e)e^{-x/\epsilon}, \quad 0 \leq x \leq 1. \qquad (13.8.11)$$

The result is sketched in Figure 13.7.

Numerous examples of this kind can be found in the two delightful books on perturbation methods by Nayfeh (1973, 1981), where extension to higher-order approximations is discussed.

13.9 Seepage flow in an aquifer with slowly varying depth

A shallow porous aquifer is sandwiched by impermeable soil layers above and below, as sketched in Figure 13.8. Along the two edges at $x = -L$

13.9 Seepage flow in an aquifer with slowly varying depth

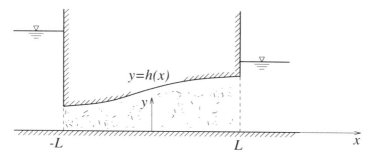

Fig. 13.8. A slowly varying aquifer.

and L the velocity potentials are known functions of local depth. Let the depth of the aquifer at any x be given by $h(x)$ with $h/L \ll 1$ and $h'(x) = 0$ at $x = \pm L$. How do we find the steady flow in the aquifer?

Recall from §1.4 on porous media that the seepage velocity $\mathbf{u}$ is related to the potential ϕ and the pore pressure p by Darcy's law

$$\mathbf{u} = \nabla \phi = -k \nabla \left(\frac{p}{\rho g} + y \right). \tag{13.9.1}$$

For constant hydraulic conductivity k, ϕ is harmonic

$$\nabla^2 \phi = 0, \quad -L < x < L, \quad 0 < y < h(x). \tag{13.9.2}$$

The boundary conditions at the two edges are

$$\phi(-L, y) = f_-(y) \tag{13.9.3}$$

and

$$\phi(L, y) = f_+(y), \tag{13.9.4}$$

where f_- and f_+ are known. On the upper and lower boundaries, the normal flux vanishes

$$\mathbf{n} \cdot \nabla \phi = 0, \quad y = h(x) \tag{13.9.5}$$

$$\frac{\partial \phi}{\partial y} = 0, \quad y = 0. \tag{13.9.6}$$

Since on the upper boundary the unit normal is

$$\mathbf{n} = \left(-\frac{dh}{dx}, 1 \right) \left[1 + \left(\frac{dh}{dx} \right)^2 \right]^{-\frac{1}{2}},$$

(13.9.5) may also be written

$$-\frac{dh}{dx}\frac{\partial\phi}{\partial x} + \frac{\partial\phi}{\partial y} = 0, \quad y = h(x). \qquad (13.9.7)$$

These equations also govern the steady two-dimensional diffusion of heat in a laterally insulated slab of variable thickness.†

Let us normalize all variables as follows:

$$X = \frac{x}{L}, \quad Y = \frac{y}{h_o}, \quad H = \frac{h}{h_o}, \quad (\Phi, F_-, F_+) = \frac{1}{kh_o}(\phi, f_-, f_+), \qquad (13.9.8)$$

where h_o is the characteristic depth of the aquifer. The normalized equations are

$$\epsilon^2 \frac{\partial^2 \Phi}{\partial X^2} + \frac{\partial^2 \Phi}{\partial Y^2} = 0, \quad -1 < X < 1, \quad 0 < Y < H(x) \qquad (13.9.9)$$

with ϵ being the shallowness parameter

$$\epsilon \equiv \frac{h_o}{L} \ll 1. \qquad (13.9.10)$$

Note that a second-order (highest) derivative is multiplied by a small parameter. The dimensionless boundary conditions are

$$\Phi(-1, Y) = F_-(Y), \quad 0 < Y < H_- \qquad (13.9.11)$$

$$\Phi(1, Y) = F_+(Y), \quad 0 < Y < H_+ \qquad (13.9.12)$$

$$\frac{\partial \Phi}{\partial Y} = \epsilon^2 H' \frac{\partial \Phi}{\partial X}, \quad Y = H(X), \quad -1 < X < 1 \qquad (13.9.13)$$

$$\frac{\partial \Phi}{\partial Y} = 0, \quad Y = 0, \quad -1 < X < 1. \qquad (13.9.14)$$

Let us introduce the outer expansion

$$\Phi = \Phi_0 + \epsilon \Phi_1 + \epsilon^2 \Phi_2 + \cdots, \qquad (13.9.15)$$

where odd powers are included to enable matching to the boundary-layer solutions. At $O(1)$ the perturbation equations are

$$\frac{\partial^2 \Phi_0}{\partial Y^2} = 0, \quad 0 < Y < H \qquad (13.9.16)$$

$$\frac{\partial \Phi_0}{\partial Y} = 0, \quad Y = H \qquad (13.9.17)$$

† The corresponding analysis for a laterally insulated body of revolution is discussed in Cole (1968) by using cylindrical coordinates and Bessel functions. The present example is just the planar fascimile of his.

13.9 Seepage flow in an aquifer with slowly varying depth

and

$$\frac{\partial \Phi_0}{\partial Y} = 0, \quad Y = 0. \tag{13.9.18}$$

These equations constitute an ordinary differential system in Y only. Clearly, Φ_0 is independent of Y, i.e.,

$$\Phi_0 = \Phi_0(X). \tag{13.9.19}$$

The task now is to determine this horizontal dependence.

At $O(\epsilon)$ Φ_1 is also governed by (13.9.16–13.9.18), hence $\Phi_1 = \Phi_1(X)$ also remains unknown. As is usual in singular perturbations information on the leading-order solution is revealed at a higher order. We examine the $O(\epsilon^2)$ outer problem

$$\frac{\partial^2 \Phi_2}{\partial Y^2} = -\frac{\partial^2 \Phi_0}{\partial X^2}, \quad 0 < Y < H \tag{13.9.20}$$

$$\frac{\partial \Phi_2}{\partial Y} = H'\frac{\partial \Phi_0}{\partial X}, \quad Y = H \tag{13.9.21}$$

$$\frac{\partial \Phi_2}{\partial Y} = 0, \quad Y = 0. \tag{13.9.22}$$

This boundary-value problem for Φ_2 is inhomogeneous; the solution satisfying the boundary condition at $Y = 0$ is

$$\Phi_2 = -\frac{\Phi_0''}{2}Y^2 + C(X),$$

where primes denote differentiation with respect to X. Applying the inhomogeneous condition at $Y = H$, we get the solvability condition for Φ_2

$$-\Phi_0'' H = H'\Phi_0' \quad \text{or} \quad (H\Phi_0')' = 0,$$

which means that the total discharge across the aquifer is constant. Hence

$$\Phi_0 = A \int_{-1}^{X} \frac{dX'}{H(X')} + B, \tag{13.9.23}$$

where A and B are pure constants. Clearly, this solution cannot satisfy the boundary conditions (13.9.11) and (13.9.12); boundary layers are needed at both ends.

Near each end, variations in both X and Y directions must be equally important. Hence, the full Laplace equation is needed and the boundary-layer thickness is $x = O(h_0)$. Accordingly, the boundary-layer coordinates are

$$X_- = \frac{L+x}{h_0} = \frac{1+X}{\epsilon} \quad \text{and} \quad X_+ = \frac{x-L}{h_0} = \frac{X-1}{\epsilon} \qquad (13.9.24)$$

for the left and right boundary layers, respectively. The corresponding pressure in the boundary layers will be denoted by $\Phi_-(X_-, Y)$ and $\Phi_+(X_+, Y)$. Since

$$\frac{\partial}{\partial X} = \frac{1}{\epsilon}\frac{\partial}{\partial X_\pm},$$

(13.9.9) becomes

$$\frac{\partial^2 \Phi_\pm}{\partial X_\pm^2} + \frac{\partial^2 \Phi_\pm}{\partial Y^2} = 0. \qquad (13.9.25)$$

The boundary condition at the left end is

$$\Phi_-(0, Y) = F_-(Y), \quad X_- = 0, \quad 0 < Y < H_-. \qquad (13.9.26)$$

On the upper surface

$$\frac{\partial \Phi_-}{\partial Y} = \epsilon \frac{dH}{dX}\frac{\partial \Phi_-}{\partial X_-}, \qquad Y = H(-1 + \epsilon X_-). \qquad (13.9.27)$$

Note that the scale of H is L everywhere, and the slope of the boundary remains $O(\epsilon)$ even near the ends. Since $\epsilon X_- = O(\epsilon)$, we approximate $H(X)$ by

$$H(-1 + \epsilon X_-) = H|_{X_-=0} + \epsilon X_- \left.\frac{dH}{dX}\right|_{X_-=0} + O(\epsilon^2).$$

It follows that

$$\frac{dH(X)}{dX} = \epsilon X_- \left.\frac{d^2 H}{dX^2}\right|_{X_-=0} + O(\epsilon^2).$$

Substituting

$$\Phi_- = \Phi_-^{(0)} + \epsilon \Phi_-^{(1)} + \epsilon^2 \Phi_-^{(2)} + \cdots$$

into (13.9.27), we get near the left end

$$\frac{\partial \Phi_-^0}{\partial Y}(X_-, H_-) = 0 \qquad (13.9.28)$$

13.9 Seepage flow in an aquifer with slowly varying depth

on top of the layer and

$$\frac{\partial \Phi^0_-}{\partial Y}(X_-, 0) = 0 \qquad (13.9.29)$$

at the bottom. The domain of the boundary-value problem for Φ_0 is now approximated by a semi-infinite rectangular strip. By separation of variables, we easily get

$$\Phi^0_- = \frac{a^-_0}{2} + \sum_{n=1}^{\infty} a^-_n \exp\left(\frac{-n\pi X_-}{H_-}\right) \cos\frac{n\pi Y}{H_-}, \qquad (13.9.30)$$

where

$$a^-_n = \frac{2}{H_-}\int_0^{H_-} F_-(Y) \cos\frac{n\pi Y}{H_-} dY, \quad n = 0, 1, 2, \ldots. \qquad (13.9.31)$$

The inner solution is completely determined. To match it to the outer solution we require

$$\lim_{X_- \gg 1} \Phi^{(0)}_- = \lim_{1+X \ll 1} \Phi_0.$$

This condition determines B,

$$B = \frac{a^-_0}{2} = \frac{1}{H_-}\int_0^{H_-} F_-(Y) dY, \qquad (13.9.32)$$

which is the average of the boundary data over the cross section.

A similar analysis for the boundary layer near $x = L$ or $X = 1$ gives

$$\Phi^{(0)}_+ = \frac{a^+_0}{2} + \sum_{n=1}^{\infty}\left\{a^+_n \exp\left(\frac{n\pi X_+}{H_+}\right) \cos\frac{n\pi Y}{H_+}\right\}, \qquad (13.9.33)$$

where

$$a^+_n = \frac{2}{H_+}\int_0^{H_+} F_+(Y) \cos\frac{n\pi Y}{H_+} dY, \quad n = 0, 1, 2, \ldots. \qquad (13.9.34)$$

Matching this result with the outer solution gives

$$A\int_{-1}^{1}\frac{dX'}{H(X')} + B = \frac{a^+_0}{2} = \frac{1}{H_+}\int_0^{H_+} F_+(Y) dY$$

or

$$A\int_{-1}^{1}\frac{dX'}{H(X')} = \frac{a^+_0 - a^-_0}{2}$$

$$= \frac{1}{2H_+}\int_0^{H_+} F_+(Y) dY - \frac{1}{2H_-}\int_0^{H_-} F_-(Y) dY, \qquad (13.9.35)$$

which determines A and the leading-order outer approximation Φ_0. The leading-order solution is now completely determined everywhere.

13.10 Water table near a cracked sheet pile

In this example, we shall illustrate how to juxtapose complex variables and perturbation methods to solve a groundwater problem approximately (Mei, 1977).

A vertical sheet pile is driven into the ground to a great depth. On the left side ($x < 0$) the ground is kept completely wet by a water layer of depth H. The side on the right ($x > 0$) is supposed to be protected from wetting. Unfortunately, the pile along the negative y axis has a horizontal line crack at some depth h, as sketched in Figure 13.9. How much wetting can be expected on the right side of the pile?

Referring to Figure 13.9, we assume that the pile is infinitesimally thin, and the crack width $2a$ is much smaller than the depth of the crack h. In the near field (inner region) of the crack the dominant scale

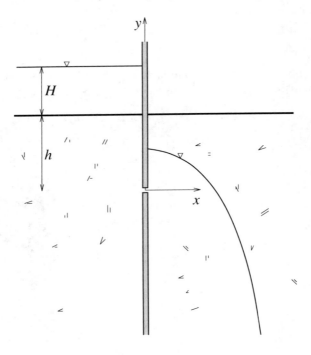

Fig. 13.9. A cracked sheet pile.

13.10 Water table near a cracked sheet pile

is $2a$; the flow is oblivious to the presence of the ground surface and leaks through the slot as if in a barrier separating two infinite regions. There are two far fields (outer regions) characterized by the length scale h. On the left side of the pile the ground is completely wet; the crack appears to be a line sink draining fluid out of the pile at the rate Q. The outer problem is just the left half of the half-space problem with a submerged line sink of strength $2Q$. On the ground surface the boundary condition is

$$\phi = \mathrm{Re}\, w = -k(H+h), \quad y = h,$$

which can be satisfied by placing an image source in the upper half plane at $y = 2h$. The complex potential is

$$w(z) = -k(H+h) + \frac{Q}{\pi}\left[-\ln z + \ln(z - 2h)\right], \qquad (13.10.1)$$

where $z = x + iy$. In the full z plane, $\ln z$ and $\ln(z - 2h)$ are rendered single-valued by horizontal cuts along $(0, \infty)$ and $(ih, ih + \infty)$. On the right side of the pile the ground is partially dry; a line source of strength Q creates a water table. The complex potential for the outer solution is given by (11.8.13), with Q changed to $2Q$,

$$w = -\frac{2Q}{\pi}\ln 2 + \frac{Q}{\pi}\ln\zeta, \qquad (13.10.2)$$

where

$$-ikz = \frac{2Q}{\pi}\ln 2 + \frac{2Q}{\pi}\left[\ln\left(\sqrt{\zeta+1} - 1\right) - \ln\zeta\right] \qquad (13.10.3)$$

maps the wetted region in the z plane onto the upper half plane of ζ, as shown in Figure 11.18.

The geometry of the near field is depicted in Figure 13.10, which can be mapped onto the upper τ plane by Joukowski transformation, also shown in the same figure,

$$\frac{2iz}{a} = \tau + \frac{1}{\tau}. \qquad (13.10.4)$$

This quadratic equation for τ can be solved

$$\tau = \frac{iz}{a} + i\sqrt{\frac{z^2}{a^2} + 1}.$$

To make the square root single-valued, we cut the z plane along $-a < y < a, x = 0$ and choose the branch so that $\sqrt{z^2/a^2 + 1} \to z/a$ for

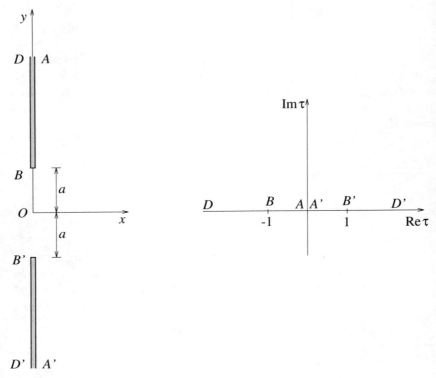

Fig. 13.10. Near field of the crack.

$|z/a| \gg 1$. Clearly, there is a sink at DD' and a source of equal strength at AA'. The most general solution, which has a source at AA' and satisfies Im w = constant along the real τ axis, is, by inspection,

$$w = C + M \ln \tau. \tag{13.10.5}$$

Now Q, C and M must be determined by asymptotic matching.

We first approximate the outer solutions for small $|z|/h$. On the side $x < 0$ the inner approximation of (13.10.1) is

$$w = -k(H+h) - \frac{Q}{\pi} \ln z + \frac{Q}{\pi}\left(\ln 2h + \frac{3\pi i}{2}\right). \tag{13.10.6}$$

To get the inner approximation of (13.10.2) for $x > 0$ we turn first

13.10 Water table near a cracked sheet pile

to (13.10.3). Since small z corresponds to small ζ, we expand

$$\ln(\sqrt{\zeta+1}-1) = \ln\left(1 + \frac{\zeta}{2} - \frac{\zeta^2}{8} + \cdots - 1\right)$$

$$= \ln\left[\frac{\zeta}{2}\left(1 - \frac{\zeta}{4} + \cdots\right)\right]$$

$$= \ln\zeta - \ln 2 + \ln\left(1 - \frac{\zeta}{4} + \cdots\right)$$

$$= \ln\zeta - \ln 2 - \frac{\zeta}{4} + O(\zeta^2).$$

It follows from (13.10.3) that

$$-ikz = -\frac{Q\zeta}{2\pi} + O(\zeta^2) \quad \text{or} \quad \zeta = \frac{2\pi i k z}{Q} + O(z^2),$$

which may be used to approximate (13.10.2)

$$w \cong -\frac{2Q}{\pi}\ln 2 + \frac{Q}{\pi}\left[\ln\frac{2\pi k}{Q} + \frac{i\pi}{2} + \ln z\right]. \tag{13.10.7}$$

The error terms are omitted for brevity.

The outer limit of the inner approximation (13.10.2) depends on the sign of x. For matching on the right we note that large $|z|/a$ implies large τ; see Figure 13.10. It follows from (13.10.4) that

$$\tau = \frac{2iz}{a}\left[1 + O\left(\frac{a^2}{z^2}\right)\right],$$

hence

$$w(z) \cong C + M\left(-\ln\frac{a}{2} + \frac{i\pi}{2} + \ln z\right), \quad x > 0. \tag{13.10.8}$$

For matching on the left we seek approximation of (13.10.5) for small τ instead; see Figure 13.10. Since

$$\tau \cong -\frac{ia}{2z}\left[1 + O\left(\frac{a^2}{z^2}\right)\right],$$

the corresponding outer approximation is

$$w(z) \cong C + M\left(\ln\frac{a}{2} + \frac{3i\pi}{2} - \ln z\right), \quad x < 0. \tag{13.10.9}$$

Now we match the logarithmic terms of (13.10.6) and (13.10.9) by requiring

$$M = \frac{Q}{\pi}.$$

The constant terms in (13.10.7) and (13.10.9) are matched if

$$C + M\left(\ln\frac{a}{2} + \frac{3i\pi}{2}\right) = -k(H+h) + \frac{Q}{\pi}\left(\ln 2h + \frac{3i\pi}{2}\right).$$

Similarly, on the right side the inner and outer solutions (13.10.8) and (13.10.7) are matched by equating the constant terms

$$C + M\left(-\ln\frac{a}{2} + \frac{i\pi}{2}\right) = -\frac{2Q}{\pi}\ln 2 + \frac{Q}{\pi}\left(\ln\frac{2\pi k}{Q} + \frac{i\pi}{2}\right).$$

From the last three equations we may solve for Q

$$\frac{Q}{\pi}\ln\frac{\pi k a^2}{16Qh} = -k(H+h).$$

Let

$$\beta = \frac{16Qh}{\pi k a^2},$$

then

$$\beta\ln\beta = \frac{16(H+h)h}{a^2}, \qquad (13.10.10)$$

which is a transcendental equation for β. Once solved numerically the discharge is found from

$$Q = \frac{\pi k(H+h)}{\ln\beta}. \qquad (13.10.11)$$

Afterward the constant C and, hence, the inner and outer solutions are determined.

Recall from (11.8.16) and (11.8.17) that the maximum width and height of the water table due to a submerged source of strength $2Q$ are given by

$$kx_{max} = \frac{2Q}{\pi} \quad \text{and} \quad ky_{max} = \frac{2Q}{\pi}\ln 2.$$

It follows that

$$\frac{x_{max}}{h} = \frac{\pi(H+h)}{h}\frac{1}{\ln\beta}, \quad \frac{y_{max}}{h} = \frac{H+h}{h}\frac{2\ln 2}{\ln\beta}. \qquad (13.10.12)$$

As the crack narrows, β increases, but $\ln\beta$ increases slowly. Hence, Q, x_{max} and y_{max} diminish slowly, implying that even a tiny crack can cause a good deal of unwanted wetting. Also x_{max} and y_{max} depend weakly on the hydraulic conductivity k through $\ln\beta$, hence change little for a wide variety of soils.

As a numerical example we take $H = h = 5m$ and $k = 10^{-5} m/sec$, which is typical of clayish sand. For a crack of width $20cm$, the theory gives $Q = 2.99\, m^3$ per day, $x_{max} = 3.46m$ and $y_{max} = 1.53m$. If the crack width is only $2cm$, then $Q = 2.04\, m^3$ per day, $x_{max} = 2.36m$ and $y_{max} = 1.04m$. Thus a ten-fold reduction of crack width does not reduce the leakage substantially.

13.11 *Vibration of a soil layer

Many physical problems are described by several unknowns and governed by a system of equations. Reduction to a single equation is often inconvenient or even infeasible, and approximations must be made directly to the original system. In general the physical reasoning for scales and perturbation analysis is more circumspect as one must consider the interplay of all related elements in the problem. The dynamics of fluid-saturated soils is such an example (Mei and Foda, 1981).

13.11.1 Formulation

We first formulate the problem for the one-dimensional vibration of a soil layer due to surface forcing. The pore fluid and the granular solid matrix will be treated as two interacting continua. Let the following symbols be used:

n = volume fraction of pores

ρ_w = density of water

ρ_s = density of solid matrix

u = seepage velocity of pore fluid

v = velocity of solid matrix

x = vertical coordinates measured from the bedrock

σ = effective stress in the solid matrix

p = pressure in the pore fluid.

Since $n\rho_w$ is the fluid mass and $(1-n)\rho_s$ the solid mass per unit volume of the soil mixture, the exact laws of mass conservation must be

$$\frac{\partial}{\partial t}(n\rho_w) + \frac{\partial}{\partial x}(n\rho_w u) = 0 \qquad (13.11.1)$$

and
$$\frac{\partial}{\partial t}\left((1-n)\rho_s\right) + \frac{\partial}{\partial x}\left((1-n)\rho_s v\right) = 0. \tag{13.11.2}$$

As equations of state, we assume that the solid density remains constant
$$\rho_s = \overline{\rho}_s = \text{constant} \tag{13.11.3}$$

and that the fluid is slightly compressible due to the presence of gas bubbles, i.e.,
$$\rho_w = \overline{\rho}_w\left(1 + \frac{p'}{\beta}\right), \tag{13.11.4}$$

where β is the coefficient of compressibility for water, due largely to entrained gas bubbles. For infinitesimal motion, we linearize all perturbation by letting
$$f = \overline{f} + f',$$

e.g.,
$$n = \overline{n} + n', \qquad \rho_w = \overline{\rho}_w + \rho'_w, \qquad p = p', \qquad \sigma = \sigma',$$

where $f'/\overline{f} \ll 1$. Since u and v are small deviations from static equilibrium, (13.11.1) and (13.11.2) can be linearized to
$$\overline{n}\frac{\partial \rho'_w}{\partial t} + \overline{\rho}_w\frac{\partial n'}{\partial t} + \overline{\rho}_w \overline{n}\frac{\partial u}{\partial x} = 0 \tag{13.11.5}$$

and
$$-\frac{\partial n'}{\partial t} + (1-\overline{n})\frac{\partial v}{\partial x} = 0. \tag{13.11.6}$$

Invoking (13.11.4), we may write (13.11.5) as
$$\frac{\overline{n}}{\beta}\frac{\partial p'}{\partial t} + \frac{\partial n'}{\partial t} + \overline{n}\frac{\partial u}{\partial x} = 0. \tag{13.11.7}$$

For infinitesimal motion, conservation of fluid momentum requires
$$\overline{n}\,\overline{\rho}_w\frac{\partial u}{\partial t} = -\overline{n}\frac{\partial p'}{\partial x} - \frac{\overline{n}^2}{k}(u-v), \tag{13.11.8}$$

while conservation of solid momentum requires
$$(1-\overline{n})\overline{\rho}_s\frac{\partial v}{\partial t} = \frac{\partial \sigma'}{\partial x} - (1-\overline{n})\frac{\partial p'}{\partial x} + \frac{\overline{n}^2}{k}(u-v). \tag{13.11.9}$$

A Darcy's resistance force proportional to the difference of velocities between two phases has been included in both momentum equations,

13.11 Vibration of a soil layer

with opposite signs. Gravity is ignored; there is no hydrostatic pressure in the pore fluid or geostatic stress in the solid.

Equations (13.11.6) and (13.11.7) may be added to eliminate n'

$$\overline{n}\frac{\partial \rho'_w}{\partial t} + \overline{\rho}_w(1-\overline{n})\frac{\partial v}{\partial x} + \overline{\rho}_w\overline{n}\frac{\partial u}{\partial x} = 0, \qquad (13.11.10)$$

which is called the *storage equation*. Equations (13.11.8) and (13.11.9) may be combined to give the momentum equation for the composite

$$\overline{n}\,\overline{\rho}_w\frac{\partial u}{\partial t} + (1-\overline{n})\overline{\rho}_s\frac{\partial v}{\partial t} = \frac{\partial}{\partial x}(\sigma' - p'). \qquad (13.11.11)$$

On the left is the inertia of the mixture per unit volume; on the right $\sigma' - p'$ is the total stress. Finally, we add Hooke's law to relate the solid stress σ' and strain rate $\partial v/\partial x$

$$\frac{\partial \sigma'}{\partial t} = E\frac{\partial v}{\partial x}. \qquad (13.11.12)$$

We shall seek the soil response due to simple harmonic stress and pressure applied at the ground surface $x = L$,

$$\sigma' = \text{Re}\left(\sigma_G e^{-i\omega t}\right), \quad x = L \qquad (13.11.13)$$

$$p = \text{Re}\left(P_G e^{-i\omega t}\right), \quad x = L, \qquad (13.11.14)$$

where σ_G and P_G are given constants. On the bedrock no motion of either phase is allowed

$$u = v = 0, \quad x = 0. \qquad (13.11.15)$$

Equations (13.11.6–13.11.9) and (13.11.12) are a system of five first-order equations for five unknowns n', p', σ', u and v. In principle they can be combined by cross differentiation and eliminated to yield a partial differential equation of fifth order. This undertaking is laborious!

Let us seek an effective approximation for sinusoidal oscillations at high frequency.

13.11.2 The outer solution

Intuitively, the strong resistance to fluid flow suggests that the relative velocity $u - v$ must be very small almost everywhere except near the unsealed ground surface. Thus the solid/water composite should behave as a single phase except in a thin boundary layer near the top. We first examine the outer region away from the boundary layer. Distinguishing all the outer unknowns by $(.)^o$, we scale the outer variables as follows:

$$x = LX, \quad t = T/\omega$$

$$\begin{pmatrix} p' \\ \sigma' \end{pmatrix} = \begin{pmatrix} p^o \\ \sigma^o \end{pmatrix} = \sigma_G \begin{pmatrix} P \\ S \end{pmatrix} \tag{13.11.16}$$

$$\begin{pmatrix} u' \\ v' \end{pmatrix} = \begin{pmatrix} u^o \\ v^o \end{pmatrix} = \frac{\sigma_G \omega L}{\beta} \begin{pmatrix} U \\ V \end{pmatrix}$$

so that the governing equations are normalized to

$$\left(\frac{\overline{\rho}_w \omega k}{\overline{n}}\right) \frac{\partial U}{\partial T} = -\left(\frac{k\beta}{\overline{n}\omega L^2}\right) \frac{\partial P}{\partial X} - (U - V) \tag{13.11.17}$$

$$\overline{n} \frac{\partial}{\partial X}(U - V) + \frac{\partial V}{\partial X} = -\overline{n} \frac{\partial P}{\partial T} \tag{13.11.18}$$

$$\frac{\omega^2 L^2}{\beta} \left[\overline{n} \rho_w \frac{\partial U}{\partial T} + (1 - \overline{n}) \overline{\rho}_s \frac{\partial V}{\partial T} \right] = \frac{\partial}{\partial X}(S - P) \tag{13.11.19}$$

$$\frac{\partial S}{\partial T} = \frac{E}{\beta} \frac{\partial V}{\partial X}. \tag{13.11.20}$$

For fine-grained soils the permeability k is very small so that the factors

$$\frac{\overline{\rho}_w \omega k}{\overline{n}} \ll 1, \qquad \frac{k\beta}{\overline{n}\omega L^2} \ll 1$$

are negligibly small. Consequently, (13.11.17) can be approximated by

$$U \cong V \tag{13.11.21}$$

as heuristically speculated before. Equation (13.11.18) becomes

$$\frac{\partial V}{\partial X} \cong -\overline{n} \frac{\partial P}{\partial T} = \frac{\beta}{E} \frac{\partial S}{\partial T}$$

because of (13.11.10), therefore,

$$\overline{P} = -\frac{\beta}{\overline{n}E} \overline{S}. \tag{13.11.22}$$

It also follows from (13.11.19) that

$$\frac{1}{C^2} \frac{\partial^2 V}{\partial T^2} = \frac{\partial^2 V}{\partial X^2}, \tag{13.11.23}$$

13.11 Vibration of a soil layer

where C denotes the dimensionless wave speed

$$C^2 = \frac{E + \beta/\bar{n}}{\omega^2 L^2 [\bar{n}\rho_w + (1-\bar{n})\rho_s]}. \tag{13.11.24}$$

Thus the outer region is a single phase elastic composite through which waves can propagate. In physical dimensions the characteristic wave velocity c is

$$c = C\omega L = \left(\frac{E + \beta/\bar{n}}{\bar{n}\rho_m + (1-\bar{n})\rho_s}\right)^{1/2}. \tag{13.11.25}$$

For sinusoidal forcing the outer solution is easily found and is given below in physical variables:

$$U^o = V^o = \frac{i\bar{n}c}{\beta} B \sin\frac{\omega x}{c} e^{-i\omega t} \tag{13.11.26}$$

$$p^o = -\frac{\beta}{nE}\sigma^o = B\cos\frac{\omega x}{c} e^{-i\omega t}. \tag{13.11.27}$$

The kinematic boundary conditions (13.11.15) at the rock bottom are satisfied. Note from (13.11.8) that on the bedrock $\partial p'/\partial x = 0$, which is satisfied by (13.11.27). But clearly the dynamic conditions on the ground surface $x = L$ are not satisfied. A boundary layer is needed.

13.11.3 The boundary-layer correction

In the boundary layer near the ground surface we assume

$$u = u^o + u^b, \qquad p = p^o + p^b, \qquad \sigma = \sigma^o + \sigma^b,$$

where $(.)^b$ represents the boundary-layer correction to the outer solution. Let δ denote the boundary-layer depth scale which is yet unknown, and renormalize the variables as follows:

$$\frac{x - L}{\delta} = \hat{x}, \qquad t = \frac{T}{\omega}$$

$$\begin{pmatrix} u^b \\ v^b \end{pmatrix} = \frac{\sigma_G \omega \delta}{\beta} \begin{pmatrix} \hat{u} \\ \hat{v} \end{pmatrix}$$

$$\begin{pmatrix} p^b \\ \sigma^b \end{pmatrix} = \sigma_G \begin{pmatrix} \hat{p} \\ \hat{\sigma} \end{pmatrix}.$$

To satisfy the dynamic boundary conditions on the ground surface, the stress and pore pressure corresponding to the boundary-layer correction must be comparable to the outer stress and outer pore pressure. Therefore, $O(\partial u^o/\partial x) = O(\partial u^b/\partial x)$, $O(\partial v^o/\partial x) = O(\partial v^b/\partial x)$, which explains the reduction of scales of (u^b, v^b) from those of (u^o, v^o) by the factor δ/L. The normalized equations are

$$\frac{\partial}{\partial \hat{x}}(\hat{u} - \hat{v}) + \frac{1}{\overline{n}}\frac{\partial \hat{v}}{\partial \hat{x}} = -\frac{\partial \hat{p}}{\partial T} \tag{13.11.28}$$

$$\left(\frac{\overline{p}_w \omega k}{\overline{n}}\right)\frac{\partial \hat{u}}{\partial T} = -\frac{1}{\overline{n}}\frac{\beta k}{\omega\delta^2}\frac{\partial \hat{p}}{\partial \hat{x}} - (\hat{u} - \hat{v}) \tag{13.11.29}$$

$$\frac{\omega^2\delta^2}{\beta}\left[\overline{n}\rho_w\frac{\partial \hat{u}}{\partial T} + (1-\overline{n})\overline{\rho}_s\frac{\partial \hat{v}}{\partial T}\right] = \frac{\partial}{\partial \hat{x}}(\hat{\sigma} - \hat{p}) \tag{13.11.30}$$

and

$$\frac{\partial \hat{v}}{\partial \hat{x}} = \frac{\beta}{E}\frac{\partial \hat{\sigma}}{\partial \hat{t}}. \tag{13.11.31}$$

For small enough δ the pressure gradient in (13.11.29) should be important and drives the relative motion between the two phases

$$0 \cong \frac{1}{\overline{n}}\frac{\beta k}{\omega\delta^2}\frac{\partial \hat{p}}{\partial \hat{x}} - (\hat{u} - \hat{v}). \tag{13.11.32}$$

Furthermore, inertia is negligible in the boundary layer ($\omega^2\delta^2/\beta \ll 1$). By integrating (13.11.30) upward from $\hat{x} \sim -\infty$ (outer edge of the boundary layer), where all corrections must vanish, we get simply

$$\hat{p} = \hat{\sigma}. \tag{13.11.33}$$

This result means that the total stress

$$\tau = (\sigma^o - p^o) + (\sigma^b - p^b) = \sigma^o - p^o \tag{13.11.34}$$

needs no correction in the boundary layer even though σ and p themselves must be corrected. Eliminating $\hat{u}, v$ and $\hat{\sigma}$ from (13.11.28, 13.11.30–13.11.32), we get

$$\frac{\partial^2 \hat{p}}{\partial \hat{x}^2} = \left(\omega\delta^2\frac{\beta + \overline{n}E}{kE\beta}\right)\frac{\partial \hat{p}}{\partial T}, \tag{13.11.35}$$

which is the diffusion equation. The implied thickness of the oscillatory boundary layer is

$$\delta = \left[\frac{kE\beta}{(\beta + \overline{n}E)\omega}\right]^{1/2}, \tag{13.11.36}$$

13.11 Vibration of a soil layer

which is smaller for higher frequency or smaller permeability. In physical variables (13.11.35) becomes

$$\frac{\partial^2 p^b}{\partial x^2} = \frac{\beta + \bar{n}E}{kE\beta} \frac{\partial p^b}{\partial t}, \qquad (13.11.37)$$

and is known as the equation of one-dimensional consolidation, due to K. Terzaghi, the founder of modern soil mechanics.

For sinusoidal motion p^b is easily found to be

$$p^b = \sigma^b = A \exp\left(\frac{1-i}{\sqrt{2}} \frac{x-L}{\delta} - i\omega t\right), \qquad (13.11.38)$$

which diminishes to zero outside the boundary layer. The coefficients A of the boundary layer correction and B of the outer solution are found by satisfying the dynamic boundary conditions on the ground surface ($x = 0$)

$$p^o + p^b = P_G = B \cos \frac{\omega L}{c} + A$$

$$\sigma^o + \sigma^b = \sigma_G = -\frac{\bar{n}E}{\beta} B \cos \frac{\omega L}{c} + A.$$

These algebraic equations are easily solved for A and B

$$A = \frac{\frac{\bar{n}E}{\beta} P_G + \sigma_G}{1 + \frac{\bar{n}E}{\beta}} \qquad (13.11.39)$$

$$B = \frac{P_G - \sigma_G}{\left(1 + \frac{\bar{n}E}{\beta}\right) \cos \frac{\omega L}{c}}. \qquad (13.11.40)$$

Note that the boundary-layer part p^b behaves like a wave propagating downward while attenuating exponentially in depth. The outer solution has the amplitude proportional to the applied total stress $\sigma_G - P_G$ on the surface and may experience very large values if

$$\frac{\omega L}{c} = \left(n + \frac{1}{2}\right)\pi, \quad n = 0, 1, 2, 3, \ldots,$$

which corresponds to resonance of the outer elastic layer.

Thus a complex problem of two-phase mechanics is solved with relatively little algebra but much insight.

In this chapter we have taken a whirlwind tour of the land of perturbations. By now the reader should realize that effective use of perturbation methods in solving physical problems requires a good deal of forethought in setting the problem up for formal expansions. The mathematics can often be tedious, but there is little cause for despair, thanks to computer algebra to be described in the last chapter.

Exercises

13.1 In an ocean of finite and constant depth a progressive surface wave of frequency ω can only exist if the wave number k is the real root of the following dispersion relation

$$\omega^2 h/g = kh \tanh kh.$$

Obtain a two-term approximation of kh for $\omega^2 h/g \ll 1$ (shallow water wave) and for $\omega^2 h/g \gg 1$ (deep water wave).

13.2 Determine a two-term expansion for the large roots of

$$x \tan x = 1.$$

13.3 Determine a two-term approximation of x if

$$\epsilon x^5 + x^3 - 1 = 0.$$

13.4 Let a solid cylinder of outer radius a be melted from within at $r = R(t)$. The surface temperatures are $T(a,t) = T_0$ and $T(R(t),t) = T_m$, where $T_m > T_0$. In terms of the normalized variables

$$T' = \frac{T - T_0}{T_m - T_0}, \qquad r' = \frac{r}{a}, \qquad R' = \frac{R}{a}, \qquad t' = \frac{kt}{a^2},$$

the present problem can be stated for $T'(R', r')$ by

$$\frac{1}{r'}\frac{\partial}{\partial r'}\left(r'\frac{\partial T'}{\partial r'}\right) = \epsilon \frac{\partial T'}{\partial R'}\frac{\partial T'}{\partial r'}\bigg|_{r'=R'}$$

with the boundary conditions

$$T'(R', r' = 1) = 0$$
$$T'(R', r' = R') = 1$$

and

$$\frac{dR'}{dt'} = \frac{\partial T'}{\partial r'}\bigg|_{r'=R'},$$

where

$$\epsilon = \frac{C(T - T_0)}{L} \ll 1.$$

Show that to the order $O(\epsilon)$,

$$T' \cong \frac{\ln r'}{\ln R'} + \epsilon \left[\frac{1 + R'^2(\ln R' - 1)}{4R'^2(\ln R')^4} \ln r' - \frac{1 + r'^2(\ln r' - 1)}{4R'^2(\ln R')^2} \right]$$

$$t' \cong \frac{1}{2}R'^2 \ln R' + \frac{1}{4}(1 - R'^2) - \frac{\epsilon}{4}(1 - R'^2)\left(1 - \frac{1}{\ln R'}\right)$$

(Asfar et al., 1979; Aziz and Na, 1984).

13.5 A nearly rectangular lake of constant depth has three straight coasts (north, south and east). The west side deviates slightly from the y axis by the amount $\epsilon f(y)$, where $\epsilon \ll 1$ and $f(y) = O(1)$. Follow the steps suggested below and study the effects of the coastal deviation on the natural frequencies of long-period oscillations in the shallow lake.

For simple harmonic oscillations the spatial factor η of the free-surface displacement $\operatorname{Re} \eta(x,y)e^{-i\omega t}$ satisfies the Helmholtz equation. Show first that the exact boundary condition on the west coast

$$\frac{\partial \eta}{\partial n} = 0 \quad \text{on} \quad x = \epsilon f(y), \quad 0 < y < a$$

can also be written as

$$\frac{\partial \eta}{\partial x} = \epsilon \frac{df}{dy} \frac{\partial \eta}{\partial y} \quad \text{on} \quad x = \epsilon f(y).$$

Use Taylor expansion to show that

$$\frac{\partial \eta(0,y)}{\partial x} + \epsilon f(y)\frac{\partial^2 \eta(0,y)}{\partial x^2} + O(\epsilon^2) = \epsilon \frac{df}{dy}\frac{\partial \eta(0,y)}{\partial x} + O(\epsilon^2).$$

Now expand

$$\eta = \eta_0 + \epsilon \eta_1 + \epsilon^2 \eta_2 + \cdots$$

and

$$\omega = \omega_0 + \epsilon \omega_1 + \epsilon^2 \omega_2 + \cdots$$

and find ω_1 either by solving the inhomogeneous problem for η_1 in the rectangle or by applying Green's formula to η_0 and η_1 over the rectangle.

Discuss the results. For the special case of $f(y) = 1 - (y/a)^2$, obtain the frequency shift explicitly.

13.6 Consider the steady diffusion of heat in a circular annulus. The temperature is T_a on the inner surface of radius a and T_b on the outer surface of radius $b(1 + \epsilon \cos\theta)$. Find the steady temperature in the annulus by perturbations to the order $O(\epsilon)$. Proceed by first expanding the boundary condition on the outer surface about $r = b$ in Taylor series up to order $O(\epsilon)$.

13.7 A thin metal wire is pulled T_1 and T_2 above the ambient atmosphere. Heat is lost from the wire to the surrounding air at the rate proportional to T. Show first that the temperature in the wire is governed by

$$U\frac{dT}{dx} = k\frac{d^2T}{dx^2} - bT,$$

where b is the coefficient of loss. Obtain the temperature in the region $0 < x < L$ exactly. Then find the approximate solution if $\epsilon = k/UL \ll 1$. Compare the approximate and exact results (Nayfeh, 1981).

13.8 Gravity induces geostatic stresses in the earth. From the equilibrium equation in elasticity

$$\frac{\partial \sigma_{ij}}{\partial x_j} - \rho g \delta_{i3} = 0,$$

verify first that

$$\sigma = \frac{\nu \rho g y}{1 - \nu}, \quad \sigma_{yy} = \rho g y, \quad \sigma_{xy} = 0$$

represents the stress field in the ground with a horizontal surface, where ρ is the density of earth and ν is the Poisson ratio.

Use the perturbation method to find the stress field in a two-dimensional mountain of small slope and the additional stresses in the ground (McTigue and Mei, 1981).

Let H be the characteristic height and L the characteristic width of the mountain. Start by normalizing the x, y coordinates by L, the mountain height $h(x)$ above the surrounding earth by H and the stresses by $\rho g H$. Show that the normalized equations for static equilibrium are

$$0 = \frac{\partial \sigma_{ij}}{\partial x_j} - \frac{\delta_{i3}}{\epsilon},$$

where $\epsilon = H/L \ll 1$. Show also that on the ground surface $y = \epsilon h(x)$ the unit normal vector is

$$\mathbf{n} = \frac{\nabla(y - \epsilon h(x))}{\sqrt{\nabla(y - \epsilon h(x)) \cdot \nabla(y - \epsilon h(x))}}.$$

Now get the perturbation equations by the following procedure:

(i) Introduce Airy's stress function F for the perturbed field

$$\sigma_{xx} = \frac{\nu y}{\epsilon(1-\nu)} + \frac{\partial^2 F}{\partial y^2}, \quad \sigma_{yy} = \frac{y}{\epsilon} + \frac{\partial^2 F}{\partial x^2}, \quad \sigma_{xy} = -\frac{\partial^2 F}{\partial x \partial y}.$$

(ii) Expand the boundary conditions about the flat ground surface $y = 0$.

(iii) Expand Airy's stress function as a power series of ϵ

$$F = F^{(0)} + \epsilon F^{(1)} + O(\epsilon^2).$$

Verify that in order to obtain the horizontal spreading effect of the uneven mountain height it is necessary to consider $O(\epsilon)$ at least.

13.9 A composite solid is made of a periodic array of identical particles imbedded in a solid matrix of different materials. Let the density, heat conductivity and specific heat of each material be ρ_a, K_a, C_a, where $a = 1$ for the grains and $a = 2$ for the matrix. Let the externally imposed temperature gradient be characterized by the length scale L, where $\ell/L = \epsilon \ll 1$. Reason physically that the time scale characterizing the global variation of temperature should be $O(L^2/(K/\rho C))$. Use the method of homogenization to show that, to the leading order in $O(\epsilon)$, T is the same in both the grains and the matrix. Prove that the global diffusion of heat in the composite is governed by

$$\langle \rho C \rangle \frac{\partial T}{\partial t} = \frac{\partial}{\partial X_i}\left(\langle K_{ij}\rangle \frac{\partial T}{\partial X_j}\right),$$

where

$$\langle \rho C \rangle = \frac{1}{\Omega} \sum_{a=1}^{2} \iiint_{\Omega_a} \rho_a C_a d\Omega,$$

and $\langle K_{ij}\rangle$ is the tensor with the components

$$\langle K_{ij}\rangle = \frac{1}{\Omega} \sum_{a=1}^{2} \iiint_{\Omega_a} K_a \left(\delta_{ij} + \frac{\partial B_{aj}}{\partial x_i}\right) d\Omega.$$

The vectors B_{ai} must be found by solving the following unit cell problem on the microscale

$$\frac{\partial}{\partial x_j}\left[K_a\left(\delta_{ij}+\frac{\partial B_{ai}}{\partial x_j}\right)\right]=0 \quad \text{in} \quad \Omega_a$$

$$B_{1i}=B_{2i} \quad \text{on} \quad \Gamma$$

$$K_1\left(\delta_{ij}+\frac{\partial B_{1i}}{\partial x_j}\right)n_j = K_2\left(\delta_{ij}+\frac{\partial B_{2i}}{\partial x_j}\right)n_j \quad \text{on} \quad \Gamma.$$

In addition B_{ai} must be Ω-periodic.

13.10 For the aquifer problem in §13.9 work out the details for the specific depth variation

$$h(x)=h_0+h_1\left[1-\left(\frac{X}{L}\right)^2\right].$$

The left end is submerged below the water surface of height H_o, while the right end is exposed to the atmosphere.

13.11 As pointed out in §12.9 the two-dimensional theory of strip loading on the surface of a half space leads to unreasonable results when the distance from the strip is much greater than the width ($2b$) of the strip. A resolution was suggested heuristically that the plane-strain problem is really just the near-field approximation of a three-dimensional problem for a long loading of finite length $2L$ with $L \gg b$. In the far field, where the scale of x, y, z is L, the problem is three-dimensional, but the strip appears to have zero width. Hence the strip is essentially as a line distribution of vertical force along $-L < z < L$ with unknown intensity $F(z)$.

In this exercise the reader is asked to confirm the anticipated two-dimensionality of the near field.

Define the near field by $\sqrt{x^2+y^2}=O(b)$, $|z\pm L|\gg b$ and the far field by $x,y,z=O(L)$. Start from the three-dimensional equations of elasticity

$$\nabla^2\mathbf{u}+\frac{1}{1-2\nu}\nabla\nabla\cdot\mathbf{u}=0,$$

where $\mathbf{u}$ is the solid displacement, and introduce dimensionless variables according to the near-field scales

$$(u,v)=U(u',v'), \quad w=\epsilon U w'$$

with $\epsilon = b/L \ll 1$, and
$$(x,y) = b(x', y'), \qquad z = Lz'.$$

Show that to the leading order in ϵ, u and v are governed by

$$\left(\frac{\partial^2}{\partial x^2} + \frac{\partial^2}{\partial y^2}\right)\binom{u}{v} + \frac{1}{1-2\nu}\binom{\frac{\partial}{\partial x}}{\frac{\partial}{\partial y}}\left(\frac{\partial u}{\partial x} + \frac{\partial v}{\partial y}\right) = O(\epsilon^2).$$

These are the equations for plane-strain problems. Next show that w can be deduced from

$$\left(\frac{\partial^2}{\partial x^2} + \frac{\partial^2}{\partial y^2}\right)w + \frac{1}{1-2\nu}\frac{\partial}{\partial z}\left(\frac{\partial u}{\partial x} + \frac{\partial v}{\partial y}\right) = O(\epsilon^2)$$

after u and v are found.

14
Computer algebra for perturbation analysis

While programming languages such as FORTRAN are designed for numerical computations, a number of them, e.g., MACSYMA, MAPLE, MATHEMATICA, REDUCE and THEORIST, have been developed to handle symbolic manipulations. With these languages one can use the computer to manipulate algebraic and trigonometric operations, differentiation and integration, Taylor series expansions, Laplace transform, and the solution of simple ordinary differential equations. When repetitive and lengthy manipulations are needed, these symbolic languages are extremely useful for speed and accuracy. In this chapter we first introduce some basic commands of MACSYMA and then illustrate their use for perturbation analysis. Only the bare essentials are introduced to get the reader started. More proficient users of the computer may very well wish to consult the MACSYMA manuals in order to find more shortcuts. There are also a few books devoted exclusively to the basics and applications of MACSYMA. *Computer Algebra in Applied Mathematics – An Introduction to MACSYMA* by R.H. Rand (1984) and *Variational Finite Element Methods – A Symbolic Computation Approach* by A.I. Beltzer (1990) are both helpful.

14.1 Getting started

MACSYMA can be installed on a variety of computers. On each the entry protocol may be slightly different. On a personal computer (PC) you simply enter WINDOWS, point the icon at the MACSYMA logo and click the mouse. After a short wait a few lines of trademark information

appear on the monitor screen, followed by the first command line,
(c1)
All the command lines are preceded by the letter c and numbered consecutively. Let us ask for the solutions of a quadratic equation and type:
(c1) solve $(a*x\wedge 2+b*x+c=0)$;
Note that the semicolon is the sign for execution. The screen gives you the solutions on the first response line, preceded by the letter d,
(d1)
$$\left[-\frac{\text{sqrt}(b^2-4ac)-b}{2a}, \frac{\text{sqrt}(b^2-4ac)-b}{2a}\right]$$
After finishing all the manipulations you quit MACSYMA by typing
(c20) quit();
or by pulling down the FILE menu and choosing EXIT with the mouse.

Comments that do not affect the calculations but appear in printouts can be added between the symbols /* and */.

In the following sections, explanations are added between input (c) and output (d) lines. MACSYMA commands used here are summarized in Appendix D.

14.2 Algebraic and trigonometric operations

14.2.1 Elementary operations

We present this entire section as a single output file. All commands have been prepared in a batch file called "lesa.txt".

(c2) BATCH("lęsa.txt");

The symbols $+, -, *, /, \wedge, !$ denote addition, subtraction, multiplication, division, exponentiation and factorial, respectively. An example to use these symbols for performing exact calculations of a group of numbers is

(c3) $2+3*4/5-6\wedge 7+8!$;
(d3)
$$-\frac{1198058}{5}$$

(c4) $x\wedge 2+2*\overset{\downarrow}{x}*y+y\wedge 2$;
(d4)
$$y^2+2xy+x^2$$

Using ":", we label the preceding output line "exp1",
(c5) exp1:(%);

(d5)
$$y^2 + 2xy + x^2$$

This labeling simplifies future operations involving the expression in (d4), thus,

(c6) $(z+3) * \exp 1$;
(d6)
$$(y^2 + 2xy + x^2)(z+3)$$

Variables can be arrays or lists. For example, let us define a subscripted array a[i]

(c7) for $i : 1$ thru 4, do $a[i] : x * i - 1$;
(d7) done
(c8) listarray(a);
(d8)
$$[x-1, 2x-1, 3x-1, 4x-1]$$

A list must always be put in square brackets $[\ldots]$.

(c9) list:$[x, y, y + z \wedge 2 = z]$;
(d9)
$$[x, y, z^2 + y = z]$$

The finite sum of a subscripted variable b_{2n-1} can be entered as

(c10) sum($b[2*n-1], n, 0, 5$);
(d10)
$$b_9 + b_7 + b_5 + b_3 + b_1 + b_{-1}$$

Algebraic operations are allowed in the summation

(c11) $(a[i] * x \wedge i, i, 0, 3)$;
(d11)
$$x^3(3x-1) + x^2(2x-1) + (x-1)x + a_0$$

The product of a subscripted variable can be written as

(c12) product$((x-i), i, 0, 5)$;
(d12)
$$(x-5)(x-4)(x-3)(x-2)(x-1)x$$

Finally, sums and products can be nested.

(c13) sum($a[i]*$product$(x-j, j, 0, 3), i, 0, 3$);

(d13)
$$\begin{aligned}&(x-3)(x-2)(x-1)x(3x-1)\\&+(x-3)(x-2)(x-1)x(2x-1)\\&+(x-3)(x-2)(x-1)^2x+a_0(x-3)(x-2)(x-1)x\end{aligned}$$

14.2.2 Functions

MACSYMA handles all the elementary transcendental functions and many special functions such as Bessel functions, Gamma function, Legendre polynomials, etc. If the argument is a number, the functions are also evaluated exactly. Certain symbols are reserved for standard constants: %pi for π, %e for e, %i for i, etc. Reserved symbols for standard functions such as log(x), sin(x), sqrt(x), ... are the same as in FORTRAN. For example,

(c14) cos(%pi) + log(%e) + cosh(%i) − sqrt(2);
(d14) cos(1) − $\sqrt{2}$.

A user-defined function is introduced by ":="
(c15) $f(x,y) := x \wedge 2 + 2*x*y + y \wedge 2$;
(d15)
$$f(x,y) := x^2 + 2xy + y^2$$

A number or an expression may replace one or all of the arguments
(c16) $f(5.5, y)$;
(d16)
$$y^2 + 11.0y + 30.25$$

(c17) $f(5.5, 3 + \%i * 2)$;
(d17)
$$(2i+3)^2 + 11.0\,(2i+3) + 30.25$$

(c18) $f(\%e, \sin(z))$;
(d18)
$$\sin^2(z) + 2e\sin(z) + e^2$$

The definition of the function can also be removed as follows:
(c19) remfunction(f);
(d19)
$$[f]$$

Let us verify it
  (c20) $f(5.5, 3 + \%i * 2)$;
  (d20) $f(5.5, 2i + 3)$
Indeed, no information is gained.

14.2.3 Algebraic reductions

Factorization and series expansion, etc., which are essential steps in perturbation analysis, can be faithfully handled by MACSYMA by using **factor**† and **expand**:
  (c21) factor($x \wedge 6 - 15*x \wedge 5 + 85*x \wedge 4 - 225*x \wedge 3 + 274*x \wedge 2 - 120*x$);
  (d21)
$$(x-5)(x-4)(x-3)(x-2)(x-1)x$$

  (c22) expand($(x+y) \wedge 3$);
  (d22)
$$y^3 + 3xy^2 + 3x^2y + x^3$$

To simplify a long algebraic expression, **ratsimp** is useful for handling rational fractions.
  (c23) ratsimp($(1 - x \wedge 3)/(x \wedge 2 - 4*x + 3)$);
  (d23)
$$\frac{x^2 + x + 1}{3 - x}$$

To simplify expressions involving log, sqrt ..., we should use **radcan** instead. Let us see the difference through the following examples:
  (c24) /* Example 1 */
  $((x-1) \wedge (3/2) - \text{sqrt}(x-1) * (x+1))/\text{sqrt}((x-1)*(x+1))$;
  (d24)
$$\frac{(x-1)^{3/2} - \sqrt{x-1}(x+1)}{\sqrt{(x-1)(x+1)}}$$

  (C25) ratsimp(%);
  (d25)
$$-\frac{2\sqrt{x-1}}{\sqrt{x^2-1}}$$

  (c26) radcan(%);

† The MACSYMA commands introduced in this chapter are listed in Appendix D.

(d26)
$$-\frac{2}{\sqrt{x+1}}$$

(c27)/* Example 2 */
eq3:log(x ∧ 2 − 1) − log(x + 1);
(d27)
$$\log(x^2 - 1) - \log(x + 1)$$

(c28) ratsimp(%);

(d28)
$$\log(x^2 - 1) - \log(x + 1)$$

No simplification is done, but
(c29) radcan(eq3);

(d29)
$$\log(x - 1)$$

14.2.4 Trigonometric reductions

The command **trigreduce** is for reducing the powers of trigonometric functions $\sin x, \cos x$, etc.; for example,
(c30) sin(x) ∧ 3 + 5 * cos(x) ∧ 5;
(d30)
$$\sin^3(x) + 5\cos^5(x)$$

(c31) trigreduce(%);
(d31)
$$5\left(\frac{\cos(5x)}{16} + \frac{5\cos(3x)}{16} + \frac{5\cos(x)}{8}\right) + \frac{3\sin(x) - \sin(3x)}{4}$$

To invoke other trigonometric identities, **trigsimp** and **trigexpand** should be used.
(c32) trigsimp(cos(x) ∧ 2+sin(x) ∧ 2);
You then see the following message before the response:
\ MACSYMA \ share \ trigsimp.fas being loaded.
(d32)
1

(c33) trigexpand(sin(2 * x + y)+cos(2 * a));

(d33)
$$\cos(2x)\sin(y) + \sin(2x)\cos(y) - \sin^2(a) + \cos^2(a)$$

14.2.5 Substitutions and manipulations

There are several ways to substitute one expression into another:**ev**, **subst** and **ratsubst**. First the protocol of **ev** is: ev(exp, arg1, arg2, ..., verb1, verb2), which means to substitute arg1 into exp, and then to expand, factor, etc. In the parentheses exp is an expression, arg1 can be an equation, or the name of an equation, a d-line; the first verb, ver1, can be **expand, factor, diff, integrate,** ....

(c34) eq3:$x \wedge 2 + 2 * x * y + y \wedge 2 = z \wedge 2$;
(d34)
$$y^2 + 2xy + x^2 = z^2$$

(c35) ev(eq3,$x = a + b$);
(d35)
$$y^2 + 2(b + a)y + (b + a)^2 = z^2$$

(c36) ev(eq3,$x = a + b$,expand);
(d36)
$$y^2 + 2by + 2ay + b^2 + 2ab + a^2 = z^2$$

To substitute a for b in the expression c, we may use **subst**,
(c37) subst($a, x + y, z + (x + y) \wedge 3$);
(d37)
$$z + a^3$$

Note that $x + y$ is in parentheses, otherwise **subst** cannot recognize the target term. To verify this we try without the parentheses in the following example:
(c38) subst($a, x + y, x + y + z$);
(d38)
$$z + y + x$$

Nothing has been changed. But the command **ratsubst** in the form ratsubst(a, b, c) is more intelligent.
(c39) ratsubst($a, x + y, x + y + z$);

(d39)
$$z + a$$

You can also make a list of substitutions in an expression by using **subst** in the form subst([arg1,arg2,...],exp);
(c40) subst([x = a, y = b],eq3);
(d40)
$$b^2 + 2ab + a^2 = z^2$$

Another useful command is **part** for extracting parts from an equation. The rule is: the first part of an equation is the left-hand side
(c41) eq6:x = 3 * y+sin(z);
(d41)
$$x = \sin(z) + 3y$$

(c42)part(eq6,1);
(d42)
$$x$$

The second part of an equation is the right-hand side
(c43) part(eq6,2);
(d43)
$$\sin(z) + 3y$$

The equality sign of an equation is the zeroth part.
(c44) part(eq6,0);
(d44)
$$=$$

Let us illustrate how to extract a part of a part of a part ... of an equation.
(c45) eq7:x − y + (z*log(x))/x ∧ 2 = 12;
(d45)
$$\frac{\log(x)z}{x^2} - y + x = 12$$

To extract −y, which is the second term on the left-hand side, we enter
(c46) part(eq7,1,2);
(d46)
$$-y$$

As an aside let us make a false move by asking for a part that does not exist
(c47) part(eq7,1,3,1);
PART fell off end.
(d47) false

We must now leave MACSYMA to correct the batch file and then re-enter to continue by typing "**batcon(true);**", the MACSYMA program is resumed.
(c48) batcon(true);
(c49) part(eq7,1,1,1);

(d49)
$$\log(x)z$$

(c50) part(eq7,1,1,2);

(d50)
$$x^2$$

To carry out a series of manipulations for an unlabeled line you may use the symbol **%th(n)** to refer to the nth previous line. For example, let the line in question be the group of expressions
(c51) $[x+1, y*z, (x+y)/z = x \wedge 2 + a]$;
(d51)
$$\left[x+1, yz, \frac{y+x}{z} = x^2 + a\right]$$

(c52) part(%,1);
(d52)
$$x+1$$

(c53) part(%th(2),2);
(d53)
$$yz$$

(c54) part(%th(3),3);
(d54)
$$\frac{y+x}{z} = x^2 + a$$

(c55) part(%th(4),3,2);

(d55)
$$x^2 + a$$
(c56) part(%th(5),3,2,1);
(d56)
$$x^2$$

14.3 Exact and perturbation methods for algebraic equations

MACSYMA has the ability to solve a system of linear equations explicitly. The tool is **linsolve**. For a system of three equations with unknowns x, y and z, the command is "**linsolve**([eq1, eq2, eq3],[x,y,z]);" after the equations have been defined. We give an example of three simultaneous equations:

(c3) eq1:3 * x + b * y + a * z − 2;
(d3)
$$az + by + 3x - 2$$
(c4) eq2:4 * x + 2 * y − a = 12;
(d4)
$$2y + 4x - a = 12$$
(c5) eq3:9 * z − 12 * y − 3 * a;
(d5)
$$9z - 12y - 3a$$

Note that the symbol "=0" is omitted but implied.
(c6) linsolve([eq1,eq2,eq3],[x,y,z]);
The solution is
(d6)
$$\left[x = \frac{(3a + 36)b + 6a^2 + 48a - 12}{12b + 16a - 18}, \ y = -\frac{4a^2 + 9a + 84}{12b + 16a - 18}, \right.$$
$$\left. z = \frac{2ab - 9a - 56}{6b + 8a - 9} \right]$$

A few nonlinear equations can also be explicitly solved by invoking **solve**, for example,
(c7) eqs:[x * y * z = 42, x + y − z = −2, 3 * x + 2 * y − 3 * z = −5];
(d7)
$$[xyz = 42, -z + y + x = -2, -3z + 2y + 3x = -5]$$

(c8) solve(eqs,[x, y, z]);

There are two sets of solutions that are complex conjugates of each other.
(d8)
$$\left[\left[x = \frac{\sqrt{167}i - 1}{2}, y = -1, z = \frac{\sqrt{167}i + 1}{2}\right],\right.$$
$$\left.\left[x = -\frac{\sqrt{167}i + 1}{2}, y = -1, z = -\frac{\sqrt{167}i - 1}{2}\right]\right]$$

There are of course many nonlinear solutions that cannot be solved by MACSYMA. For some linear equations, the explicit solutions, though possible by MACSYMA, are too lengthy to be of practical use. A case in point is the standard cubic equation, whose output takes up three pages of printout.

We illustrate the perturbation solution of a cubic equation with a small parameter c:

$$x^3 - \frac{1}{2}x^2 + (c-1)x + cd + \frac{1}{2} = 0.$$

(c3) eq0:$x \wedge 3 - x \wedge 2/2 + (c-1)*x + c*d + 1/2 = 0$;
(d3)
$$x^3 - \frac{x^2}{2} + (c-1)x + cd + \frac{1}{2} = 0$$

Let us expand

$$x = \sum_{0}^{4} c^i x[i]$$

(c4) $x = \text{sum}(c \wedge i * x[i], i, 0, 4)$;
(d4)
$$x = x_4 c^4 + x_3 c^3 + x_2 c^2 + x_1 c + x_0$$

and substitute the expansion into eq0.
(c5) eqa:eq0,%;
(d5)
$$cd + (x_4 c^4 + x_3 c^3 + x_2 c^2 + x_1 c + x_0)^3$$
$$- \frac{(x_4 c^4 + x_3 c^3 + x_2 c^2 + x_1 c + x_0)^2}{2}$$
$$+ (c-1)(x_4 c^4 + x_3 c^3 + x_2 c^2 + x_1 c + x_0) + \frac{1}{2} = 0$$

14.3 Exact and perturbation methods for algebraic equations

The zeroth order perturbation equation is found by using **ratcoef** to extract the coefficient of c^0 in the expansion:
(c6) ratcoef(eqa,c,0);
(d6)
$$\frac{2x_0^3 - x_0^2 - 2x_0 + 1}{2} = 0$$

The set of three roots for x_0 to be labeled as "soln0" is therefore easily found by **solve**.
(c7) soln0:solve(%, x[0]);
(d7)
$$\left[x_0 = \frac{1}{2}, x_0 = -1, x_0 = 1 \right]$$

For later convenience, let us christen the three roots as $r[1], r[2]$ and $r[3]$.
(c8) For $i : 1$ thru 3 do $r[i]$:rhs(part(soln0, i));
(d8)

$$\text{done}$$

(c9) $r[1]$;
(d9)

$$\frac{1}{2}$$

(c10) $r[2]$;
(d10)

$$-1$$

(c11) $r[3]$;
(d11)

$$1$$

Now the first-order problem.
(c12) ratcoef(eqa,c,1);
(d12)
$$d + (3x_0^2 - x_0 - 1)x_1 + x_0 = 0$$

This is a first-order equation that can be solved in terms of $x[0]$ and factorized.
(c13) solve(%, x[1]),factor;
(d13)
$$\left[x_1 = -\frac{d + x_0}{3x_0^3 - x_0 - 1} \right]$$

The procedure is repeated for $x[2]$.
(c14) ev(ratcoef(eqa,c,2),factor);
(d14)
$$\frac{6x_0^2 x_2 - 2x_0 x_2 - 2x_2 + 6x_0 x_1^2 - x_1^2 + 2x_1}{2} = 0$$

which is solved in terms of $x[0]$ and $x[1]$.
(c15) solve(%, $x[2]$),factor;
(d15)
$$\left[x_2 = -\frac{x_1(6x_0 x_1 - x_1 + 2)}{2(3x_0^2 - x_0 - 1)} \right]$$

We next use line d13 to eliminate $x[1]$ from the above result
(c16) ev(%,d13);
(d16)
$$\left[x_2 = \frac{(d+x_0)\left(-\frac{6x_0(d+x_0)}{3x_0^2 - x_0 - 1} + \frac{d+x_0}{3x_0^2 - x_0 - 1} + 2\right)}{2(3x_0^2 - x_0 - 1)^2} \right]$$

(c17) ratsimp(%);
(d17)
$$\left[x_2 = -\frac{(6x_0 - 1)d^2 + (6x_0^2 + 2)d + x_0^2 + 2x_0}{54x_0^6 - 54x_0^5 - 36x_0^4 + 34x_0^3 + 12x_0^2 - 6x_0 - 2} \right]$$

Let us collect the first three orders to get the perturbation solution in terms of x_0 and call it xx.
(c18) xx=sum($c \wedge i * x[i], i, 0, 2$);
(d18)
$$xx = x_2 c^2 + x_1 c + x_0$$

Corresponding to each of the three roots of x_0, there is an xx. For the first root, we ask
(c19) ev(d18,d13,d17,$x[0] = r[1]$,factor);
(d19)
$$xx = \frac{128c^2 d^2 + 224c^2 d + 72cd + 80c^2 + 36c + 27}{54}$$

Similarly, for the second root,
(c21) ev(d18,d13,d17,$x[0] = r[2]$,factor);
(d21)
$$xx = \frac{7c^2 d^2 - 8c^2 d - 18cd + c^2 + 18c - 54}{54}$$

Finally, for the third root,
(c23) ev($d18, d13, d17, x[0] = r[3]$,factor);
(d23)
$$xx = -\frac{5c^2d^2 + 8c^2d + 2cd + 3c^2 + 2c - 2}{2}$$

The procedure may obviously be extended to very high orders.

14.4 Calculus

First we introduce the command for Taylor expansion **taylor**, which is always needed for perturbation analyses. For example, to get the Taylor expansion of $(\sin x)/x$ in powers of x up to x^4, we enter
(c2) taylor($\sin(x)/x, x, 0, 4$);
(d2)/T/
$$1 - \frac{x^2}{6} + \frac{x^4}{120} + \cdots$$

We now turn to differentiation.

14.4.1 Differentiation

The ordinary derivative of $f(x)$ with respect to x is denoted by **diff**(f,x). Be sure to declare first the functional dependence before differentiation is requested, otherwise f is treated as a constant.
(c3) diff(f,x);
(d3)
$$0$$

To declare functional dependence we use **depends**.
(c4) depends(f, x);
(d4)
$$[f(x)]$$

The function can be explicit
(c5) $f : x \wedge 3 + \text{sqrt}(x) + \log(x)$;
(d5)
$$\log(x) + x^3 + \sqrt{x}$$

Now the first derivative may be taken
(c6) diff(f, x);

(d6)
$$3x^2 + \frac{1}{2\sqrt{x}} + \frac{1}{x}$$

For second and higher derivatives, a third argument indicating the order of differentiation is added

(c7) diff($f, x, 2$);
(d7)
$$6x - \frac{1}{4x^{3/2}} - \frac{1}{x^2}$$

(c8) diff($f, x, 3$);
(d8)
$$\frac{3}{8x^{5/2}} + \frac{2}{x^3} + 6$$

It is possible to withhold the evaluation of the differentiation by adding an apostrophe

(c9) 'diff($x \wedge 3 +$ sqrt(x) + log(x), x);
(d9)
$$\frac{d}{dx}(\log(x) + x^3 + \sqrt{x})$$

(x10) ev(%,diff);
(d10)
$$3x^2 + \frac{1}{2\sqrt{x}} + \frac{1}{x}$$

A function of several variables is defined by

(c11) depends($a, [x1, x2, x3]$);
(d11)
$$[a(x1, x2, x3)]$$

The second partial derivative of a with respect to $x2$ is denoted by

(c12) diff($a, x2, 2$);
(d12)
$$\frac{d^2 a}{dx2^2}$$

For example, let us write the heat equation.

(c13) depends($c, [x, t]$);
(d13)
$$[c(x, t)]$$

(c14) diff(c, t) = $k*$diff($c, x, 2$);

(d14)
$$\frac{dc}{dt} = \frac{d^2c}{dx^2}k$$

To remove the functional dependence we simply use **remove**.
(c15) remove(c,dependency);
(d15)
$$\text{done}$$

Now c is reduced to a constant. As a check,
(c16) diff(c, t) = diff(c, x, 2);
(d16)
$$0 = 0$$

14.4.2 Integration

Consider the following integrand, which is a function of x:
(c17) x ∧ 2 + 1/x + sin(x);
(d17)
$$\sin(x) + x^2 + \frac{1}{x}$$

To get an indefinite integral we use **integrate** without specifying the limits
(c18) integrate(%,x);
(d18)
$$\log(x) - \cos(x) + \frac{x^3}{3}$$

If the integrand is not specified, the output is just a formal expression; for example,
(c19) integrate(g(x), x);
(d19)
$$\int g(x)dx$$

But if g is defined,
(c20) g(x) := x ∧ 2 + 1/x + sin(x);
MACSYMA tries its best to evaluate it.
(d20)
$$g(x) := x^2 + \frac{1}{x} + \sin(x)$$

Integration can be withheld by adding an apostrophe,
(c21) 'integrate($g(x), x$);
(d21)
$$\int \left(\sin(x) + x^2 + \frac{1}{x} \right) dx$$

(c22) ev(%,integrate);
(d22)
$$\log(x) - \cos(x) + \frac{x^3}{3}$$

The evaluation of a definite integral by the computer is very time consuming. To see this, let us turn on the clock,†

(c23) showtime:true$

Time = 0 msecs

(c24) 'integrate($\sin(x), x, 0, \%pi$);

Time = 0 msecs

$$\int_0^\pi \sin(x) dx$$

(c25) ev(%,integrate);
Time = 5000 msecs

(d25)
$$2$$

The way to evaluate a multiple integral is shown in the following example:

(c26) integrate(integrate(integrate($x \wedge 3 * y * z, x, 0, 1), y, 0, 2), z, 0, 3$);
Time = 6209 msecs
(d26)
$$\frac{9}{4}$$

It is much faster to perform the indefinite integration first and then evaluate the limits.

(c27) integrate($\sin(x), x$);
Time = 0 msecs

---

† The clock speed varies with the computer. The times recorded below are for MAC-SYMA WINDOWS version, Release 417.25 delta 2, on a 386 PC at 25 MHz with 16 MB RAM.

(d27)
$$-\cos(x)$$

(c28) ev(%, $x = $ %pi) $-$ ev(%, $x = 0$);
Time $= 59$ msecs
(d28)
$$2$$

The saving of time is evident. As another example, the definite integral
$$\int_0^\pi \sin x^2 \cos x \, dx$$
would cause a PC 386 to crash. Hence we first evaluate the indefinite integral

(c29) 'integrate(sin($x \wedge 2$)*cos(x), x);
Time $= 0$ msecs
(d29)
$$\int \cos(x) \sin(x^2) dx$$

(c30) ev(%,integrate);
Time $= 8889$ msecs
(d30)
$$\frac{\sqrt{\pi}}{16} \begin{pmatrix} ((\sqrt{2i}-\sqrt{2})\sin\left(\tfrac{1}{4}\right)+(\sqrt{2i}+\sqrt{2})\cos\left(\tfrac{1}{4}\right)) * \\ \operatorname{erf}\left(\tfrac{1}{4}((2\sqrt{2i}+2\sqrt{2})x+\sqrt{2i}+\sqrt{2})\right) \\ +((\sqrt{2i}-\sqrt{2})\sin\left(\tfrac{1}{4}\right)+(\sqrt{2i}+\sqrt{2})\cos\left(\tfrac{1}{4}\right)) * \\ \operatorname{erf}\left(\tfrac{1}{4}((2\sqrt{2i}+2\sqrt{2})x-\sqrt{2i}-\sqrt{2})\right) \\ +((\sqrt{2i}+\sqrt{2})\sin\left(\tfrac{1}{4}\right)+(\sqrt{2i}-\sqrt{2})\cos\left(\tfrac{1}{4}\right)) * \\ \operatorname{erf}\left(\tfrac{1}{4}((2\sqrt{2i}-2\sqrt{2})x+\sqrt{2i}-\sqrt{2})\right) \\ +((\sqrt{2i}+\sqrt{2})\sin\left(\tfrac{1}{4}\right)+(\sqrt{2i}-\sqrt{2})\cos\left(\tfrac{1}{4}\right)) * \\ \operatorname{erf}\left(\tfrac{1}{4}((2\sqrt{2i}-2\sqrt{2})x-\sqrt{2i}+\sqrt{2})\right) \end{pmatrix}$$

The limits can then be substituted to evaluate the definite integral. We omit the lengthy output and only clock the process.

(c31) ev(%, $x = $ %pi) $-$ ev(%, $x = 0$)$
Time $= 1810$ msecs

If the integrand contains nonnumeric parameters, it is necessary to specify their signs. For example,

(c32) integrate(exp($-a * x \wedge 2$), x, 0, inf);

Is a positive, negative or zero?
pos;
Time = 42509 msecs
(d32)
$$\frac{\sqrt{\pi}}{2\sqrt{a}}$$

14.5 Ordinary differential equations

Linear differential equations of the first and second order can be solved by the package **ode2** with the commands **bc2** for boundary conditions and **ic2** for initial conditions. Consider the example

$$\frac{d^2y}{dx^2} + 4y = \sin x,$$

subject to the boundary conditions

$$y(0) = 2, \quad y(1) = 3.$$

(c1) depends (y,x);
(d1)
$$[y(x)]$$
(c2) eq1:diff(y, x, 2) + 4 * y = sin(x);
(d2)
$$\frac{d^2y}{dx^2} + 4y = \sin(x)$$

To solve this inhomogeneous equation we invoke **ode2**.
(c3) ode2(eq1,y,x);
\ macsyma \ ode \ ode2.fas being loaded.
(d3)
$$y = \%k1 \sin(2x) + \%k2 \cos(2x) + \frac{\sin(x)}{3}$$

where %k1 and %k2 are the constants of integration. Now we apply the boundary conditions to the general solution above.
(c4) bc2(%, $x = 0, y = 2, x = 1, y = 3$);
The constants are then determined to yield the final solution.
(d4)
$$y = -\frac{(6\cos(2) + \sin(1) - 9)\sin(2x)}{3\sin(2)} \\ + 2\cos(2x) + \frac{\sin(x)}{3}$$

If instead the initial conditions are imposed, then **ic2** may be used.
(c5) ic2(d3, x = 0, y = 1, diff(y, x) = 2);
(d5)
$$y = \frac{5\sin(2x)}{6} + \cos(2x) + \frac{\sin(x)}{3}$$

Note that the argument x of y need not be shown in the inputs.

The Laplace transform package in MACSYMA can also be used to solve initial-value problems governed by ordinary differential equations. For simple cases the inverse transform can be given explicitly. The interested reader should consult MACSYMA manuals for such information.

We now turn to an example of regular perturbations.

14.6 Pipe flow in a vertical thermal gradient

In §8.9 the problem of a pipe flow through a vertical thermal gradient was formulated and solved exactly for all Rayleigh numbers ϵ. We now derive an approximate result, also given by Morton (1960), for small ϵ by perturbation analysis. The principle is to expand the solution in ascending powers of ϵ,

$$U = U_0 + \epsilon U_1 + \epsilon^2 U_2 + \cdots \tag{14.6.1}$$

$$\theta = \theta_0 + \epsilon \theta_1 + \epsilon^2 \theta_2 + \cdots. \tag{14.6.2}$$

At the leading order $O(1)$, the perturbation equations are

$$\frac{d^2 U_0}{dR^2} + \frac{1}{R}\frac{dU_0}{dR} = -P$$

and

$$\frac{d^2 \theta_0}{dR^2} + \frac{1}{R}\frac{d\theta_0}{dR} = -U_0$$

with the boundary conditions

$$\frac{dU_0}{dR} = \frac{d\theta_0}{dR} = 0, \quad R = 0$$

and

$$U_0 = \theta_0 = 0, \quad R = 1.$$

The perturbation equations at any order $O(\epsilon^n), n = 1, 2, \ldots$, are

$$\frac{d^2 U_n}{dR^2} + \frac{1}{R}\frac{dU_n}{dR} = \theta_{n-1}\mathrm{sgn}\tau$$

and
$$\frac{d^2\theta_n}{dR^2} + \frac{1}{R}\frac{d\theta_n}{dR} = -U_n$$

with the boundary conditions
$$\frac{dU_n}{dR} = \frac{d\theta_n}{dR} = 0, \quad R = 0$$

and
$$U_n = \theta_n = 0, \quad R = 1.$$

The solutions to these linear equations are straightforward although the algebra becomes tedious for $n = 2$ and higher. Let us employ computer algebra.

In this MACSYMA program, the following notations will be used: u = axial velocity, t = temperature, r = radial coordinate, q the prescribed total pressure gradient, and e = Rayleigh number signifying buoyancy. First we introduce the perturbation quantities and define their dependence in r.

(c1) depends($[u, u0, u1, u2, t, t0, t1, t2], r$);
(d1)
$$[u(r), u0(r), u1(r), u2(r), t(r), t0(r), t1(r), t2(r)]$$

The equation for fluid momentum reads
(c2) mom:diff($u, r, 2$) + $(1/r)*$diff(u, r) = $-q + e*t$;
(d2)
$$\frac{d^2u}{dr^2} + \frac{\frac{du}{dr}}{r} = et - q$$

where the ambient temperature is assumed to be increasing with height. The law of energy conservation reads
(c3) heat:diff($t, r, 2$) + $(1/r)*$diff(t, r) = $-u$;
(d3)
$$\frac{d^2t}{dr^2} + \frac{\frac{dt}{dr}}{r} = -u$$

The boundary conditions for the fluid velocity are
(c4) /*Symmetry along pipe axis $r = 0$. */
bc1:diff(u, r) = 0;
(d4)
$$\frac{du}{dr} = 0$$

(c5) /*No slip on the pipe wall $r = 1$. */

14.6 Pipe flow in a vertical thermal gradient

bc2:$u = 0$;
(d5)
$$u = 0$$

The boundary conditions for the temperature are
(c6) /*Symmetry along the pipe axis on $r = 0$. */
bc3:diff(t, r) = 0;
(d6)
$$\frac{dt}{dr} = 0$$

(c7) /*On $r = 1$, the temperature is given. */
bc4: $t = 0$;
(d7)
$$t = 0$$

Now we use $u0, u1, u2, \ldots$ to mean $u_0, u_1, u_2, \ldots$, respectively, and $t1, t2, t3, \ldots$ to mean $t_1, t_2, t_3, \ldots$, respectively, and assume the following perturbation expansions up to the second power of e
(c8) uu:$u = u0 + e*u1 + e \wedge 2 * u2$;
(d8)
$$u = e^2 u2 + e u1 + u0$$

(c9) tt:$t = t0 + e*t1 + e \wedge 2 * t2$;
(d9)
$$t = e^2 t2 + e t1 + t0$$

and substitute the expansions into the momentum equation and carry out the differentiation. The expanded momentum equation is also labeled "mom".
(c10) mom:ev(mom,uu, tt,diff);
(d10)
$$e^2 \frac{d^2 u2}{dr^2} + \frac{e^2 \frac{du2}{dr} + e \frac{du1}{dr} + \frac{du0}{dr}}{r}$$
$$+ e \frac{d^2 u1}{dr^2} + \frac{d^2 u0}{dr^2} = e(e^2 t2 + e\, t1 + t0) - q$$

The perturbation equations at various orders can now be extracted by the command **ratcoef**:
(c11) mom0:ratcoef(mom,e,0);

(d11)
$$\frac{r\frac{d^2 u0}{dr^2} + \frac{du0}{dr}}{r} = -q$$

(c12) mom1:ratcoef(eq3,e,1);
(d12)
$$\frac{r\frac{d^2 u1}{dr^2} + \frac{du1}{dr}}{r} = t0$$

(c13) mom2:ratcoef(mom,e,2);
(d13)
$$\frac{r\frac{d^2 u2}{dr^2} + \frac{du2}{dr}}{r} = t1$$

Similarly, the expansions are substituted into the energy equation.
(c14) heat:ev(heat,uu,tt,diff);
(d14)
$$e^2\frac{d^2 t2}{dr^2} + \frac{e^2\frac{dt2}{dr} + e\frac{dt1}{dr} + \frac{dt0}{dr}}{r} + e\frac{d^2 t1}{dr^2} + \frac{d^2 t0}{dr^2} = -e^2 u2 - e\, u1 - u0$$

The following perturbation equations are obtained:
(c15) heat0:ratcoef(heat,e,0);
(d15)
$$\frac{r\frac{d^2 t0}{dr^2} + \frac{dt0}{dr}}{r} = -u0$$

(c16) heat1:ratcoef(heat,e,1);
(d16)
$$\frac{r\frac{d^2 t1}{dr^2} + \frac{dt1}{dr}}{r} = -u1$$

(c17) heat2:ratcoef(heat,e,2);
(d17)
$$\frac{r\frac{d^2 t2}{dr^2} + \frac{dt2}{dr}}{r} = -u2$$

14.6 Pipe flow in a vertical thermal gradient

The assumed expansions are inserted in the boundary conditions for u. First, along the axis,
(c18) bc1:ev(bc1,uu,diff);
(d18)
$$e^2 \frac{du2}{dr} + e\frac{du1}{dr} + \frac{du0}{dr} = 0$$
from which the boundary conditions at various orders can be trivially deduced.
(c19) [ratcoef(bc1,e,0),ratcoef(bc1,e,1),ratcoef(bc1,e,2)];
(d19)
$$\left[\frac{du0}{dr} = 0, \frac{du1}{dr} = 0, \frac{du2}{dr} = 0\right]$$
(c20) bc2:ev(bc2,UU);
(d20)
$$e^2 u2 + e\, u1 + u0 = 0$$
(c21) [ratcoef(bc2,e,0),ratcoef(bc2,e,1),ratcoef(bc2,e,2)];
(d21)
$$[u0 = 0, u1 = 0, u2 = 0]$$

The time is ripe for solving the perturbation problems. We first solve the differential equation for $u0$,
(c22) vel0:ode2(mom0,u0,r);
\ macsyma \ ode \ ode2.fas being loaded.
(d22)
$$u0 = \%k1 \log(r) - \frac{qr^2}{4} + \%k2$$
where %k1 and %k2 are integration constants. Now apply the boundary condition at $r = 0$ for $u0$(d(19)).
(c23) diff(%,r);
(d23)
$$\frac{du0}{dr} = \frac{\%k1}{r} - \frac{qr}{2}$$
For finiteness at $r = 0$ we require $\%k1 = 0$.
(c24) vel0:subst([%k1= 0],vel0);
(d24)
$$u0 = \%k2 - \frac{qr^2}{4}$$

Next, the first of the boundary conditions in (d21) is applied.
(c25) solve(ev(rhs(vel0),$r = 1$),%k2);
(d25)
$$\left[\%k2 = \frac{q}{4} \right]$$

(c26) vel0:subst(%,vel0);
(d26)
$$u0 = \frac{q}{4} - \frac{qr^2}{4}$$

The solution for $u0$ is now complete and may be inserted in the zeroth order equation for t

(c27) heat0:subst([vel0],heat0);

with the result

(d27)
$$\frac{r \frac{d^2 t0}{dr^2} + \frac{dt0}{dr}}{r} = \frac{qr^2}{4} - \frac{q}{4}$$

The general solution is found by

(c28) temp0:ode2(heat0,t0,r);
(d28)
$$t0 = \frac{\%k1 \log(r)}{4} + \frac{qr^4 - 4qr^2}{64} + \%k2$$

To apply the boundary condition at $r = 0$ we need

(c29) diff(%,r);
(d29)
$$\frac{dt0}{dr} = \frac{4qr^3 - 8qr}{64} + \frac{\%k1}{4r}$$

Clearly, %k1 must vanish to ensure finiteness.

(c30)temp0:subst(%k1= 0,temp0);
(d30)
$$t0 = \frac{qr^4 - 4qr^2}{64} + \%k2$$

Let us apply the boundary condition on the pipe wall

(c31) subst($r = 1, \%$);
(d31)
$$t0 = \%k2 - \frac{3q}{64}$$

(c32) rhs(%)= 0;

(d32)
$$\%k2 - \frac{3q}{64} = 0$$

(c33) solve(%,%k2);
(d33)
$$\left[\%k2 = \frac{3q}{64} \right]$$

(c34) temp0:subst(%,temp0);
(d34)
$$t0 = \frac{qr^4 - 4qr^2}{64} + \frac{3q}{64}$$

The solution for $t0$ is complete. We now begin the solution for $u1$.

(c35) subst([et0],eq31);
(d35)
$$\frac{r\frac{d^2u1}{dr^2} + \frac{du1}{dr}}{r} = \frac{qr^4 - 4qr^2}{64} + \frac{3q}{64}$$

(c36) vel1:ode2(%,u1,r);
(d36)
$$u1 = \frac{\%k1\log(r)}{64} + \frac{qr^6 - 9qr^4 + 27qr^2}{2304} + \%k2$$

(c37) vel1: subst([%k1 = 0], %);
(d37)
$$u1 = \frac{qr^6 - 9qr^4 + 27qr^2}{2304} + \%k2$$

(c38) subst($r = 1$, %);
(d38)
$$u1 = \frac{19q}{2304} + \%k2$$

(c39) solve(rhs(%) = 0, %k2);
(d39)
$$\left[\%k2 = -\frac{19q}{2304} \right]$$

(c40) vel1:subst(%,vel1);
(d40)
$$u1 = \frac{qr^6 - 9qr^4 + 27qr^2}{2304} - \frac{19q}{2304}$$

The first-order velocity is found.
(c41) subst([vel1],heat1);
(d41)
$$\frac{r\frac{d^2 t1}{dr^2} + \frac{dt1}{dr}}{r} = \frac{19q}{2304} - \frac{qr^6 - 9qr^4 + 27qr^2}{2304}$$

(c42) temp1:ode2(%,t1,r);
(d42)
$$t1 = \frac{\%k1 \log(r)}{2304} - \frac{qr^8 - 16qr^6 + 108qr^4 - 304qr^2}{147456} + \%k2$$

(c43) temp1:subst(%k1= 0, %);
(d43)
$$t1 = \%k2 - \frac{qr^8 - 16qr^6 + 108qr^4 - 304qr^2}{147456}$$

(c44) subst($r = 1$, %);
(d44)
$$t1 = \frac{211q}{147456} + \%k2$$

(c45) solve(rhs(%) = 0, %k2);
(d45)
$$\left[\%k2 = -\frac{211p}{147456} \right]$$

(c46) temp1:subst(%,temp1);
(d46)
$$t1 = -\frac{qr^8 - 16qr^6 + 108qr^4 - 304qr^2}{147456} - \frac{211q}{147456}$$

The first-order temperature is also found.

We stop here and summarize the approximate solution with $O(e)$ accuracy.

(c47) u = rhs(vel0) + $e*$rhs(vel1);
(d47)
$$u = e\left(\frac{qr^6 - 9qr^4 + 27qr^2}{2304} - \frac{19q}{2304}\right) - \frac{qr^2}{4} + \frac{q}{4}$$

(c48) t = rhs(temp0) + $e*$rhs(temp1);
(d48)
$$t = e\left(-\frac{qr^8 - 16qr^6 + 108qr^4 - 304qr^2}{147456} - \frac{211q}{147456}\right) + \frac{qr^4 - 4qr^2}{64} + \frac{3q}{64}$$

14.7 Duffing problem by multiple scales

Using the symbols defined in §8.9, we summarize the results accurate to the order $O(\epsilon^2)$ as follows:

$$U_0 = \frac{P}{4}(1-R^2) \tag{14.6.3}$$

$$\theta_0 = \frac{P}{64}(1-R^2)(1-3R^2) \tag{14.6.4}$$

$$U_1 = \frac{P}{2,304}\left(R^6 - 9R^4 + 27R^2 + 19\right) \tag{14.6.5}$$

$$\theta_1 = \frac{P}{147,456}\left(R^8 + 16R^6 - 108R^4 + 304R^2 - 211\right) \tag{14.6.6}$$

$$U_2 = -\frac{P}{14,745,600}\left(R^{10} - 25R^8 + 300R^6 - 1,900R^4 \right.$$
$$\left. + 5,275R^2 - 3,651\right) \tag{14.6.7}$$

$$\theta_2 = \frac{P}{2,123,366,400}\left(R^{12} - 36R^{10} + 675R^8 - 7,600R^6 \right.$$
$$\left. +47,475R^4 - 131,436R^2 + 9,0921\right). \tag{14.6.8}$$

The formulas above are written in the form given by Aziz and Na (1984), who also compared them with the exact solution of Morton. The agreement is remarkably good even for quite large ϵ.

14.7 Duffing problem by multiple scales

Computer algebra is particularly suited for the multiple-scale analysis of §13.5. We demonstrate the procedure by seeking the periodic solution to the Duffing equation

$$\frac{d^2u}{dt^2} + \omega^2 u + \epsilon u^3 = 0,$$

which describes an anharmonic oscillator with a weak nonlinear restoring force. Anticipating multiple-scale expansions, we first declare the dependence on fast and slow coordinates by using the command **depends**.
(c3) depends([u, u0, u1, u2], [t, t1, t2]);
(d3)

$$[u(t, t1, t2), u0(t, t1, t2), u1(t, t1, t2), u2(t, t1, t2)]$$

In the program we shall denote the small parameter by e.
(c4) eqa: diff(u, t, 2) + omega ∧ 2 * u + e * u ∧ 3 = 0;

(d4)
$$u\omega^2 + \frac{d^2u}{dt^2} + e u^3 = 0$$

We now define slow coordinates $t1 = et, t2 = e^2 t, \ldots$ with the help of the command **gradef**. In particular we define the gradient of f with respect to $x_1, x_2, \ldots$ to be $g_1, g_2, \ldots$ and write **gradef**$(f(x_1, \ldots, x_n), g_1, \ldots, g_n)$; thus

(c5) gradef($t1, t, e$);
(d5)
$$t1$$

The slow coordinate $t2 = e^2 t$ is defined similarly.

(c6) gradef($t2, t, e \wedge 2$);
(d6)
$$t2$$

With these definitions the differentiation in (eqa) can now be carried out.

(c7) eqa:eqa,diff;
(d7)
$$u\omega^2 + e^2\left(e^2\frac{d^2u}{dt2^2} + e\frac{d^2u}{dt1\,dt2} + \frac{d^2u}{dt\,dt2}\right)$$
$$+ e\left(e\frac{d^2u}{dt1^2} + e^2\frac{d^2u}{dt1\,dt2} + \frac{d^2u}{dt\,dt1}\right) + \frac{d^2u}{dt^2}$$
$$+ e^2\frac{d^2u}{dt\,dt2} + e\frac{d^2u}{dt\,dt1} + e u^3 = 0$$

Now we must treat $t, t1$ and $t2$ as independent variables by removing the dependence of $t1, t2$ on t, i.e., by declaring that the derivatives of $t1$ and $t2$ with respect to t are zero.

(c8) gradef(t1,t,0)$
(c9) gradef(t2,t,0)$

The following perturbation series is then introduced.

(c10) ev(eqa,$u = u0 + e * u1 + e \wedge 2 * u2$,diff);

14.7 Duffing problem by multiple scales

(d10)

$$\left(e^2 u2 + e\, u1 + u0\right)\omega^2$$

$$+ e^2 \begin{pmatrix} e^2\left(\frac{d^2u2}{dt2^2} + e\frac{d^2u1}{dt2^2} + \frac{d^2u0}{dt2^2}\right) \\ + e\left(e^2\frac{d^2u2}{dt1\,dt2} + e\frac{d^2u1}{dt1\,dt2} + \frac{d^2u0}{dt1\,dt2}\right) \\ + e^2\frac{d^2u2}{dt\,dt2} + e\frac{d^2u1}{dt\,dt2} + \frac{d^2u0}{dt\,dt2} \end{pmatrix}$$

$$+ e \begin{pmatrix} e\left(e^2\frac{d^2u2}{dt1^2} + e\frac{d^2u1}{dt1^2} + \frac{d^2u0}{dt1^2}\right) \\ + e^2\left(e^2\frac{d^2u2}{dt1\,dt2} + e\frac{d^2u1}{dt1\,dt2} + \frac{d^2u0}{dt1\,dt2}\right) \\ + e^2\frac{d^2u2}{dt\,dt1} + e\frac{d^2u1}{dt\,dt1} + \frac{d^2u0}{dt\,dt1} \end{pmatrix} + e^2\frac{d^2u2}{dt^2}$$

$$+ e^2\left(e^2\frac{d^2u2}{dt\,dt2} + e\frac{d^2u1}{dt\,dt2} + \frac{d^2u0}{dt\,dt2}\right)$$

$$+ e\left(e^2\frac{d^2u2}{dt\,dt1} + e\frac{d^2u1}{dt\,dt1} + \frac{d^2u0}{dt\,dt1}\right)$$

$$+ e(e^2 u2 + e\, u1 + u0)^3 + e\frac{d^2u1}{dt^2} + \frac{d^2u0}{dt^2} = 0$$

Terms of equal powers of e will now be grouped.

(c11) eqb:taylor(%,e,0,2);

(d11) /T/

$$\frac{d^2u0}{dt^2} + \omega^2 u0 + \left(\frac{d^2u1}{dt^2} + u0^3 + \omega^2 u1 + 2\frac{d^2u0}{dt\,dt1}\right)e$$

$$+ \left(3\,u1\,u0^2 + \omega^2 u2 + 2\frac{d^2u1}{dt\,dt1} + 2\frac{d^2u0}{dt\,dt2} + \frac{d^2u2}{dt^2} + \frac{d^2u0}{dt1^2}\right)e^2 + \cdots$$

$$= 0 + \cdots$$

Now we separate the powers of e:

(c12) /* Order $O(1)$. */
eq0:coeff(eqb,e,0);

(d12) /R/

$$\frac{d^2u0}{dt^2} + \omega^2 u0 = 0$$

(c13) /*Order $O(e)$. */
eq1:coeff(eqb,e,1);

(d13) /R/
$$\frac{d^2 u1}{dt^2} + u0^3 + \omega^2 u1 + 2\frac{d^2 u0}{dt\, dt1} = 0$$

(c14) /* Order $O(e^2)$. */
eq2:coeff(eqb,e,2);
(d14) /R/
$$3\, u1\, u0^2 + \omega^2 u2 + 2\frac{d^2 u1}{dt\, dt1} + 2\frac{d^2 u0}{dt\, dt2} + \frac{d^2 u2}{dt^2} + \frac{d^2 u0}{dt1^2} = 0$$

The periodic solution $u0$ to (eq0) in line (d12) is obviously
$$a(t_1, t_2) e^{i\omega t} + a_s(t_1, t_2) e^{-i\omega t},$$
where a_s is the complex conjugate of a. Let us denote a_s by as in the program and introduce the abbreviation $x \equiv e^{i\omega t}$

(c15) x : exp(%i * omega * t);
(d15)
$$e^{it\omega}$$

(c16) depends([a, as], [t1, t2]);
(d16)
$$[a(t1, t2), as(t1, t2)]$$

(c17) ev(eq1,u0 = a * x + as/x,diff);
(d17) /R/
$$\frac{1}{(e^{it\omega})^3} \begin{pmatrix} (e^{it\omega})^3 \frac{d^2 u1}{dt^2} + (e^{it\omega})^3 \omega^2 u1 \\ + \left(2i\frac{da}{dt1}(e^{it\omega})^4 - 2i\frac{das}{dt1}(e^{it\omega})^2\right)\omega \\ + a^3 (e^{it\omega})^6 + 3a^2 as (e^{it\omega})^4 \\ + 3a\, as^2 (e^{it\omega})^2 + as^3 \end{pmatrix} = 0$$

(c18) eq1:expand(%);
(d18)
$$a^3 e^{3it\omega} + 2i\frac{da}{dt1}\omega\, e^{it\omega} + 3a^2\, as\, e^{it\omega} - 2i\frac{das}{dt1}\omega\, e^{it\omega}$$
$$+ 3a\, as^2\, e^{-it\omega} + as^3\, e^{-3it\omega} + u1\,\omega + \frac{d^2 u1}{dt^2} = 0$$

The solvability conditions are now enforced by equating to zero the secular terms proportional to x and x^{-1} in (eq1), currently in line (d18).
(c19) sc1:coeff(eq1,x,1);

14.7 Duffing problem by multiple scales

(d19)
$$2i\frac{da}{dt1}\omega + 3\,a^2\,as = 0$$

(c20) sc2:coeff(eq1,x,−1);
(d20)
$$3\,a\,as^2 - 2i\frac{das}{dt1}\omega = 0$$

The remaining parts of eq1 then yield the governing equation for $u1$.

(c21) eq1:part(lhs(eq1),[1,6,7,8]) = 0;
(d21)
$$a^3\,e^{3it\omega} + as^3\,e^{-3it\omega} + u1\,\omega^2 + \frac{d^2u1}{dt^2} = 0$$

This inhomogeneous equation contains only the third harmonics and can be solved by
$$c e^{3i\omega t} + c_s e^{-3i\omega t}.$$

(c22) expand(ev(eq1,u1 = c*x^3 + cs/x^3,diff));
(d22)
$$-8\,c\,\omega^2 e^{3it\omega} + a^3\,e^{3it\omega} - 8\,cs\,\omega^2\,e^{-3it\omega} + as^3\,e^{-3it\omega} = 0$$

The coefficients of $x \wedge 3$ give the equation for c.

(c23) coeff(%, $x \wedge 3$);
(d23)
$$a^3 - 8\,c\,\omega^2 = 0$$

(c24) solve(%, c);
(d24)
$$\left[c = \frac{a^3}{8\,\omega^2}\right]$$

(c25) coeff(%th(3), $x \wedge (-3)$);
(d25)
$$as^3 - 8\,cs\,\omega^2 = 0$$

(c26) solve(%, cs);
(d26)
$$\left[cs = \frac{as^3}{8\,\omega^2}\right]$$

The coefficients c and $cs(= c*)$ are now known in terms of a and $as(= a*)$ and may be substituted in $u1$.

(c27) ev(u1,[%,%th(2)],diff);
(d27)
$$u1$$

Next we write
$$a = \frac{b}{2}e^{i\phi} \qquad a_s = \frac{b}{2}e^{-i\phi},$$
where b denotes the magnitude and ϕ the phase of $u0$.

(c28) depends([b, phi], [t1, t2]);
(d28)
$$[b(t1,t2), \phi(t1,t2)]$$

(c29) a:a = (1/2)*b*exp(%i*phi);
(d29)
$$a = \frac{b\,e^{i\phi}}{2}$$

(c30) as:as = (1/2)*b*exp(-%i*phi);
(d30)
$$as = \frac{b\,e^{-i\phi}}{2}$$

(c31) expand(ev(sc1,[%,%th(2)],diff));
(d31)
$$-b\,e^{i\phi}\frac{d\phi}{dt1}w + i\frac{db}{dt1}e^{i\phi}w + \frac{3\,b^3 e^{i\phi}}{8} = 0$$

Let us remove the factor $e^{i\phi}$.

(c32) coeff(%,exp(%i*phi));
(d32)
$$-b\frac{d\phi}{dt1}w + i\frac{db}{dt1}w + \frac{3\,b^3}{8} = 0$$

The real and imaginary parts of these equations yield

(c33) imagpart(%);
(d33)
$$\frac{db}{dt1}\omega = 0$$

(c34) realpart(%th(2));
(d34)
$$\frac{3\,b^3}{8} - b\frac{d\phi}{dt1}\omega = 0$$

(c35) depends([b, phi0], t2);
(d35)
$$[b(t2), \phi(t2)]$$
(c36) phi:phi = 3 * b ∧ 2 * t1/(8 w) + phi0;
(d36)
$$\phi = \frac{3 b^2 t1}{8 w} + \phi 0$$

The amplitude of the zeroth-order solution is, therefore,
$$a = \frac{|a|}{2} \exp\left(\frac{i3|a|^2}{8\omega} t_1 + \phi_0\right),$$
where $|a|$ and ϕ_0 are only functions of t_2.

14.8 Evolution of wave envelope on a nonlinear string

In §13.7 the slow evolution of sinusoidal waves propagating along an elastically supported string was shown to be governed by a Schrödinger equation. Here we consider how a nonlinear addition to the restoring force affects the evolution equation for the envelope. Specifically, we start from the following wave equation that has been suitably normalized,
$$\frac{\partial^2 y}{\partial t^2} - \frac{\partial^2 y}{\partial x^2} + y + \epsilon^2 \sigma y^3 = 0, \quad -\infty < x < \infty, \qquad (14.8.1)$$
where $\epsilon \ll 1$, and σ is a constant of order unity. We now apply the method of multiple scales. Since the mathematical reasoning has been discussed in §13.7, we shall only display the input and output statements with little explanation.

/* Declare dependence on multiple-scale variables. */
(c3) depends(y, [x, x1, x2, t, t1, t2])$
(c4) depends(y0, [x, x1, x2, t, t1, t2])$
(c5) depends(y1, [x, x1, x2, t, t1, t2])$
(c6) depends(y2, [x, x1, x2, t, t1, t2])$
/* The governing equation. */
(c7) eq1:diff(y, t, 2) − diff(y, x, 2) + y + sigma * e ∧ 2 * y ∧ 3 = 0;
(d7)
$$-\frac{d^2 y}{dx^2} + \frac{d^2 y}{dt^2} + e^2 \sigma y^3 + y = 0$$
/* Define multiple-scale variables.*/
(c8) gradef(x1, x, e)$

(c9) gradef($x2, x, e \wedge 2$)\$
(c10) gradef($t1, t, e$)\$
(c11) gradef($t2, t, e \wedge 2$)\$
/* A perturbation series is introduced in the governing equation, but the lengthy result will not be displayed. */
(c12) eq1:ev(eq1,$y = y0 + e*y1 + e \wedge 2*y2$, diff, expand)\$
/* Pretend from here on that the multiple-scale variables are independent. */
(c13) gradef($x1, x, 0$)\$
(c14) gradef($x2, x, 0$)\$
(c15) gradef($t1, t, 0$)\$
(c16) gradef($t2, t, 0$)\$
(c17) eq1:taylor(eq1,e,0,2);
(d17) /T/

$$y0 + \frac{d^2 y0}{dt^2} - \frac{d^2 y0}{dx^2} + \left(2 \frac{d^2 y0}{dt\,dt1} - 2 \frac{d^2 y0}{dx\,dx1} + y1 + \frac{d^2 y1}{dt^2} - \frac{d^2 y1}{dx^2}\right) e$$

$$+ \left(\sigma\, y0^3 + 2 \frac{d^2 y0}{dt\,dt2} + \frac{d^2 y0}{dt1^2} - 2 \frac{d^2 y0}{dx\,dx2} - \frac{d^2 y0}{dx1^2}\right.$$

$$\left. + 2 \frac{d^2 y1}{dt\,dt1} - 2 \frac{d^2 y1}{dx\,dx1} + y2 + \frac{d^2 y2}{dt^2} - \frac{d^2 y2}{dx^2}\right) e^2 + \cdots = 0 + \cdots$$

(c18) eq10:coeff(eq1,e,0);
(d18) /R/

$$y0 + \frac{d^2 y0}{dt^2} - \frac{d^2 y0}{dx^2} = 0$$

(c19) eq11: coeff(eq1,e,1);
(d19) /R/

$$2 \frac{d^2 y0}{dt\,dt1} - 2 \frac{d^2 y0}{dx\,dx1} + y1 + \frac{d^2 y1}{dt^2} - \frac{d^2 y1}{dx^2} = 0$$

(c20) eq12:coeff(eq1,e,2);
(d20) /R/

$$\sigma\, y0^3 + 2 \frac{d^2 y0}{dt\,dt2} + \frac{d^2 y0}{dt1^2} - 2 \frac{d^2 y0}{dx\,dx2} - \frac{d^2 y0}{dx1^2}$$

$$+ 2 \frac{d^2 y1}{dt\,dt1} - 2 \frac{d^2 y1}{dx\,dx1} + y2 + \frac{d^2 y2}{dt^2} - \frac{d^2 y2}{dx^2} = 0$$

/* Denote $z = e^{ikx - i\omega t}$, a = wave amplitude, and as = the complex conjugate of a. */

14.8 Evolution of wave envelope on a nonlinear string

(c21) gradef($z, x, \%i * k * z$)\$
(c22) gradef($z, t, -\%i * w * z$)\$
(c23) y0 : y0 = $a * z + as/z$\$
(c24) depends($[a, as], [x1, x2, t1, t2]$)\$
(c25) eq10: ev(eq10, y0, diff, expand);
(d25) /R/

$$-\frac{(a\omega^2 - ak^2 - a)z^2 + as\,\omega^2 - as\,k^2 - as}{z} = 0$$

/* The dispersion relation. */
(c26) disp: coeff(eq10, $A * z$);
(d26) /R/

$$-\omega^2 + k^2 + 1 = 0$$

/* As a check we should get the same dispersion relation from the coefficient of the conjugate as/z. */
(c27) disp:coeff(eq10, as/z);
(d27) /R/

$$-\omega^2 + k^2 + 1 = 0$$

(c28) eq11:ev(eq11, y0, diff, expand);
(d28) /R/

$$y1 + \frac{d^2y1}{dt^2} - \frac{d^2y1}{dx^2} + \left(-2i\frac{da}{dt1}\omega - 2i\frac{da}{dx1}k\right)z = 0$$

/* Let us kill the secular terms proportional to $a/z, as/z$. */
(c29) evo1:coeff(eq11, $-2 * \%i * z, 1$);
(d29)

$$\frac{da}{dt1}\omega + \frac{da}{dx1}k = 0$$

/* Define $c_g = k/\omega$ = group velocity. */
(c30) evo1:diff($a, t1$) + c[g]*diff($a, x1$) = 0;
(d30)

$$c_g\frac{da}{dx1} + \frac{da}{dt1} = 0$$

/* The preceding equation governs the evolution in the range $(x_1, t_1) = O$. Now y1 can be set to zero. */
(c31) eq12:ev(eq12,y0, y1 = 0, diff, expand);

(d31) /R/

$$\left(\left(a^3 z^6 + 3\, a^2\, as\, z^4 + 3\, a\, as^2\, z^2 + as^3\right) \sigma + z^3\, y2 + z^3 \frac{d^2 y2}{dt^2} \right.$$

$$\left. - z^3 \frac{d^2 y2}{dx^2} + \left(-2i \frac{da}{dt2} w - 2i \frac{da}{dx2} k - \frac{d^2 a}{dx1^2} + \frac{d^2 a}{dt1^2} \right) z^4 \right) \frac{1}{z^3} = 0$$

(c32) eq12:taylor(eq12,z)$
(c33) eq22:coeff(eq12,z);
(d33)

$$-2i \frac{da}{dt2} \omega + 3\, a^2\, as\, \sigma - 2i \frac{da}{dx2} k - \frac{d^2 a}{dx1^2} + \frac{d^2 a}{dt1^2} = 0$$

(c34) evo2:ev(eq22, diff$(a, t1, 2) = c[g] \wedge 2*$diff$(a, x1, 2)$);
(d34)

$$-2i \frac{da}{dt2} \omega + 3\, a^2\, as\, \sigma - 2i \frac{da}{dx2} k + \frac{d^2 a}{dx1^2} c_g^2 - \frac{d^2 a}{dx1^2} = 0$$

Equation (d34) describes the evolution of a wave envelope over a much larger domain $(x_2, t_2) = O(1)$ or $(x_1, t_1) = O(1/\epsilon)$. Equations (d30) and (d34) can be combined to give

$$-2i \left(\frac{\partial a}{\partial t_1} + c_g \frac{\partial a}{\partial x_1} \right) + \epsilon \left(-(1 - c_g^2) \frac{\partial^2 a}{\partial x_1^2} + 3\sigma a^2 as \right) = 0. \quad (14.8.2)$$

By changing to the coordinate system moving at the group velocity, we can transform this equation to the nonlinear Schrödinger equation, which is ubiquitous in the mechanics and physics of nonlinear dispersive waves. Many analytical and numerical studies have been devoted to (14.8.2).

To see what happens in a still larger domain or longer time, say $(x_3, t_3) = O(1)$ or $(x_1, t_1) = O(\epsilon^2)$, we need to extend the analysis to the order $O(\epsilon^3)$. This extension was first made by Dysthe (1979) for waves on deep water. The derivation is a daunting task if it is done with just pencil and paper. As an exercise on computer algebra the reader should make such an extension for our example of weakly nonlinear spring.

While we have chosen MACSYMA to demonstrate the elements of symbolic computation, the reader may find that any of the other commercial softwares to be his or her personal favorite. The essence is to take advantage of computer algebra when the analysis appears too cumbersome for pencil and paper. In the age of word processors there is little reason to remain loyal to the typewriter!

Exercises

14.1 Verify the $O(\epsilon^2)$ results in (14.6.7) and (14.6.8) by computer algebra.

14.2 Solve the following initial-value problem by Taylor expansion

$$\frac{d^2y}{dx^2} - 6y^2 = 0, \quad x > 0$$

with

$$y(0) = p \quad \text{and} \quad \frac{dy(0)}{dx} = q.$$

Show that the solution with the first seven terms is

$$y = p + qx + 3p^2x^2 + 2pqx^3 + \frac{1}{2}(q^2 + 6p^3)x^4$$

$$+ 3p^2qx^5 + (pq^2 + 3p^4)x^6 + \frac{1}{7}(q^3 + 24p^3q)x^7$$

$$+ \cdots.$$

Complete the series up to $O(x^{10})$. Suggestion: assume

$$y = \text{sum}(x \wedge i * a[i], i, 0, 10);$$

and use **taylor** (eqa,x,0,10), **ratcoef** (eqb,x,n) and **solve** (eqc,a[i]).

14.3 Solve the following boundary-value problem† by matched asymptotics

$$\epsilon \frac{d^2y}{dx} + (1+x)\frac{dy}{dx} + y = 0$$

with

$$y(0) = y(1) = 1.$$

Show first that the outer approximation satisfying $y(1) = 1$ is

$$yout = y0 + e * y1 + e \wedge 2 * y2 + \cdots,$$

where e stands for ϵ, and

$$y0 = \frac{2}{x+1}, \quad y1 = -\frac{(x-1)(x+3)}{2(x+1)^3}$$

and

$$y2 = -\frac{(x-1)(x+3)(x^2+2x+7)}{4(x+1)^5}.$$

† From Bender and Orszag (1978), p. 431.

Near the end at $x = 0$ use the inner (boundary layer) coordinate $xx = x/e$, **gradef** and **depend**, and assume the inner expansion

$$yin = yy0 + e * yy1 + e \wedge 2 * yy2 + \cdots.$$

Solve the inner problems and determine the constants of integration by matching with the outer solutions order by order. Show, in particular,

$$yy0 = 2 - e^{-xx}.$$

Find $yy1$ and $yy2$.

14.4 Assume $A = A(t_1, t_2, t_3)$ and extend the analysis of §14.8 to $O(\epsilon^3)$ to get the governing equation for the variation of A over the range $(x_3, t_3) = O(1)$.

Appendices

A. Gauss' theorem

Let $\mathbf{q} = u\mathbf{i} + v\mathbf{j} + w\mathbf{k}$ be any vector, and $\mathbf{n}$ be the outward unit normal to the surface element $d\mathcal{S}$, as sketched in Figure A.1. Then

$$\iint_{\mathcal{S}} \mathbf{q} \cdot \mathbf{n} d\mathcal{S} = \iiint_{\mathcal{V}} \nabla \cdot \mathbf{q} d\mathcal{V} \tag{A.1}$$

or, in expanded form,

$$\iint_{\mathcal{S}} (u\mathbf{i} + v\mathbf{j} + w\mathbf{k}) \cdot \mathbf{n} d\mathcal{S} = \iiint_{\mathcal{V}} \left(\frac{\partial u}{\partial x} + \frac{\partial v}{\partial y} + \frac{\partial w}{\partial z} \right) d\mathcal{V}. \tag{A.2}$$

To prove it, we refer to Figure A.1 and consider just one component

$$\iiint_{\mathcal{V}} \frac{\partial w}{\partial z} d\mathcal{V} = \iiint_{\mathcal{V}} \frac{\partial w}{\partial z} dxdydz$$
$$= \iint_{\mathcal{S}} [w(\mathcal{S}_2) - w(\mathcal{S}_1)] dxdy,$$

where $(w(\mathcal{S}_1), w(\mathcal{S}_2))$ are the values of w on the (bottom, top) part of the bounding surface; the two parts are divided by a closed curve touching the smallest vertical cylinder surrounding the volume. As shown in Figure A.1, on the bottom half of the surface $\mathcal{S}$, i.e., $\mathcal{S}_1$, we have $\mathbf{k} \cdot \mathbf{n} d\mathcal{S} = -dxdy$. On the top half of the surface $\mathcal{S}$, i.e., $\mathcal{S}_\in$, we have $\mathbf{k} \cdot \mathbf{n} d\mathcal{S} = dxdy$. Hence the last surface integral can be written

$$\iint_{\mathcal{S}_2} w\mathbf{k} \cdot \mathbf{n} d\mathcal{S} + \iint_{\mathcal{S}_1} w\mathbf{k} \cdot \mathbf{n} d\mathcal{S} = \iint_{\mathcal{S}} w\mathbf{k} \cdot \mathbf{n} d\mathcal{S}.$$

Similarly,

$$\iiint_{\mathcal{V}} \frac{\partial u}{\partial x} d\mathcal{V} = \iint_{\mathcal{S}} u\mathbf{i} \cdot \mathbf{n} d\mathcal{S}$$

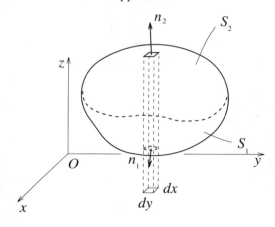

Fig. A.1. Control volume and its bounding surface.

and

$$\iiint_V \frac{\partial v}{\partial y} dV = \iint_S v\mathbf{j}\cdot\mathbf{n} dS.$$

Adding the last three equations, we get (A.2) and the Gauss theorem.

B. Stokes' theorem

Let us denote the components of the two-dimensional vector $\mathbf{q}$ by (q_x, q_y) so that

$$\nabla \times \mathbf{q} = \left(\frac{\partial q_y}{\partial x} - \frac{\partial q_x}{\partial y}\right),$$

which is normal to the x, y plane. For any area $\mathcal{S}$ enclosed by the curve $\mathcal{C}$,

$$\iint_\mathcal{S} \frac{\partial q_y}{\partial x} dx dy = \int_{y_{min}}^{y_{max}} \left([q_y]_R - [q_y]_L\right) dy, \qquad (B.1)$$

where $q_y(L), q_y(R)$ denote the value of q_y taken on the (left, right) end of a horizontal slice at y, see Figure B.1a. As the horizontal slice moves up from y_{min} to y_{max} the right-end R moves counterclockwise along the right part of $\mathcal{C}$, while the left-end L moves clockwise along the left part of $\mathcal{C}$. Hence the right-hand side of (B.1) is

$$\oint_\mathcal{C} q_y\, dy,$$

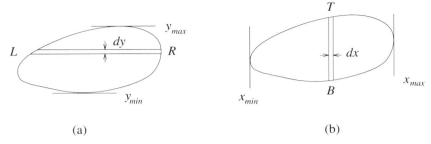

Fig. B.1. Two ways of integrating over an area.

where the line integral covers the entire closed curve in the counterclockwise direction. Similarly, consider

$$\iint_S \frac{\partial q_x}{\partial y} dx dy = \int_{x_{min}}^{x_{max}} ([q_x]_T - [q_x]_B) \, dx, \tag{B.2}$$

where (T, B) denote, respectively, the (top, bottom) ends of a vertical slice at x, as shown in Figure B.1b. As the vertical slice moves from left (x_{min}) to right (x_{max}), B moves counterclockwise along the bottom part of the curve, and T moves clockwise along the top curve. Hence the right-hand side of (B.2) is

$$-\oint_C q_x \, dx.$$

The difference of (B.1) and (B.2) gives

$$\iint_S \left(\frac{\partial q_y}{\partial x} - \frac{\partial q_x}{\partial y}\right) = \oint_C (q_x dx + q_y dy) = \oint_C (q_x, q_y) \cdot d\mathbf{s},$$

which is Stokes' theorem.

C. Gamma function

The Gamma function is defined by the integral

$$\Gamma(x) = \int_0^\infty t^{x-1} e^{-t} \, dt, \tag{C.1}$$

which is convergent for $x > 0$. By changing the argument and partial integration, we get

$$\Gamma(x+1) = \int_0^\infty t^x e^{-t} \, dt = \left[-t^x e^{-t}\right]_0^\infty + x \int_0^\infty t^{x-1} e^{-t} \, dt.$$

The integrated terms vanish, therefore,
$$\Gamma(x+1) = x\Gamma(x), \qquad (C.2)$$
which is a recursion relation.

From the defining equations (C.1) and (C.2), we get
$$\Gamma(1) = 0! = \int_0^\infty e^{-t}\, dt = 1. \qquad (C.3)$$
Repeated use of the recursion relation gives
$$\Gamma(2) = 1! = 1 \cdot 1$$
$$\Gamma(3) = 2! = 2 \cdot \Gamma(2) = 2 \cdot 1$$
$$\Gamma(4) = 3! = 3 \cdot \Gamma(3) = 3 \cdot 2 \cdot 1$$
$$\Gamma(n+1) = n! = n(n-1) \cdots 3 \cdot 2 \cdot 1. \qquad (C.4)$$
For any $x > 0$ we have
$$\Gamma(x) = \frac{\Gamma(x+1)}{x} = \frac{\Gamma(x+2)}{x(x+1)} = \frac{\Gamma(x+3)}{x(x+1)(x+2)}$$
$$= \frac{\Gamma(x+n)}{x(x+1)(x+2)\cdots(x+n-1)}. \qquad (C.5)$$
For negative x, $\Gamma(x)$ is not defined by (C.1). Nevertheless, (C.5) can be used to extend the definition of $\Gamma(x)$ because it is always possible to find a large enough n so that $x + n > 0$, and $\Gamma(x+n)$ is defined by the integral. The recursion formula (C.5) then defines $\Gamma(x)$, from which it is seen that
$$\Gamma(x) \sim \infty \quad \text{at} \quad x = 0, -1, -2, -3, \ldots.$$
Figure C.1 shows the variation of $\Gamma(x)$ with x.

D. Glossary of MACSYMA commands

We list below the few commands introduced in Chapter 14. For hundreds more it is necessary to go to the MACSYMA *Reference Manual* (1988) and MACSYMA *User's Guide* (1988).

bc2(*solution, xvalue1, yvalue1, xvalue2, yvalue2*): Solves a second-order ordinary differential equation by **ode2**. **bc2** then applies the boundary conditions $y(x_1) = y_1$ and $y(x_2) = y_2$, where $x_i = xvaluei$, $y_i = yvaluei$, $i = 1, 2$.

D. Glossary of MACSYMA commands

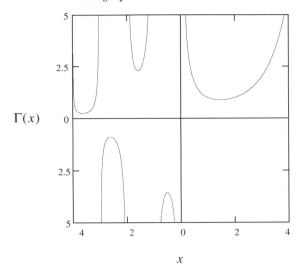

Fig. C.1. Gamma function.

depends($funlist_1, \ldots, varlist_1, \ldots, funlist_k, varlist_k$): Declares that the list of functions in $funlist_1$ depends on the variables in the list $varlist_1$.

diff($exp, var_1, n_1, \ldots var_k, n_k$): Differentiates the expression exp with respect to var_1 n_1 times,... and with respect to var_k n_k times.

ev($exp, arg_1, \ldots arg_n$): Evaluates the expression exp by carrying out the commands specified in the arguments $arg_1, \ldots, arg_n$.

expand(exp): Expands exp by multiplying out products of sums and exponential sums.

factor(exp): Factors exp.

gradef($f(x_1, \ldots, x_n), g_1, \ldots, g_n$): Defines the gradients for each derivative of a function with respect to the function's arguments. For i from 1 to n, $df/dx_i = g_i$.

ic2(*solution, xvalue, yvalue, derivative value*): Solves a second-order ordinary differential equation by **ode2** and **ic2**, then applies the initial conditions that $y(x_0) = y_0$ and $y'(x_0) = y_1$, where $x_0 = xvalue$, $y_0 = yvalue$ and $y_1 = derivative\ value$.

integrate (exp, var): Indefinite integral of exp with respect to var.

integrate(exp, var, a, b): Definite integral of exp with respect to var from a to b.

linsolve($[exp_1, \ldots, exp_n], [var_1, var_n]$): Solves the list of simultaneous linear equations for the list of variables.

ode2($diffeqn, depvar, indepvar$): Solves an ordinary differential equation for $depvar$, as a function of the independent variable $indepvar$. If successful, the integration constants are denoted by % c for first-order equations, and by % k1 and % k2 for second-order equations.

part($exp, n_1, \ldots, n_k$): Selects in succession the parts of the expression exp. In exponentiation, the base is part 1 and the exponent part 2. In a sum or product, the ith term of factor is part i. In any expression, the main operator is part 0.

radcan(exp): Simplifies expressions involving radicals, logarithms, and exponentials.

ratcoef(exp, var, n): Gives the coefficient of var^n.

ratsimp(exp): Rationally simplifies the expression exp to nonrational functions.

ratsubst(a, b, c): Substitutes a for b in c. It is more intelligent than **subst**.

remfunction($func$): Removes the user-defined function properties of $func$.

remove($name$, FUNCTION): Removes the function property of $name$ if one exists.

solve(exp, var): Solves an algebraic equation $exp = 0$ for var. The equation can be nonlinear.

subst(a, b, c): Substitutes a for all occurrences of b in c.

sum($exp, 'ind, lo, hi$): Sums the values of exp with respect to the index $'ind$ from lo to hi.

taylor(exp, var, pt, pow): Taylor expansion of exp with respect to var about the pt up to the power pow.

trigexpand(exp): Expands the trigonometric or hyperbolic functions of sums of angles or multiple angles occurring in exp.

trigreduce(exp, var): Combines products and powers of trigonometric and hyperbolic sines and cosines of var into those of multiples of var.

trigsimp(exp): Employs the identities $\cos^2 x + \sin^2 x = 1$ or $\cosh^2 x - \sinh^2 x = 1$ to simplify exp.

%th(n): n(c) lines before.

Bibliography

Abramov, V.M. (1937). The problem of the contact of an elastic half-plane with an absolutely rigid base for the determination of the frictional force. *Doklady SSSR*. **II**, 173–178.

Abramowitz, M. and Stegun, I.A. (1964). *Handbook of Mathematical Functions with Formulas, Graphs and Mathematical Tables*. National Bureau of Standards, Applied Mathematics Series 55, New York.

Asfar, O.R., Aziz, A. and Soliman, M.A. (1979). A uniformly valid solution for inward cylindrical solidification. *Mech. Res. Comm.* **6**, 325–332.

Aziz, A. and Na, T.Y. (1984). *Perturbation Methods in Heat Transfer*. Hemisphere Publishing Co., Washington.

Bakhavlov, N. and Pansenko, G. (1989). *Homogenization: Averaging Processes in Periodic Media*. Kluwer Academic, Dordrecht.

Beltzer, A.I. (1990). *Variational Finite Element Methods – A Symbolic Computation Approach*. Springer-Verlag, Berlin.

Bender, C.M and Orszag, S.A. (1978). *Advanced Mathematical Methods for Scientists and Engineers*. McGraw-Hill, New York.

Bensoussan, A., Lions, J.L. and Papanicolaou, G. (1978). *Asymptotic Analysis for Periodic Structures*. North-Holland, Amsterdam.

Birkhoff, G. and Zarantonello, E.H. (1957). *Jets, Wakes and Cavities*. Academic Press, New York.

Boussinesq, J. (1885). *Applications des Potentials à l'Etude de l'Equilibre et du Mouvement des Solides Elastique*. Gauthiers Villars, Paris.

Carrier, G.F., Krook, M. and Pearson, C.E. (1966). *Functions of a Complex Variable – Theory and Technique*. McGraw-Hill, New York.

Carrier, G.F. and Pearson, G.E. (1976). *Partial Differential Equations – Theory and Technique*. Academic Press, New York.

Carslaw, H.S. and Jaeger, J.C. (1959). *Conduction of Heat in Solids*. Oxford University Press, London, U.K.

Carslaw, H.S. and Jaeger, J.C. (1963). *Operational Methods in Applied Mathematics*. Dover, New York.

Cheng, H. (1984). An exactly solvable nonlinear partial differential equation with solitary-wave solutions. *Studies in Applied Mathematics*. Elsevier Science Publishing Co., Inc., 183–187.

Cheng, H. and Wu, T.T. (1970). An aging spring. *Studies in Applied Mathematics*. **49**, 183–185.

Cole, J.D. (1968). *Perturbation Methods in Applied Mathematics.* Blaisdell Publishing Co., Waltham, MA.
Copson, E.T. (1935). *An Introduction to the Theory of Functions of a Complex Variable.* Oxford University Press, London, U.K.
Courant, R. and Friedrichs, K.O. (1948). *Supersonic Flows and Shock Waves.* Wiley-Interscience, New York.
Dysthe, K.B. (1979). Note on a modification to the nonlinear Schrödinger equation for application to deep water waves. *Proc. R. Soc. Lond.* **A 369**, 105–114.
Ene, H.I. and Sanchez-Palencia, E. (1975). Equations et phénomène de surface pour l'écoulement dans un modèle de milieux poreux. *J. Mécanique.* **14(1)**, 73–108.
Erdelyi, A. (1953a). *Higher Transcendental Functions.* Vol II. McGraw-Hill, New York.
Erdelyi, A. (1953b). *Tables of Integral Transforms.* Vol I. McGraw-Hill, New York.
Fock, V.A. (1960). *Electromagnetic Diffraction and Propagation Problems.* Macmillan, New York.
Fung, Y.C. (1965). *Foundation of Solid Mechanics.* 525 pp. Prentice-Hall, Englewood Cliffs, NJ.
Gilbarg, D. (1960). Jets and Cavities. *Encyclopedia of Physics.* Vol. 9, 311–445. Ed. S. Flügge. Springer-Verlag, Berlin.
Gipson, G.S. (1987). *Boundary Elements Fundamentals.* Computational Mechanics, Southampton, U.K.
Graf, K.F. (1975). *Wave Motion in Elastic Solids.* Ohio University Press.
Haberman, R. (1977). *Mathematical Models: Mechanical Vibrations, Population Dynamics and Traffic Flow: An Introduction to Applied Mathematics.* Prentice-Hall, Englewood Cliffs, NJ.
Hildebrand, F.B. (1964). *Advanced Calculus for Engineers.* Prentice-Hall, Englewood Cliffs, NJ.
Holmes, P. (1979). A nonlinear oscillator with a strange attractor. *Philo. Trans.* Royal Society, London, **A 292**, 419–48.
Ince, E.L. (1956). *Ordinary Differential Equations.* Dover, New York.
Jeffreys, H. and Jeffreys, B.S. (1950). *Methods of Mathematical Physics.* 2nd ed. Cambridge University Press, Cambridge, U.K.
Jones, D.S. (1964). *The Theory of Electromagnetism.* Pergamon Press, Oxford, U.K
Junger, M.C. and Feit, D. (1972). *Sound, Structures and Their Interaction.* MIT Press, Cambridge, MA.
Kajiura, K. (1968). A model of the bottom boundary layer in water waves. *Bull Earthq. Res. Inst.,* University of Tokyo, **46**, 75–123.
Kalker, J.J. (1977). A survey of the mechanics of contact between solids. *Zeit. Ang. Mat. Mech.* **57**, T3–T17.
Kanninen, M.F. and Popelar, C.H. (1985). *Advanced Fracture Mechanics.* Oxford University Press, London, U.K.
Kantorovich, L.V. and Krylov, V.I. (1964). *Approximate Methods of Higher Analysis.* Wiley-Interscience, New York.
Kármán, Th. v. and Biot, M.A. (1944). *Mathematical Methods in Engineering.* McGraw-Hill, New York.
Keller, J.B. (1980). Darcy's law for flow in porous media and the two-space method. *Nonlinear Partial Differential Equation and Applied Science.*

Ed. R.L. Sternberg, A.J. Kalinowski and J.S. Papadakis, 429–413, Dekker, New York.
Kober, H. (1957). *Dictionary of Conformal Representations*. Dover, New York.
Koshlyakov, N.S., Smirnov, M.M., and Gliner, E.B. (1964). *Differential Equations of Mathematical Physics*. North-Holland Publishers, Amsterdam.
Kynch, G.J. (1952). A theory of sedimentation. *Trans. Faraday Soc.* **48**, 106–176, London.
Levine, H. and Rodemich, E. (1958). Scattering of surface waves on an ideal fluid. *Appl. Math. Statis. Lab. Tech. Report* **78**, 64 pp., Stanford University, CA.
Liepman, H.W. and Roshko, A. (1957). *Elements of Gas Dynamics*. Wiley, New York.
Lighthill, M.J. and Whitham, G.B. (1955). On kinematic waves: I . Flood movement in long rivers; II. Theory of traffic flow on long crowded roads. *Proc. Roy. Soc. A* **229**, 281–345.
Lin, C.C. and Segal, L.A. (1988). *Mathematics Applied to Deterministic Problems in the Natural Sciences*. Soc. Ind. Appl. Math., Philadelphia.
Longuet-Higgins, M.S. (1967). On the trapping of wave energy round islands. *J. Fluid Mech.* **29**, 781–821.
Lunardini, V.J. (1981). *Heat Transfer in Cold Climates*. Van Nostrand Reinhold, New York.
MACSYMA (1988). *Reference Manual*. Version 13. Symbolics, Burlington, MA.
MACSYMA (1988). *User's Guide*. Symbolics, Burlington, MA.
McLachlan, N.W. (1955). *Bessel Functions for Engineers*. Clarendon Press, Oxford, U.K.
McTigue, D.F. and Mei, C.C. (1981). Gravity induced stresses near topography of small slope. *J. Geophysics Res.* **86**, 9268–9278.
Mei, C.C. (1966). Radiation and scattering of transient gravity waves by vertical plates. *Quart. Appl. Math. Mech.* **XIX.** 417–446.
Mei, C.C. (1977). Leakage of groundwater through a slot in a sheet pile. *J. Hydraulic Res.* **15**, 253–259.
Mei, C.C. (1989). *The Applied Dynamics of Ocean Surface Waves*. World Scientific, Singapore.
Mei, C.C. and Foda, M.A. (1981). Wave-induced responses in a fluid-filled poro-elastic solid with a free surface – a boundary layer theory. *Geophys. J. R. Astro Soc.* **66**, 597–631.
Milne-Thomson, L.M. (1955). *Theoretical Hydrodynamics*. 3rd ed. Macmillan, New York.
Morse, P.M. and Feshbach, H. (1953). *Methods of Theoretical Physics*. Parts I and II. McGraw-Hill, New York.
Morton, B.R. (1960). Laminar convection in uniformly heated vertical pipes. *J. Fluid Mechanics.* **8**, 227–240.
Muskhelishvili, N.I. (1962). *Some Basic Problems of the Mathematical Theory of Elasticity*. Noordhoff, Leyden.
Nayfeh, A.H. (1973). *Perturbation Methods*. Wiley-Interscience, New York.
Nayfeh, A.H. (1981). *Introduction to Perturbation Techniques*. Wiley-Interscience, New York.

Polubarinova-Kochina, P. Ya. (1962). *Theory of Ground Water Movement.* Princeton University Press, Princeton, NJ.

Press, W.H., Flannery, B.P., Teukolsky, S.A. and Vetterling, W.T. (1985). *Numerical Recipes: The Art of Scientific Computing.* Cambridge University Press, Cambridge, U.K.

Rand, R.H. (1984). *Computer Algebra in Applied Mathematics – An Introduction to MACSYMA.* Pitman Advanced Publishing, Boston, MA.

Rice, J.R. (1968). Mathematical analysis in the mechanics of fraction. *Fracture - An Advanced Treatise.* Chapter 3, 191–311. Ed. H. Liebowitz, Academic Press, New York.

Sanchez-Palencia, E. (1980). *Non Homogeneous Media and Vibration Theory.* Lecture Notes in Physics 127. Springer-Verlag, Berlin.

Sih, G.C. (1973). *Mechanics of Fracture.* Vol. 1. Noordhoff, Leyden.

Stoker, J.J. (1957). *Water Waves, the Mathematical Theory with Applications.* Wiley-Interscience, New York.

Strack, O.D.L. (1989). *Groundwater Mechanics.* Prentice-Hall, Englewood Cliffs, NJ.

Tolstov, G.P. (1962). *Fourier Series.* Dover, New York.

Tricomi, F.G. (1957). *Integral Equations.* Wiley-Interscience, New York.

Tuck, E.O. and Mei, C.C. (1983). Contact of one or more slender bodies with an elastic half space. *Int. J. Solids and Structures.* **19**, 1–23.

Tulin, M.P. (1953). Some two-dimensional cavity flows about slender bodies. *David Taylor Model Basin.* Report 834.

Turian, R.M. and Bird, R.B. (1963). Viscous heating in the cone-and-plate viscometer-II. *Chem. Eng. Sci.* **18**, 689–696.

Watson, G.N. (1958). *A Treatise on the Theory of Bessel Functions.* Cambridge University Press, Cambridge, U.K.

Wehausen, J.V. and Laitone, E.V. (1960). Surface Waves. *Encyclopedia of Physics.* **IX**, 446–776. Ed. S. Flügge. Springer-Verlag, Berlin.

Whitham, G.B. (1976). *Linear and Nonlinear Waves.* Wiley-Interscience, New York.

Wu, T.Y.-T. (1968). Inviscid cavity and wave flows. *Basic Development in Fluid Dynamics.* **2**, 1–116. Ed. M. Holt. Academic Press, New York.

Yih, C.S. (1965). *Dynamics of Nonhomogeneous Fluids.* Macmillan, New York.

Zauderer, E. (1983). *Partial Differential Equations of Applied Mathematics.* Wiley-Interscience, New York.

Index

δ function, 105–109, 111, 113, 122, 123

Abramov, V.M., 337
Abramowitz, M., 165, 178, 181, 203, 279
Airy function, 203
Airy's stress function, 147, 330, 405
analytic continuation, 318, 328, 330, 340
analytic function, 289, 290, 300, 310, 320, 324, 329, 334, 339, 342
argument, 211
Asfar, O.R., 403
Aziz, A., 349, 352, 355, 357, 403, 435

Bakhavlov, N., 370
base vectors, 65, 66
Beltzer, A.I., 408
Bender, C.M., 445
Bensoussan, A., 370, 371, 375
Bernoulli's equation, 227, 258, 323
Bernoulli's theorem, 258
Bessel equation, 167, 170, 171, 174, 177, 178, 182, 186, 192, 205
   canonical form of, 167
Bessel function, 165, 199, 204
   with complex argument, 194
   of first kind, 167, 169
   with a large argument, 176
   modified, 193, 195
   recurrence relation, 175
   of second kind, 171, 193
   with a small argument, 176
Bessel's inequality, 93
biharmonic equation, 147
biharmonic function, 318, 330
Biot, M.A., xiv
Bird, R.B., 352
Birkhoff, G., 322
Borda's mouth piece, 317

boundary integral method, 128
boundary layer, 382, 383, 387–389, 397, 399–401, 446
   technique, 345, 381
Boussinesq, J., 145
branch cut, 215, 216, 243, 246, 256, 257, 273, 282, 300
branch point, 215, 216, 218, 219, 238, 256, 268, 273, 282, 302

Carrier, G.F., 161, 280, 285
Carslaw, H.S., 270, 285, 357
Cauchy initial condition, 34, 35
Cauchy problem, 44
Cauchy's inequality, 237, 252
Cauchy's integral formula, 235, 236, 239, 248, 250, 253, 254, 263, 266, 271, 272, 274, 318
Cauchy's theorem, 229–231, 233, 236, 239, 272, 329
Cauchy–Riemann conditions, 220–223, 226, 231, 251, 329
cavitation number, 323
cavity flow, 318
   linearized theory of, 322, 327
characteristic triangle, 35, 38, 41–43
characteristics, 21, 23–25, 27, 29–35, 37, 38, 40, 45–47, 50–52, 54
Cheng, H., 206
circular domain, 165, 180
circular pond, 184
classification of equations, 20
Cole, J.D., 386
compatibility condition, 146, 147
complex functions, 210, 212, 220, 221, 230
complex number, 210–212
complex variables, 210, 215

457

computer algebra, 402, 428, 435, 444, 445
concentrated load, 108, 117, 120
conduction equation, 11
conformal mapping, 210, 289–291, 297, 311, 314, 317
conservation, 1
  of energy, 10, 11, 428
  of mass, 12, 18, 19, 296, 311, 395
  of momentum, 1, 4, 5, 8, 13, 19, 114, 396
conservation law, 1, 7
  of mass, 1
  of momentum, 1
consolidation, 401
contour integral, 243, 245, 253, 254, 257, 262, 266, 282
control volume, 1
convolution theorem, 135, 263–265
  of Fourier transform, 134
  of Laplace transform, 264
Courant, R., 24

d'Alembert's solution, 35, 38, 41, 43, 143
dam/reservoir problem, 19
Darcy's law, 1, 8, 11, 18, 89, 143, 223, 370, 375, 376, 385
differential approach, 1, 12
diffraction, 275, 279
diffusion, 10, 63, 65, 67, 78, 84, 87, 135, 138, 267, 272, 286, 349, 386, 403, 405
  equation, 11, 25, 27, 65–67, 75, 85, 135, 287, 354, 400
Dirichlet condition, 9, 123, 129
dispersive wave, 377
domain of dependence, 33, 35, 39, 43
Duffing problem, 435
Dysthe, K.B., 444

earthquake, 88
eigenfrequencies, 181
eigenfunction, 57, 59, 62, 64–66, 70, 72–74, 76, 77, 79, 87, 91, 105, 124, 125, 180, 182, 183
eigensolutions, 79, 112
eigenvalue, 57, 59, 62, 64, 70–74, 79, 182, 183, 204
  condition, 59, 64, 69, 71, 206
elastic beam, 114
elasticity, 330, 339
elliptic, 23, 27–30
Ene, H.I., 374
equipotential lines, 144
Erdelyi, A., 161, 165, 178, 203, 282, 285
error function, 140

essential singularity, 238
estuary, 301, 313
Euler constant, 174
even and odd functions, 94, 95

Feit, D., 192
Feshbach, H., xiv, 80, 128, 161
finite domains, 57
first-order equation, 20, 22, 29, 31
  system of, 22
Fock, V.A., 277
Foda, M.A., 395
Fourier coefficient, 92
Fourier integral, 241, 249, 258
Fourier integral theorem, 132, 133, 153
Fourier series, 91–102, 104, 132
  acceleration of convergence, 104
  convergence of, 99
  cosine, 95, 102
  double, 125
  exponential, 98, 99
  full, 93
  sine, 95
  trigonometric, 93, 94, 98
Fourier theorem, 94, 97, 99
Fourier transform, 132–148, 161, 162, 246–250, 258, 260, 263, 272, 377
  cosine, 132, 153, 154
  exponential, 132, 133, 143, 259
  inverse, 133, 210
  of derivatives, 134
  sine, 132, 153, 159
Fourier's law, 1, 11
free-boundary problem, 10
Fresnel integrals, 279, 288
Friedrichs, K.O., 24
Frobenius method, 167
fundamental solution, 122–124, 129
Fung, Y.C., 145, 152

Gamma function, 169, 411, 449
Gauss' theorem, 8, 10, 127, 231, 375, 448
Gaussian function, 106
geothermal gradient, 195, 427
Gibson, G.S., 128
Gilbarg, D., 322
Graf, K.F., 55
Green's function, 105, 108–114, 117, 120–131
  symmetry of, 110, 114
Green's identity, 72
Green's theorem, 105, 113, 114, 121, 123, 126, 130

Haberman, R., 53
Hankel functions, 185–188, 191

harmonic conjugates, 222, 226
heat conduction, 82, 84
heat equation, 280, 281, 355, 356
Heaviside step function, 260
Helmholtz equation, 191, 275, 277, 403
higher dimensions, 78
Hildebrand, F.B., 81, 174, 202, 203
Holmes, P., 257
homogenization, 367, 370, 376, 405
   method of, 345
Hooke's law, 1, 2, 4, 17, 89, 146, 151, 397
   method of, 345
hydrofoil, 322, 323, 327
hyperbolic, 23, 26, 28, 29

Ince, E.L., 62
indicial equation, 168, 170, 171
inertia
   moment of, 115
   rotatory, 116, 117, 129
infinite channel, 295–297
influence function, 110
initial-boundary-value problem, 40, 74–76, 86, 184, 267, 272
inner approximation, 382–384, 392, 393
inner expansion, 382
inner solution, 382, 383, 389, 394
integral approach, 5
integral equations, 159
irrotational flow, 226, 227, 290, 313

Jacobi symbol, 61, 179, 185
Jaeger, J.C., 270, 285, 357
Jeffreys, B.S., 161
Jeffreys, H., 161
Jones, D.S., 192
Jordan's lemma, 246–249, 258, 262, 263, 266, 268, 273, 282
Joukowski transformation, 340, 391
Junger, M.C., 192

Kármán, Th. v. , xiv
Kajiura, K., 209
Kalker, J.J., 339
Kanninen, M.F., 339
Kantorovich, L.V., 102
Keller, J.B., 375
Kelvin functions, 195, 199, 201, 203
kinematic condition, 9, 10
kinetic energy, 44
Kober, H., 291
Koshlyakov, N.S., 31, 54, 127
Krook, M., 161, 285
Krylov, V.I., 102
Kynch, G.J., 55

l'Hopital's rule, 174
Laitone, E.V., 342
Lamé constants, 330
Laplace equation, 8, 25, 82, 123–125, 129, 143, 161, 207, 222, 327, 328, 340, 388
Laplace transform, 132, 159, 210, 247, 260–265, 267, 270, 272, 276, 278, 281, 286–288, 408, 427
   inverse, 210, 260–262
Laurent expansion, 255, 326
   theorem, 252, 254
Legendre polynomials, 165
Liepman, H.W., 22
Lighthill, M.J., 6
Lin, C.C., xiv
linear second-order equation, 25
linearization, 12, 15
Liouville's theorem, 237, 238
   extended, 252
localized source, 137
long and taut string, 38
Longuet-Higgins, M.S., 207
Lunardini, V.J., 312

macroscale, 345, 370–372, 375, 376
MACSYMA, 345, 408, 409, 411, 412, 416–418, 424, 427, 428, 444, 450
magnitude, 211
matching, 382, 383, 386, 389, 392, 393
McLachlan, N.W., 178, 201, 203
McTigue, D.F., 404
mean-square error, 92
Mei, C.C., 24, 339, 342, 390, 395, 404
method of images, 123, 129
method of multiple scales, 345, 358, 377, 435, 441
microscale, 345, 368, 369, 371, 372
Milne-Thomson, L.M., 311
mixed-boundary-value problem, 210, 256, 318
Mohr's circle, 150
Morse, P.M., xiv, 80, 128, 161
Morton, B.R., 196, 201, 427, 435
multiple-scale coordinates, 373
Muskhelishvili, N.I., 330, 337

Na, T.Y., 349, 352, 355, 357, 403, 435
natural mode, 112, 180, 182
Navier-Stokes equations, 371
Nayfeh, A.H., 161, 285, 381, 384, 404
Neuman condition, 9, 123, 129

one-dimensional waves, 33
operator
   adjoint, 125–127
   non-self-adjoint, 125, 126

self-adjoint, 125
Sturm–Liouville, 113, 125
Orszag, S.A., 445
orthogonal curvilinear coordinates, 80
orthogonality, 57, 65, 75, 77, 78, 80, 83, 84, 91, 92, 94, 99, 125, 165, 179, 182, 185, 188
outer approximation, 382, 384, 390
outer expansion, 386
outer limit, 393
outer solution, 382, 383, 389, 391, 392, 394, 397, 399, 401

Panasenko, G., 370
parabolic, 27–30, 277
parabolic approximation, 275, 288
Parseval theorem, 94
partial wave expansion, 179, 191
Pearson, C.E., 161, 280, 285
perfect fluid, 225
perturbation equations, 344, 346–348, 364, 365, 368, 369, 405
perturbation methods, 343–417
pipe flow, 195, 208
plane elasticity, 145
Plemelj's formulas, 318, 320, 341
point vortex, 228
Poisson equation, 125, 130
Poisson ratio, 146
polar coordinates, 122, 151, 211
cylindrical, 57, 80, 122, 166, 174
spherical, 124
polar form, 211–214, 217, 227, 228
pole, 238, 240–250, 252, 254, 257–259, 262, 263, 266, 268, 269, 271–273, 282, 284
Polubarinova-Kochina, P.Ya., 311
Popelar, C.H., 339
porous medium, 7, 8, 223, 311, 371
Polubarinova-Kochina, P.Ya., 311
Popelar, C.H., 339
potential energy, 44
potential flow, 227, 290, 296, 303, 313
Prandtl, L., 381
Press, W.H., 176
principal value, 212
principal-valued integral, 319

quasilinear equation, 20

radiation condition, 120, 121, 187, 191, 247, 250, 258, 342
Rand, R.H., 408
range of influence, 33, 35
Rayleigh number, 198, 201
recursion relation, 168, 173, 175, 176
reflection, 38, 54

regular perturbation, 344, 345, 349, 358
Reid, W.H., 352
residue, 266, 269, 272, 282
theorem, 254, 255, 258, 319, 326
Rice, J.R., 339
Riemann invariant, 24, 31
Riemann surface, 215–218, 301
Riemann–Hilbert problem, 318, 335, 339, 342
solution of, 320
Riemann–Hilbert technique, 210, 318–342
Roshko, A., 22

Sanchez-Palencia, E., 371, 374, 375
Schwarz's principle, 318, 327, 328
Schwarz–Christoffel transformation, 291, 295, 301
secular term, 367, 379
sedimentation, 18
seepage, 1, 7, 9, 11, 18, 143, 223, 302, 367, 370, 375, 384
Segal, L.A., xiv
self-adjoint, 72
semi-infinite channel, 298
semi-infinite domain, 153, 154
separation of variables, 57, 61, 63, 69, 71, 78, 80, 82, 84, 87, 88, 91, 132, 136, 160, 161, 166, 180, 258, 267, 269, 286, 389
shallow water waves, 12, 15, 23, 24, 57, 186
shocks, 50
Sih, G.C., 339
singular perturbation, 344, 345, 347, 358
singularity, 236, 238, 241, 243, 268, 274, 300, 326, 327
sink, 227, 291, 297, 327, 328
smoothness, 100
solvability condition, 379, 387
source, 227, 291, 296, 297, 307, 310, 311, 313, 322, 327, 328
space-like condition, 40
spherical boundaries, 165
static equilibrium, 145
Stegun, I.A., 165, 178, 181, 203, 279
Stoker, J.J., 24
Stokes' theorem, 41, 448, 449
Strack, O.D.I., 311
stream function, 223, 226, 227
streamline, 224, 227, 229, 306, 307, 314
strip footing, 335, 336
Sturm–Liouville problem, 57, 71–74, 112–114, 182

Taylor expansion, 2, 255, 268, 421, 445, 452
theorem, 250, 252

Terzaghi, K., 401
theory of homogenization, 367
thermal diffusivity, 11
time-like condition, 40
Tolstov, G.P., 100
traffic congestion, 50
traffic flow, 6, 20, 33, 45, 48, 53
traffic jam, 50
Tricomi, F.G., 159
Tuck, E.O., 339
Tulin, M.P., 322
Turien, R.M., 352

uniformly valid approximation, 384
uniqueness, 44

velocity potential, 8, 9
vibration, 1, 17, 163, 165
  in a rod, 4, 17
  of a soil layer, 395
  of a taut string, 1, 376
vortex, 291, 305
vorticity, 225, 226

water table, 307, 311
Watson, G.N., 165
water table, 307, 311
wave envelope, 376, 441, 444
wave equation, 3, 5, 16, 19, 25, 27, 180
  of canonical form, 34
wave propagation, 165, 179, 185–187
wave scattering, 190
Weber function, 174, 180
weighting function, 91
Wehausen, J.V., 342
Whitham, G.B., 6, 24
Wronskian, 178, 192
Wu, T.T., 206
Wu, T.Y.-T., 322, 327

Yih, C.S., 311
Young's modulus, 2, 4, 115, 128

Zarantonello, E.H., 322
Zauderer, E., 128
Zhukovsky function, 224, 308